AF358690

Innate Immunity to Virus Infection 2023

Innate Immunity to Virus Infection 2023

Editors

Caijun Sun
Feng Ma

Basel • Beijing • Wuhan • Barcelona • Belgrade • Novi Sad • Cluj • Manchester

Editors
Caijun Sun
School of Public Health
Sun Yat-Sen University
Shenzhen
China

Feng Ma
Chinese Academy of Medical
Sciences & Peking Union
Medical College
Suzhou
China

Editorial Office
MDPI
St. Alban-Anlage 66
4052 Basel, Switzerland

This is a reprint of articles from the Special Issue published online in the open access journal *Viruses* (ISSN 1999-4915) (available at: www.mdpi.com/journal/viruses/special_issues/1VSN050620).

For citation purposes, cite each article independently as indicated on the article page online and as indicated below:

Lastname, A.A.; Lastname, B.B. Article Title. *Journal Name* **Year**, *Volume Number*, Page Range.

ISBN 978-3-7258-0406-1 (Hbk)
ISBN 978-3-7258-0405-4 (PDF)
doi.org/10.3390/books978-3-7258-0405-4

Contents

About the Editors vii

Preface ix

Congcong Wang, Feng Ma and Caijun Sun
Special Issue: "Innate Immunity to Virus Infection, 1st Edition"
Reprinted from: *Viruses* 2023, *15*, 2060, doi:10.3390/v15102060 1

Felipe de Andrade Vieira Alves, Priscila Conrado Guerra Nunes, Laíza Vianna Arruda, Natália Gedeão Salomão and Kíssila Rabelo
The Innate Immune Response in DENV- and CHIKV-Infected Placentas and the Consequences for the Fetuses: A Minireview
Reprinted from: *Viruses* 2023, *15*, 1885, doi:10.3390/v15091885 4

Youwen Zhang, Yulin Xu, Sen Jiang, Shaohua Sun, Jiajia Zhang and Jia Luo et al.
Multiple Porcine Innate Immune Signaling Pathways Are Involved in the Anti-PEDV Response
Reprinted from: *Viruses* 2023, *15*, 1629, doi:10.3390/v15081629 23

Thomas Charles Bisom, Hope Smelser, Jean-Marc Lanchy and J. Stephen Lodmell
Alternative Splicing of RIOK3 Engages the Noncanonical NFκB Pathway during Rift Valley Fever Virus Infection
Reprinted from: *Viruses* 2023, *15*, 1566, doi:10.3390/v15071566 36

Xinyue Yao, Wanwan Dai, Siyu Yang, Zhaoli Wang, Qian Zhang and Qinghui Meng et al.
Synergistic Effect of Treatment with Highly Pathogenic Porcine Reproductive and Respiratory Syndrome Virus and Lipopolysaccharide on the Inflammatory Response of Porcine Pulmonary Microvascular Endothelial Cells
Reprinted from: *Viruses* 2023, *15*, 1523, doi:10.3390/v15071523 54

Meilin Li, Dingkun Peng, Hongwei Cao, Xiaoke Yang, Su Li and Hua-Ji Qiu et al.
The Host Cytoskeleton Functions as a Pleiotropic Scaffold: Orchestrating Regulation of the Viral Life Cycle and Mediating Host Antiviral Innate Immune Responses
Reprinted from: *Viruses* 2023, *15*, 1354, doi:10.3390/v15061354 70

Ziyu Wen, Yue Yuan, Yangguo Zhao, Haohang Wang, Zirong Han and Minchao Li et al.
Enhancement of SARS-CoV-2 N Antigen-Specific T Cell Functionality by Modulating the Autophagy-Mediated Signal Pathway in Mice
Reprinted from: *Viruses* 2023, *15*, 1316, doi:10.3390/v15061316 90

Eugenia Quiros-Roldan, Alessandra Sottini, Simona Giulia Signorini, Federico Serana, Giorgio Tiecco and Luisa Imberti
Autoantibodies to Interferons in Infectious Diseases
Reprinted from: *Viruses* 2023, *15*, 1215, doi:10.3390/v15051215 102

Emna Benzarti, Kristy O. Murray and Shannon E. Ronca
Interleukins, Chemokines, and Tumor Necrosis Factor Superfamily Ligands in the Pathogenesis of West Nile Virus Infection
Reprinted from: *Viruses* 2023, *15*, 806, doi:10.3390/v15030806 123

Jie Min, Wenjun Liu and Jing Li
Emerging Role of Interferon-Induced Noncoding RNA in Innate Antiviral Immunity
Reprinted from: *Viruses* 2022, *14*, 2607, doi:10.3390/v14122607 141

Fengwen Xu, Geng Wang, Fei Zhao, Yu Huang, Zhangling Fan and Shan Mei et al.
IFITM3 Inhibits SARS-CoV-2 Infection and Is Associated with COVID-19 Susceptibility
Reprinted from: *Viruses* **2022**, *14*, 2553, doi:10.3390/v14112553 . . . . **157**

About the Editors

Caijun Sun

Professor Caijun Sun's work mainly focus on the field of vaccine development, anti-viral drugs, virus–host interactions, and epidemiology research against infectious diseases. In recent years, he has published more than 100 academic papers in *Advanced Science, Plos Pathogens, J Virology, J Infection*, etc. He serves as Editorial board for *Viruses, Frontiers in Immunology, Human Vaccines & Immunotherapeutics*, etc.

Feng Ma

Professor Feng Ma's work mainly focus on the field of antiviral immunity, immunometabolism, infectious diseases, and inflammation. In recent years, he has published more than 30 papers in *Nature Immunology, Immunity, Nature Communications, Cell reports, EMBO report*, and *The Journal of Immunology*.

Preface

Emerging and re-emerging outbreaks of highly pathogenic viruses are becoming a severe crisis for global public health. As the first line of host defense against viral infections, the interferon (IFN)-mediated innate immunity and a variety of IFN-stimulated genes (ISGs) are well-known to play a critical role in interfering with virus entry and replication. Thus, it is of great importance to deeply understand the comprehensive interplay between innate immunity and virus infection, which will provide insights into developing novel antiviral therapeutics and vaccines. To deeply understand this complex interplay, we launched this Special Issue to gather novel knowledge about innate immunity and viral infections, including but not limited to interferon (IFN)-stimulated genes (ISGs), IFN signaling pathways, antiviral immunity, inflammation, immunometabolism, vaccine-related innate immunity, and broadly antiviral drugs. We hope that these latest studies can provide insights for the subsequent development of novel approaches to prevent and control viral infections.

Caijun Sun and Feng Ma
Editors

viruses

Editorial

Special Issue: "Innate Immunity to Virus Infection, 1st Edition"

Congcong Wang [1,2], Feng Ma [3,*] and Caijun Sun [1,2,*]

1 School of Public Health (Shenzhen), Shenzhen Campus of Sun Yat-sen University, Shenzhen 518107, China; wangcc5@mail2.sysu.edu.cn
2 Key Laboratory of Tropical Disease Control, Sun Yat-sen University, Ministry of Education, Guangzhou 510080, China
3 National Key Laboratory of Immunity and Inflammation, and CAMS Key Laboratory of Synthetic Biology Regulatory Elements, Suzhou Institute of Systems Medicine, Chinese Academy of Medical Sciences & Peking Union Medical College, Suzhou 215123, China
* Correspondence: maf@ism.pumc.edu.cn (F.M.); suncaijun@mail.sysu.edu.cn (C.S.)

Citation: Wang, C.; Ma, F.; Sun, C. Special Issue: "Innate Immunity to Virus Infection, 1st Edition". *Viruses* 2023, 15, 2060. https://doi.org/10.3390/v15102060

Received: 21 September 2023
Accepted: 28 September 2023
Published: 7 October 2023

Frequent outbreaks of emerging and re-emerging pathogenic viruses have become one of the major challenges for global public health. As the first line of defense, the innate immune system plays a vital role in fighting the invasion of pathogenic microorganisms. In response to viral entry into the host cell, pattern recognition receptors (PRRs) recognize the pathogen-associated molecular patterns (PAMPs) of viruses and then activate innate immune signaling pathways, which subsequently trigger the expression of numerous interferon-stimulated genes (ISGs) to exert direct antiviral effects [1,2]. Meanwhile, many viruses have developed various strategies to escape the innate immunity [3]. To deeply understand this complex interplay, we launched this Special Issue to gather novel knowledge about innate immunity and viral infections, and we hope that these latest studies can provide insights into developing antiviral therapeutics and vaccines.

Porcine epidemic diarrhea virus (PEDV) is a positive-sense single-stranded RNA virus that belongs to a coronavirus family. Many studies have shown that several PEDV proteins, including nsp1, nsp3, nsp5, nsp8, nsp14, nsp15, nsp16, E, M, and N, can restrict host IFN signaling. The research article by Zhang et al. investigated multiple PRR-mediated signaling pathways involved in the anti-PEDV responses. The innate immune signaling adaptors TRIF, MAVS, and STING exhibit blatant anti-PEDV activity, according to the authors' screening of porcine innate immune signaling adaptors' antiviral activity using transfected Vero cells. To further confirm it, knockdown or knockout of endogenous TRIF, MAVS, and STING promoted PEDV replication via siRNA and CRISPR approaches [4]. These results show that multiple porcine PRR-mediated signaling pathways are involved in PEDV recognition and defense, expanding our understanding of innate immunity responses to PEDV infection.

Recently, it has been important to study how the noncanonical NF-κB pathway participates in innate immunity. Bisom et al. conducted research to investigate the function of RIOK3 during Rift Valley Fever virus (RVFV) infection. They found that RVFV infection activated the noncanonical NF-κB pathway to weaken the antiviral IFN signaling response due to the production of the alternatively spliced RIOK3 X2 isoform, which encodes a truncated RIOK3 [5]. This finding will be helpful for deeply understanding the pathogenesis of RVFV through the regulation of the noncanonical NF-κB pathway to enhance viral replication.

Yao et al. reported their data on the crucial role of pulmonary microvascular endothelial cells (MVECs) in regulating inflammation during highly pathogenic porcine reproductive and respiratory syndrome virus (HP-PRRSV) infections. They reported that HP-PRRSV primarily induced virus-associated innate immune responses, whereas bacterial lipopolysaccharide (LPS) was responsible for the inflammatory response. HP-PRRSV infection exacerbated the inflammatory response due to secondary bacterial infections [6].

These results supported the importance of pulmonary MVECs in lung inflammation injury by primary HP-PRRSV infection and secondary bacterial infection.

Wen et al. focused on fusing the autophagosome-associated LC3b protein to the nucleocapsid (N) antigen, which is expected to improve the SARS-CoV-2-specific T cell functionality for developing the next-generation vaccine against SARS-CoV-2 variants. They concluded that the N-LC3b protein group can simultaneously secrete multiple cytokines (IFN-γ+/IL-2+/TNF-α+), improving T cell proliferation, especially for CD8+ T cell responses. In addition, their strategy was also induced a robust humoral immune response against the N antigen [7].

The role of IFITM3 in the SARS-CoV-2 pandemic is still controversial. Xu et al. reported their data on the association between IFITM3 and the risk of acquiring a SARS-CoV-2 infection. They demonstrated that IFITM3 inhibited SARS-CoV-2 infection by preventing virus entry, which is dependent on the first 21 amino acids of IFITM3. In addition, they also found that the rs12252 CC genotype of IFITM3 increased the risk of acquiring a SARS-CoV-2 infection and the decreased level of neutralizing antibodies against SARS-CoV-2 [8].

Another five review publications summarized the state-of-the-art research on the interplay between viruses and host innate immunity. Alves et al. summarized how placental cells engaged in innate immune responses play roles in response to Dengue virus (DENV) and chikungunya (CHIKV) infections [9]. Li et al. reported the roles of various well-known viruses in hijacking cytoskeletal structures and the accompanying antiviral responses [10]. Roldan et al. described the comprehensive understanding of the possible mechanisms of anti-cytokine autoantibody production, which could improve the approach to treating some infections, not only targeting pathogens but as a treatment for some autoimmunity patients [11]. Benzarti et al. discussed the complicated roles and expression patterns of interleukins, chemokines, and tumor necrosis factor superfamily ligands associated with West Nile virus (WNV) infection and pathogenesis [12]. Min et al. discussed the regulatory role of IFN-induced noncoding RNA (ncRNA) in antiviral innate immunity, aiming to improve our understanding of ncRNAs and provide insights for the basic research of antiviral innate immunity [13].

In conclusion, these ten articles published in this Special Issue should improve our understanding on the complex interactions between viral infections and host innate immune responses. These findings provide a summary of the most updated findings on PEDV, RVFV, PRRSV, SARS-CoV-2, DENV, CHIKV, and WNV, which are crucial for the subsequent development of novel approaches to prevent and control viral infections.

Author Contributions: Conceptualization and writing by C.W. and C.S., review and editing by C.S. and F.M. All authors have read and agreed to the published version of the manuscript.

Funding: This research was funded by the National Key R&D Program of China (2022YFE0203100, 2021YFC2300103, 2018YFA0900803), the National Natural Science Foundation of China (82271786, 32170880, 81971927), Science and Technology Planning Project of Guangdong Province, China (2021B1212040017), the Natural Science Foundation of Jiangsu Province (BK20221256 and BK20200004), Shenzhen Science and Technology Program (JCYJ20190807155009482, JSGG20200225152008136), High-Level Project of Medicine in Nanshan, Shenzhen (SZSM202103008), and Non-Profit Central Research Institute Fund of CAMS (2019PT310028).

Conflicts of Interest: The authors declare no conflict of interest.

References

1. Thaiss, C.A.; Zmora, N.; Levy, M.; Elinav, E. The microbiome and innate immunity. *Nature* **2016**, *535*, 65–74. [CrossRef] [PubMed]
2. Zhao, J.; Chen, J.; Wang, C.; Liu, Y.; Li, M.; Li, Y.; Li, R.; Han, Z.; Wang, J.; Chen, L.; et al. Kynurenine-3-monooxygenase (KMO) broadly inhibits viral infections via triggering NMDAR/Ca^{2+} influx and CaMKII/ IRF3-mediated IFN-beta production. *PLoS Pathog.* **2022**, *18*, e1010366. [CrossRef] [PubMed]
3. Minkoff, J.M.; tenOever, B. Innate immune evasion strategies of SARS-CoV-2. *Nat. Rev. Microbiol.* **2023**, *21*, 178–194. [CrossRef] [PubMed]
4. Zhang, Y.; Xu, Y.; Jiang, S.; Sun, S.; Zhang, J.; Luo, J.; Cao, Q.; Zheng, W.; Meurens, F.; Chen, N.; et al. Multiple Porcine Innate Immune Signaling Pathways Are Involved in the Anti-PEDV Response. *Viruses* **2023**, *15*, 1629. [CrossRef] [PubMed]

5.   Bisom, T.C.; Smelser, H.; Lanchy, J.M.; Lodmell, J.S. Alternative Splicing of RIOK3 Engages the Noncanonical NFkappaB Pathway during Rift Valley Fever Virus Infection. *Viruses* **2023**, *15*, 1566. [CrossRef] [PubMed]

6.   Yao, X.; Dai, W.; Yang, S.; Wang, Z.; Zhang, Q.; Meng, Q.; Zhang, T. Synergistic Effect of Treatment with Highly Pathogenic Porcine Reproductive and Respiratory Syndrome Virus and Lipopolysaccharide on the Inflammatory Response of Porcine Pulmonary Microvascular Endothelial Cells. *Viruses* **2023**, *15*, 1523. [CrossRef] [PubMed]

7.   Wen, Z.; Yuan, Y.; Zhao, Y.; Wang, H.; Han, Z.; Li, M.; Yuan, J.; Sun, C. Enhancement of SARS-CoV-2 N Antigen-Specific T Cell Functionality by Modulating the Autophagy-Mediated Signal Pathway in Mice. *Viruses* **2023**, *15*, 1316. [CrossRef] [PubMed]

8.   Xu, F.; Wang, G.; Zhao, F.; Huang, Y.; Fan, Z.; Mei, S.; Xie, Y.; Wei, L.; Hu, Y.; Wang, C.; et al. IFITM3 Inhibits SARS-CoV-2 Infection and Is Associated with COVID-19 Susceptibility. *Viruses* **2022**, *14*, 2553. [CrossRef] [PubMed]

9.   de Andrade Vieira Alves, F.; Nunes, P.C.G.; Arruda, L.V.; Salomão, N.G.; Rabelo, K. The Innate Immune Response in DENV- and CHIKV-Infected Placentas and the Consequences for the Fetuses: A Minireview. *Viruses* **2023**, *15*, 1885. [CrossRef] [PubMed]

10.   Li, M.; Peng, D.; Cao, H.; Yang, X.; Li, S.; Qiu, H.-J.; Li, L.-F. The Host Cytoskeleton Functions as a Pleiotropic Scaffold: Orchestrating Regulation of the Viral Life Cycle and Mediating Host Antiviral Innate Immune Responses. *Viruses* **2023**, *15*, 1354. [CrossRef] [PubMed]

11.   Quiros-Roldan, E.; Sottini, A.; Signorini, S.G.; Serana, F.; Tiecco, G.; Imberti, L. Autoantibodies to Interferons in Infectious Diseases. *Viruses* **2023**, *15*, 1215. [CrossRef] [PubMed]

12.   Benzarti, E.; Murray, K.O.; Ronca, S.E. Interleukins, Chemokines, and Tumor Necrosis Factor Superfamily Ligands in the Pathogenesis of West Nile Virus Infection. *Viruses* **2023**, *15*, 806. [CrossRef] [PubMed]

13.   Min, J.; Liu, W.; Li, J. Emerging Role of Interferon-Induced Noncoding RNA in Innate Antiviral Immunity. *Viruses* **2022**, *14*, 2607. [CrossRef] [PubMed]

Review

The Innate Immune Response in DENV- and CHIKV-Infected Placentas and the Consequences for the Fetuses: A Minireview

Felipe de Andrade Vieira Alves [1,2,†], Priscila Conrado Guerra Nunes [3,†], Laíza Vianna Arruda [1,2], Natália Gedeão Salomão [2,3,*] and Kíssila Rabelo [1,2,*]

1   Laboratório de Ultraestrutura e Biologia Tecidual, Universidade do Estado do Rio de Janeiro/UERJ, Rio de Janeiro 20550170, RJ, Brazil; favalves@id.uff.br (F.d.A.V.A.); viannalaiza3@gmail.com (L.V.A.)
2   Laboratório Interdisciplinar de Pesquisas Médicas, Instituto Oswaldo Cruz, Fundação Oswaldo Cruz, Rio de Janeiro 21040900, RJ, Brazil
3   Laboratório de Imunologia Viral, Instituto Oswaldo Cruz, Fundação Oswaldo Cruz, Rio de Janeiro 21040900, RJ, Brazil; priscila.nunes87@gmail.com
*   Correspondence: natgsalomao@gmail.com (N.G.S.); kissilarabelo91@gmail.com (K.R.)
†   These authors contributed equally to this work.

Abstract: Dengue virus (DENV) and chikungunya (CHIKV) are arthropod-borne viruses belonging to the *Flaviviridae* and *Togaviridae* families, respectively. Infection by both viruses can lead to a mild indistinct fever or even lead to more severe forms of the diseases, which are characterized by a generalized inflammatory state and multiorgan involvement. Infected mothers are considered a high-risk group due to their immunosuppressed state and the possibility of vertical transmission. Thereby, infection by arboviruses during pregnancy portrays a major public health concern, especially in countries where epidemics of both diseases are regular and public health policies are left aside. Placental involvement during both infections has been already described and the presence of either DENV or CHIKV has been observed in constituent cells of the placenta. In spite of that, there is little knowledge regarding the intrinsic earlier immunological mechanisms that are developed by placental cells in response to infection by both arboviruses. Here, we approach some of the current information available in the literature about the exacerbated presence of cells involved in the innate immune defense of the placenta during DENV and CHIKV infections.

Keywords: innate immunity; dengue; chikungunya; immune cells; interferon

check for
updates

Citation: de Andrade Vieira Alves, F.; Nunes, P.C.G.; Arruda, L.V.; Salomão, N.G.; Rabelo, K. The Innate Immune Response in DENV- and CHIKV-Infected Placentas and the Consequences for the Fetuses: A Minireview. *Viruses* **2023**, *15*, 1885. https://doi.org/10.3390/v15091885

Academic Editors: Caijun Sun and Feng Ma

Received: 28 June 2023
Revised: 25 August 2023
Accepted: 30 August 2023
Published: 6 September 2023

1. Introduction

Over the last 40 years, the emergence and re-emergence of dengue have posed a considerable threat to global health, with the last 10 years seeing consecutive outbreaks of the equally severe chikungunya [1]. Throughout these years, many studies have been conducted in order to understand infection control, pathogenesis, and the host immune response to these diseases and much has evolved in this knowledge [2–6]. In light of this information, for a long time, there has been a considerable gap in the knowledge and understanding of these infections in pregnant patients, namely, about the claim that there really is vertical transmission, the effects on the development of the pregnancy and fetus, and the immunological effects of these infections. We know that the innate immune response plays an extremely relevant role in viral infections, acting systemically and, also, locally [4–6]. Thus, in this review, we will investigate the already-known aspects of the innate response to these infections in a specific organ, the placenta, in order to compile and better clarify its role in the consequences and resolution of the infection.

1.1. The Dengue Virus

Although the history of dengue is uncertain, the earlier registers of a disease consistent with dengue fever date back to the period of the Chinese dynasty, on the territory of the present-day People's Republic of China [7]. Later on, between 1779 and 1780, the illness affected the continents of Africa, Asia, and North America, causing the first well-known epidemics of dengue [8]. In spite of that, the isolation of the dengue virus was performed in 1943. Between this period and nowadays, large outbreaks occurred worldwide [2,9,10].

The etiological agent of the disease, dengue virus (DENV), is an arthropod-borne virus (arbovirus) belonging to the *Flaviviridae* family and *Flavivirus* genus, comprising four major antigenically distinct serotypes (DENV 1–4), each one capable of causing the sickness [11]. All serotypes circulate mostly in tropical and subtropical areas of the globe due to the temperature and rainy seasons, factors that are favorable to the life cycle of mosquitoes of the genus *Aedes*, remarkable vectors of arboviruses [12,13]. According to the World Health Organization (WHO), it is estimated that 25,000 deaths occur per year and over 2 billion people live in endemic areas [14].

DENV, which shares similarities with other flaviviruses, such as Zika virus (ZIKV), Japanese encephalitis virus (JEV), and yellow fever virus (YFV), is an icosahedral enveloped virus of approximately 40–50 nm in size, composed of a lipid bilayer where the structural proteins of the membrane (M) and envelope (E) are inserted [15]. Inside the lipid bilayer, there is the nucleocapsid (N), a structure composed of the viral genome surrounded by multiple copies of the capsid protein (C) [15–18]. The virus genome consists of a single positive-strand RNA of about ~11 kb in length with a $5'$ cap end and lack of polyadenylated tail at its $3'$ end. This genome has only one open reading frame (ORF) that is translated into a single large polyprotein that, later on, is cleaved by cellular and viral proteases in another ten distinct proteins: three structural proteins (C, prM, and E) that constitute the viral particle and seven non-structural proteins (NS1, NS2A, NS2B, NS3, NS4A, NS4B, and NS5) related to both the viral replication process and the assembly of the virions [18–20].

1.2. The Chikungunya Virus

Chikungunya virus (CHIKV) is an arbovirus that belongs to the *Togaviridae* family and *Alphavirus* genus. It is classified as Old-World alphavirus, due to its geographical origin, and is more associated with the predominance of polyarthralgia [21,22]. Its first isolation was in 1953 in Tanzania (East Africa); it was obtained from a fevered man's blood once it was found to be responsible for causing a febrile illness known as chikungunya fever (CHIKF) [23]. Since then, the virus has been identified in more than 60 countries in Asia, Africa, Europe, and the Americas. As of now, three genotypes have been identified: West African, East-Central-Southern African (ECSA) and Asian, and the Indian Ocean lineage, originating from ECSA [24].

The CHIKV genome is a single-strand positive-sense RNA molecule, with 11.8-kbp, encoding 2472 amino acid nonstructural and 1244 amino acid structural polyproteins [25], which gives rise to four nonstructural (nsP1-4) and five structural proteins (C, E3, E2, 6K, and E1) [26,27]. The nonstructural proteins are responsible for the viral replication; meanwhile, structural proteins shape the viral particle with a 60–70 nm diameter, which is enveloped with icosahedral nucleocapsid [28].

2. Transmission and Clinical Manifestations

2.1. In Dengue

The transmission of dengue occurs through the bites of female hematophagous mosquitoes of the genus *Aedes*, mainly *Aedes aegypti*; although, other species, such as *Aedes albopictus*, are also important vectors of the disease [7,29]. Expansion of these vectors, especially *Aedes aegypti*, which is more adapted to the urban environment, is in close association with the exponential increase in urbanization, climatic changes, and socio-economic factors [30]. Transplacental transmission, organ transplantation, and blood transfusion are also types of dengue transmission reported in the literature; although, they are rare and unusual [31–33].

At the end of the 1990s, dengue was classified according to the parameters of the WHO, which included undifferentiated fever, dengue fever (DF), dengue hemorrhagic fever (DHF), and dengue shock syndrome (SCD) [34]. In general, the incubation period of the virus lasts between 4 to 7 days and the infection by any DENV serotype can cause a wide variety of symptoms and clinical manifestations, from a mild illness with undifferentiated fever to a life-threatening hemorrhagic fever [35,36].

Dengue fever (DF) was characterized by the presence of common symptoms, such as fever, arthralgia, headache, emesis, myalgia, and cutaneous rash [35]. A small number of patients tended to progress to a more severe clinical condition called dengue hemorrhagic fever (DHF), in which hemorrhagic manifestations, homeostasis abnormalities, and increased vascular permeability features could be noticed. Therefore, DHF was classified into four degrees of severity, with the latter (III and IV) coinciding with dengue shock syndrome (SCD), characterized by hypovolemic shock, with slight arterial pulse and hypotension [34]. However, the criteria used were outdated for applying during large outbreaks and difficult to meet, which led the WHO to create a new classification scheme [37]. This new consensus introduced the concept of classifying dengue into dengue without warning signs, dengue with warning signs, and severe dengue [14].

In dengue without warning signs, symptoms such as rash, nausea, vomiting, myalgia, arthralgia, and leukopenia, among others are common; meanwhile, the warning signs include abdominal pain, persistent vomiting, accumulation of fluid in the cavities, mucosal bleeding, and liver enlargement. Usually, hepatomegaly precedes plasma leakage, being an indicator of the evolution of the severity of the disease. On the other hand, clinical manifestations characteristic of severe dengue include severe plasma leakage, severe hemorrhage, and severe organ involvement [14,38,39].

The liver appears to be the central target during dengue infections and its involvement seems to be a usual complication [40]. This is supported by the presence of the dengue virus in this organ already being demonstrated in several studies, as well as hepatic injury due to the infection [41–45]. In contrast, atypical manifestations during infection, such as the commitment of the central nervous and skeletal muscle systems, heart, and lungs were also reported [46–48]. Nonetheless, previous studies showed the presence of the virus in the kidneys, pancreas, spleen, and even the placenta, which are unusual sites of the infection [46,49–52]. Thus, today we consider dengue to be a broad disease that affects the entire body and can cause systemic damage.

2.2. In Chikungunya

CHIKV is transmitted mainly by infected mosquitoes from the *Aedes* species, such as *Aedes aegypti* and *Aedes albopictus*, and is prevalent in urban and peri-urban areas, respectively [53]. The virus had already been detected in semen and vaginal secretions; however, sexual transmission was not confirmed [54]. Usually, the first infection occurs in the skin: in fibroblast, keratinocytes, and endothelial cells from blood vessels. Upon reaching the bloodstream, the virus disseminates to various organs, such as the lymphoid tissues, liver, muscle, spleen, heart, and brain [55].

The incubation period, which is the time between infection and the onset of symptoms, lasts between 3 to 12 days. In the acute phase, the most common symptoms are high fever (>38.5 °C), rash, and intense polyarthralgia; this gives the name of the disease, which originated from the Makonde language meaning "that which bends up" [56], due to hunched posture of infected individuals and it being a disease of high morbidity [57,58]. In addition, headache, discomfort in the throat, abdominal pain, constipation or diarrhea, persistent conjunctivitis, vomiting, and lymphadenopathy (cervical or generalized) may also occur [59]. It is not rare to observe dermal manifestations, mainly on the face, trunk, and extremities [60]. A maculopapular rash is the most common cutaneous manifestation in adults and vesiculobullous lesions are predominant in children [61–65]. During the post-acute phase, individuals may present with arthritis; rheumatic disorders, such as tenosynovitis; bursitis; enthesitis; periostitis; and tendonitis. Clinical manifestations could persist, evolving into a chronic disease for months or years, including joint pain and swelling varying in intensity and frequency [66].

Although it is not common, some individuals develop severe forms of the disease, with multiple organ dysfunction characterized by vascular congestion, edema, and hemorrhage [67] or culminating in death. Atypical manifestations, such as respiratory disorders, arterial hypertension, hepatitis, myocarditis with sinus tachycardia, cardiomegaly, ectopic ventricular beats, abnormal electrocardiograms, and congestive heart failure, were reported [68–70]. Age and comorbidities (such as diabetes; cardiovascular, respiratory, renal, and autoimmune diseases; and hypertension) seem to be important factors for such; however, they may occur in low-risk populations [67,68,71]. Regarding asymptomatic individuals, the percentage is between 3 and 28% [72]. A total of 123,000 severe cases of CHIKV infection were reported in an important outbreak in the 2005–2006 period located on Reunion Island, in which about a third of the population was affected [68,71,73]. It was associated with the E1-A226V mutation, a single nucleotide change at E1 glycoprotein position 226 of the ECSA genotype resulting in an alanine (Ala) to a valine (Val) substitution. This mutation was identified in more than 90% of the isolates in the Reunion Island outbreak [26]; it seems to improve CHIKV infectivity and replication in *Aedes albopictus* and, consequently, its dissemination to humans [74].

3. The Placenta

Previous evidence of DENV and CHIKV outbreaks has demonstrated that pregnant women are at high risk of experiencing pregnancy complications during viral infection [52,75–78]. In addition, there are some reports of the vertical transmission of these microorganisms, raising awareness of the importance of better understanding the role of the placenta in DENV and CHIKV infections [79–81]. Established in the third week of gestation, the placenta is characterized as a temporary and chimerical organ, formed by maternal and fetal tissue, that plays an essential role in the development and support of pregnancy. This organ supplies essential oxygen, nutrients, and hormones to the fetus, as well as carrying out the elimination of toxic waste [82].

The maternal portion of the placenta is called the decidua basalis, a tissue derived from the endometrium. On the other hand, the fetal portion includes several types of embryo-derived trophoblastic cells. These cells are specialized epithelial cells that are essential for the establishment and continuation of pregnancy. The fetal portion projects the chorionic villi, the functional unit of the placenta. They are characterized as an arboreal structure that can be anchored in the decidua or float in the intervillous space. The villi have an apical layer of syncytiotrophoblasts, which comprises the first barrier of placental defense against invading pathogens, followed by a layer of cytotrophoblastic progenitor cells and villous stroma that contain stromal fibroblasts, Hofbauer cells, and fetal vascular endothelium cells [83,84]. From the second semester, the chorionic villi are bathed by maternal blood, derived from vessels of the decidua basalis, in the intervillous space. Therefore, the human placenta is said to be hemochorial, meaning maternal blood is in contact with trophoblastic cells of fetal origin [85].

In this way, maternal and fetal blood do not mix, except for the rupture of capillary walls, which rarely occurs outside of the delivery situation. The separation between fetal and maternal blood is called the placental barrier, which is composed of syncytiotrophoblast, cytotrophoblast, connective tissue (containing mesenchymal cells and fibroblasts), and fetal endothelium. However, as pregnancy advances, the cytotrophoblast layer thins and disperses, making the placental barrier thinner, optimizing the exchange of substances [85].

4. Placental Immune Cells

The proper development of a pregnancy requires a series of physiological adaptations and a highly dynamic balance in the maternal immune response [86,87]. This is because the fetus and placenta consist of a semi-allogeneic graft and, for this reason, adaptations are necessary in the maternal immune system, which is aimed at immune regulation and fetal tolerance parallel to an effective immune defense [88]. So, maternal immune cells are subject to constant modifications in subpopulations [89], with the upregulation of those involved with innate immunity [90].

In early pregnancy, the pro-inflammatory environment, rich in dendritic cells and natural killer (NK) cells, supports tissue remodeling and trophoblastic invasion, essential for placental establishment [91]. Natural killer cells make up about 70% of decidual leukocytes in early pregnancy [92]. These cells contain a distinct phenotype of peripheral natural killer cells and secrete several growth factors, as well as angiogenic factors and cytokines that contribute to remodeling the decidua and spiral arteries [93,94]. On the other hand, dendritic cells make up only 2% of decidual leukocytes and participate in the early stages of implantation by secreting stromal cell-derived factor 1 (SDF-1), which aids in vascular expansion and decidual angiogenesis [92,95].

In addition, the decidual immunity cell population is also composed of decidual macrophages (20–25%) [92]. Decidual macrophages are the major antigen-presenting cells (APCs) at the maternal–fetal interface in early gestation; these cells are thought to also participate in vascular remodeling, trophoblastic invasion, and immune tolerance [96–99]. Most decidual leukocytes are recruited primarily by chemokines, such as CXCL12, CXCL8, TGF-β, and CCL2, secreted by trophoblast cells and decidual cells [85,92].

As pregnancy advances, placental growth slows and the peripheral environment becomes anti-inflammatory, with Hofbauer cells and regulatory T cells secreting anti-inflammatory cytokines that aid fetal immune tolerance and rapid fetal growth [91]. In general, it can be said that fetal immune tolerance is regulated by the restriction and modulation of some leukocytes present in the maternal–fetal interface. Despite the high density of natural killer cells, the number of dendritic cells and effector T cells is relatively small. In addition, the dendritic cells present in the decidua have a unique behavior: after exposure to the fetal antigen, these cells are retained in the decidual stroma and, therefore, are not able to migrate toward the maternal lymphatic vessels [88,91]. Thus, fetal antigens reach maternal lymph nodes only by passive transport and are presented to T cells by lymph-node-resident dendritic cells, a paradigm that does not trigger an effective immune response [88,100].

In the last stage of pregnancy, the maternal immune system shifts again to a pro-inflammatory state that will be essential at the time of delivery since the uterine musculature will have to contract and expel the fetus in addition to releasing the placenta [101].

The innate immune response is responsible for controlling the viral spread during the early stages of infection [102]. The effectiveness of the innate immune system is especially important during pregnancy since vertical viral transmission can lead to developmental anomalies, intrauterine growth restriction, and premature delivery/stillbirth [90]. The role of decidual innate immune cells in the defense against viral infections and their role in vertical transmission is an emerging field; but, it is still little explored. Later, we will discuss what is known about the involvement of these cells during viral infection by DENV and CHIKV.

5. Vertical Transmission in Dengue

Despite the high incidence of the disease, studies related to the maternal/fetal consequences of DENV infection during pregnancy are still limited. In addition, there is still no consensus regarding the effects of the infection on pregnant women and/or newborns; however, some studies indicate that vertical transmission can occur and present severe outcomes, such as premature births and maternal/fetal death [103–112].

Although pregnancy is considered a risk factor for the clinical course of the disease, previous studies have not found an association between the severity of maternal infection and neonatal disease [113,114]. However, it is suggested that maternal natural immunosuppression during pregnancy may favor the occurrence of more severe infections, causing damage to the health of the mother and fetus [115].

In Brazil, a study carried out by Paixão et al. (2018) reported a risk of maternal death three times higher in cases of dengue and four-hundred-and-fifty times higher when the pregnant woman had DHF [107]. In addition, a study by our group showed that the severity of dengue fever led to the death of a pregnant patient, with an intense inflammation profile in the placental and fetal tissues analyzed [52].

A recent study in India carried out by Brar et al. (2021) observed that the average gestation period was 31.89 ± 7.31 weeks. The incidence of maternal systemic complications was high: 52.3% of pregnant women had thrombocytopenia, 25% developed postpartum hemorrhage, 18.2% of pregnant women developed acute kidney injury, 4.5% required hemodialysis support, 18.2% developed acute respiratory distress syndrome (ARDS), 15.9% required ventilatory support, 9.1% developed acute liver failure, 40.9% had evidence of shock, and 15.9% of women died. With regard to the fetus, it was observed that 4.5% of pregnancies suffered spontaneous abortion, 9% were stillbirths, and 4.5% evolved to neonatal deaths. In addition, they reported that premature babies were born in 34.1% of cases and 29.5% of women had low birth weight babies [116].

In Mexico, of the pregnant women infected with DENV in 2013, 65.9% were classified as being without warning signs of dengue (WWSD), 18.3% with warning signs of dengue (WSD), and 15.9% with severe dengue (SD). Pregnant women with SD (38.5%) had fetal distress and underwent emergency cesarean sections; this condition was associated with obstetric hemorrhage (30.8%), pre-eclampsia (15.4%), and eclampsia (7.7%). Pregnant women who did not have SD had full-term pregnancies, delivered vaginally, and had apparently healthy babies with normal birth weights [117].

In Vietnam, an investigation of pregnant women infected with DENV in 2015 showed that 90% were positive for the NS1 antigen and primary infection, 20% had premature births, and 5% had stillbirths. All neonates born alive were discharged uneventfully and no maternal death was reported [118].

During pregnancy, the fetus may be susceptible to DENV infection, especially during the critical period of organogenesis or in late pregnancy [119–123].

A recent study evaluated pregnant women during an epidemic in French Guiana and reported a vertical transmission rate of 18.5%, with viral transmission, both at the beginning and at the end of pregnancy. It was possible to verify that it is more frequent when maternal infection occurs late during pregnancy, close to delivery, and that newborns may present neonates with warning signs of dengue that require platelet transfusion. Furthermore, it points out that if there is a fever during the 15 days prior to delivery, the cord blood and placenta should be sampled and tested for the virus and the newborn should be closely monitored during the postpartum period [124].

Viral transmission to the fetus via the placenta can occur via the movement of the maternal vascular endothelium to trophoblasts by infected maternal monocytes, which transmit the infection to placental trophoblasts; they also do so via paracellular pathways from maternal blood to the fetal capillaries [125,126]. It has recently been reported that DENVs preferentially infect the decidua; the intensity of the decidual infection appears to be associated with the risk of fetal infection. Viral infection in the decidua in early pregnancy may modulate decidual roles in arterial remodeling and placentation that eventually influence the placental barrier balance [127].

Potential mechanisms by which a maternal infection could result in fetal death include direct fetal infection and organ damage, placental infection resulting in decreased transmission of nutrients and oxygen, and increased production of cytokines and chemokines [128].

In the histopathological evaluation of pregnant women with dengue during pregnancy carried out by Ribeiro et al., (2017), signs of hypoxia, choriodeciduitis, deciduitis, and intervillitis were observed and viral antigens were found in the trophoblast cytoplasm, villous stroma, and decidua. In this study, two possible mechanisms of fetal and neonatal morbidity were proposed: the presence of hemodynamic changes during pregnancy that could affect the placenta and cause fetal hypoxia or the direct effect of the infection on the fetus [129].

6. Vertical Transmission in Chikungunya

CHIKV-infected pregnant women usually present the same clinical presentation as non-pregnant women. Basurko and collaborators carried out a study in French Guiana between June 2012 and June 2015 in which the median term of CHIKV infection was 30.7 weeks; the appearance of symptoms occurred mainly in the third trimester, with fever, arthralgia, and headache being the most common symptoms [130]. The hospitalization rate for maternal CHIKV was greater than 50%, mainly within 24 h of symptom onset; they did not observe differences in the frequency of pregnancy and neonatal outcomes when comparing to the control group (pregnant women who had no fever, no dengue, and no CHIKV infection at gestation) [130]. Similar results were found by Foeller in Grenada (August–December 2014); however, they found intense arthralgia and myalgia but with shorter durations in women who became infected with CHIKV during gestation [131]. Both authors found that the frequency of newborns who need intensive care unit admission seems to be higher when the women are exposed to CHIKV within 1 week before delivery, as well as pregnancy complications [130,131]. In contrast, a study conducted in India between August and October 2016 enrolled 150 CHIKV-infected pregnant women with a mean period of gestation of 25.62 ± 13.475 weeks. Of these women, 30 developed adverse pregnancy outcomes, mainly during the third trimester (80%), such as preterm delivery (7.33%), premature\rupture of membranes (3.33%), decreased fetal movements (2.67%), intrauterine death (2.67%), and oligohydramnios and preterm labor pains (2%) [132]. In the same way, AbdelAziem and collaborators reported cases of miscarriage (19.4%), preterm birth (13.9%), and stillbirth (4.3%) in a total of 93 women [133].

Although rare, vertical transmission (mother-to-child) has already been reported in CHIKV infection. The first report was conducted in June 2005 on the Reunion Island epidemic, which occurred between March 2005 and December 2006 [134]. In this outbreak, the rate of vertical transmission was close to 50% in mothers with high viremia during the intrapartum period [135]. Most authors believe the infection occurs by microtransfusions at the placental barrier or the breakdown of the syncytiotrophoblast due to uterine contractions [136,137]. The role of the placenta in CHIKV transmission is not fully understood; however, even after postponing normal birth or performing a cesarean delivery, the transmission of the virus to the baby is not avoided [135,136]. CHIKV antigens were detected in the placenta, such as the decidual, trophoblast, endothelial, Hoffbauer cells, and inside fetal capillaries [77,78,138].

During the intrapartum period, when the mother presents high viremia, the risk of the occurrence of CHIKV vertical transmission is increased; however, early maternal–fetal transmission of the virus has also been reported. Three cases of CHIKV infection before 16 weeks of gestation were reported, culminating in spontaneous abortions, with viral genome detection in the amniotic fluid, chorionic villi, and fetal brain [139]. Our group reported that spontaneous abortions occurred during the first and second trimesters, which exhibited microscopical and ultrastructural alterations and CHIKV antigen detection in abortion material [138]. The pregnant women infected with CHIKV in the studies cited were aged between 24 and 40 years old. In general, they denied smoking, alcohol use, or comorbidities. In most cases, infections in the first or second semester were symptomatic and led to miscarriage. The placentas of pregnant women who became infected with CHIKV during the second and third trimesters also exhibited histopathological alterations, CHIKV antigen detection, and an increase of cellularity and cytokines (pro- and anti-inflammatories) [77]. Several studies demonstrate the presence of CHIKV in the placenta [140], newborn cerebrospinal fluid, amniotic fluids [141], serum [79,142], and urine [79]. Although RNA CHIKV was detected in breast milk, transmission to infants was not reported [143].

Some of the obstetric complications already reported in CHIKV infection were: spontaneous abortion, preeclampsia, postpartum hemorrhage, premature birth, intrauterine death, oligohydramnios, and sepsis [76,132]. It is recommended to observe, for 7 days, the newborns of mothers who are suspected of having CHIKV infection as symptoms in infected neonates usually appear between the 3rd and 7th day of life [144]; these symptoms include fever, refusal to breastfeed, rash, swollen extremities, skin hyperpigmentation, thrombocytopenia, and irritability. However, neurological involvement may occur, leading to cases of meningoencephalitis, cerebral edema, intracranial hemorrhage, seizures, postnatal microcephaly, cerebral palsy, and neurodevelopmental delay [134,144–149]. It is important to emphasize that asymptomatic pregnant women could transmit the virus to the fetus [137].

7. Dendritic Cells, Macrophages, and Natural Killer Cells in Vertical Transmission

Vertical transmission of the dengue and chikungunya viruses has already been shown in previous studies [33,77,114,134,138,150]. However, little is known about the intrinsic mechanisms and cells involved in this event.

Dendritic cells (DC), alongside macrophages and natural killer (NK) cells, are essential cell subpopulations in placental homeostasis, participating in the regulation of implantation events and the success of pregnancy [151]. The first ones, in particular, are abundant cells located in the basal/parietal decidua, where both CD83+ (mature dendritic cells) and DC-SIGN+ (immature dendritic cells) contribute to the homeostasis in the placental tissue and modulate the cytokine expression and function of NK cells at the maternal–fetal interface [152,153]. Furthermore, these major subpopulations of cells are considered sentinels, responsible for the dissemination and amplification of both DENV and CHIKV infection [154,155]. Even though previous works have already shown that the dendritic cells of placental tissues are permissive to ZIKV infection [156,157], in another *Flavivirus*, the exact role of these cells in the vertical transmission of DENV and CHIKV is yet to be further investigated.

Macrophages, another type of immune cell found in maternal decidua, are highly associated with several important events, including the secretion of angiogenic molecules, remodeling of spiral arteries, and clearance of apoptotic cell remains in the placental bed [158,159]. These immune cells, alongside Hofbauer cells (HC), a type of chorionic villi-resident macrophage, represent an important barrier against pathogens and play a critical role in vertical transmission [160]. Therefore, infected maternal macrophages are thought to be crucial for vertical transmission events as they could interact with the placental trophoblast cells and transmit the infection [126]. In DENV cases, Hofbauer cells and macrophages appear to be pivot cells in the pathogenesis of the disease in placental tissues as the NS3 protein, implicated in dengue virus replication, was observed in the cytoplasm of both immune cells and in several organs of aborted fetuses, as well as in the maternal and fetal region of placentas [52]. The expression of TNF-α, IFN-γ, and RANTES was also found in DENV-infected placentas, revealing the maintenance of a pro-inflammatory environment in these cases [52]. Additionally, in an immunocompromised animal model, DENV vertical transmission was observed in the early stages of pregnancy and associated with an increased antibody-dependent enhancing (ADE) condition, which makes it conceivable that Hofbauer cells and macrophages at the maternal portion expressing Fc-gamma receptors could play an important role in inducing an ADE condition and, consequently, fetal infection [161]. Regarding CHIKV infections, several virus antigens were found in

Hofbauer cells in the placentas of infected pregnant women, evidencing the permissiveness of these cells to infection [78,138]. The presence of pro-inflammatory mediators was also noticed [77].

Decidual NK cells compose the majority of decidual cells (dNK) during early pregnancy and are specifically located around expanding extravillous trophoblasts [162]. These specialized maternal cells differ both in phenotype and function when compared to peripheral NK (pNK) cells and play a critical role during trophoblast invasion and placentation [163–165]. They also display distinct cytotoxic responses as dNK cells seem to produce high levels of cytokines and be less cytotoxic during trophoblast infection [166,167]. Therefore, dNK cells tend to preserve the placental trophoblasts during the development of an immune response against some pathogens, evidencing the fact that the placenta is considered a highly privileged organ [168–171]. Despite the fact that dNK cells are an immunotolerant subpopulation of cells, the gaining of a cytotoxic phenotype can occur regarding some specific infections [172,173].

8. IFN-I Response to Dengue and Chikungunya Placental Infection

Type I interferons (IFN-I) are the main cytokine mediators of the innate immune response and constitute a key defense mechanism against viral infections [173–175]. Soon after a viral infection, IFN-I synthesis is rapidly induced upon detection of viral RNA by pattern recognition receptors (PRRs) and consequent activation of interferon regulatory factor (IRF) [173]. Once synthesized, these cytokines act in a paracrine fashion to induce a peripheral antiviral state [176,177]. To complete this action, different subtypes of IFN-I, including IFN beta and IFN alpha, interact with the heterodimeric IFNAR receptor (IFNAR1/FNAR 2) to trigger a JAK-STAT-mediated signaling cascade that culminates in the transcription of hundreds of genes stimulated by interferons (ISGs) that have antiviral and immunomodulatory activities [99,178]. ISG molecules can act by several mechanisms in order to repress viral replication, including the inhibition of virus entry into the cell, inhibition of viral protein synthesis, degradation of essential viral components, and changes in cell metabolism [179,180]; they even play a regulatory and immunomodulatory role [181]. As much as they act through a shared receptor, it is noteworthy that the IFN-I subtypes have different properties [175,177,181,182].

In general, it is known that DENV is capable of inhibiting IFN-I signaling by two mechanisms: directly interfering in ISG synthesis pathways in parallel with the evasion of innate immune receptors. The non-structural proteins of DENV, especially NS2, NS4 (NS4A/NS4B), and NS5, have the ability to inhibit the activation of tyrosine kinase 2 (Tyk2), inhibit the phosphorylation of STAT1, and decrease the expression and inhibit the phosphorylation of STAT 2 (essential intermediates in the ISG synthesis cascade). Furthermore, the NS5 protein induces STAT2 degradation through a mechanism involving the cellular proteasome. On the other hand, the evasion of cell receptors would be related to the site of viral replication. Like other flaviviruses, the dengue virus induces the formation of intracellular vesicles from the membrane of the endoplasmic reticulum, which functions as a viral replication site. These vesicles resemble cellular organelles and, for this reason, are not recognized by components of the innate response [183–185].

With regard to in vitro studies, Luo and collaborators performed infection tests with flaviviruses, such as ZIKV, YFV, and DENV, in first-trimester human extravillous trophoblast cells (HTR8). DENV-RNA levels in the infected HTR8 cells were significantly enhanced on day 1 and continued to increase on day 4 and day 6 pi. On day 4 pi, IL-6, TNF-α, IL-8, and CCL2 production was augmented in ZIKV-infected HTR8 cells compared to YFV and DENV; however, DENV-infected cells produced more of these cytokines compared with the YFV-infected cells. Meanwhile, CCL3 (macrophage inflammatory protein-1 α, MIP-1α) and RANTES/CCL5 production were higher in DENV-infected cells. The IFN-alpha response was low in DENV-infected cells and the IFN-beta response was higher in DENV-infected cells compared to ZIKV-infected cells in each of the three infection times; this was also the case when compared to YFV-infected cells 6dpi [186]. The cytokine profile of DENV-infected HTR8 cells was characterized by high levels of IL-6, IL-10, IL-15, CCL2, CCL3, IL-8, VEGF, IFN-gamma, and IFN-alpha 2 [187]. In addition, DENV was shown to be able to infect other trophoblast cell lines, such as JEG3 and JAR, and promote the expression of IFNλ1 better than IFNλ2 [188]. In experiments with mice infected with DENV, the decidua exhibited a higher number of genes being upregulated, including caspase (2, 6, 8, and 9), IRF1, and NOS2. In the fetal placenta, there were expressions of complements, such as C4A, C6, and CFB [161].

Although some studies have already shown that CHIKV is able to inhibit the phosphorylation of the intermediates of the JAK-STAT cascade and, therefore, interfere with the IFN-I-mediated response [189], it is already well established that the response mediated by IFN-I has a critical role in limiting the replication and pathogenesis of CHIKV in human and mouse models and that the different subtypes of IFN-I (IFN alpha and IFN beta) play a protective role via different mechanisms [174,190,191]. While IFN alpha acts by limiting viral replication and spread, IFN beta acts by modulating neutrophil density at the site of infection, regulating inflammation during acute infection [174,192]. Furthermore, it is believed that IFN alpha somehow interferes with the chronic version of the pathology. Locke and collaborators demonstrated that early IFN alpha activity is able to limit persistent viral RNA, as well as the number of surviving immune cells, suggesting that the IFN alpha-mediated response plays a central role in the development of chronic chikungunya [177]. However, further studies are needed to clarify the role of each subtype of IFN-I in the chronic condition of the pathology.

Despite the high incidence of DENV and CHIKV infection in pregnant women, the role of IFN-I during placental viral infection is a gap in the knowledge. Studies investigating the impact of IFN-I during placental infection with DENV or CHIKV are extremely scarce. It is noteworthy that IFN-I is an essential molecule for the proper development of a pregnancy since these cytokines act in the placenta by regulating inflammation, protecting against viral infections, and contributing to fetal immunity [193,194]. Loss of an IFN-I-mediated response in the placenta can lead to a number of events, including exacerbated viral replication, fetal infection, and other factors that contribute to pregnancy complications [194–196]. Thus, the need for and urgency of carrying out studies evaluating the role of IFN-I in placental infection by DENV and CHIKV is evident.

9. Conclusions

The occurrence of arboviruses during pregnancy is an additional concern, due to the possibility of vertical transmission and fetal involvement. The various placental immune cells play a role in viral dissemination and may contribute to vertical transmission. IFN-I proteins are the main cytokine mediators of the innate immune response and constitute a key defense mechanism against viral infections. Despite the high incidence of DENV and CHIKV infections in pregnant women, the role of IFN-I during placental viral infection is a gap in knowledge and must be better studied (Figure 1). Most of the studies reported here were case studies of patients who had infections during pregnancy, some of which led to serious outcomes, such as miscarriage or maternal and fetal death. We therefore believe that infection with these arboviruses in pregnancy can be very dangerous and should be studied further.

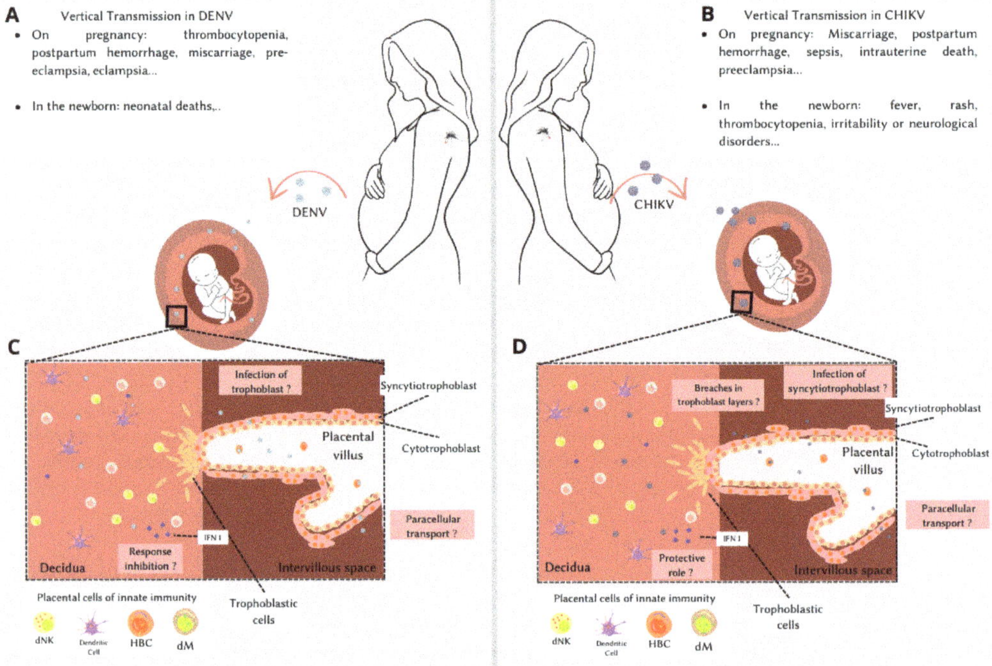

Figure 1. Schematic representation of the human maternal–fetal interface during DENV or CHIKV infection. DENV or CHIKV infection has immense potential to affect both maternal and fetal health. (**A**) During pregnancy, DENV infection can lead to thrombocytopenia, postpartum hemorrhage, miscarriage, and preeclampsia, in addition to representing an increased risk of neonatal death. (**B**) On the other hand, CHIKV infection can cause spontaneous abortion, postpartum hemorrhage, sepsis, intrauterine death, and preeclampsia and can also cause thrombocytopenia, fever, rash, irritability, and neurological disorders in the newborn. In the basal decidua are cells of the immune system: decidual natural killer (dNK) cells, dendritic cells, and maternal macrophages (dM). Chorionic villi contain trophoblast cells, Hofbauer cells (HBC), and fetal capillaries surrounded by a layer of cytotrophoblasts and multinucleated syncytiotrophoblast cells. The chorionic villus is floating in the intervillous space, bathed in maternal blood. So far, the mechanism involved in the vertical transmission of both viruses remains unclear. It is believed that vertical transmission can occur via the direct infection of trophoblasts (**C**) or syncytiotrophoblasts (**D**), as well as from breaches on the trophoblast layer (**D**) or via paracellular transport (**C,D**) from maternal blood to the fetal capillaries. The role of decidual immune system cells during DENV or CHIKV infection is not well established and nor is the IFN-I-mediated response, representing a gap in knowledge.

Author Contributions: Conceptualization, K.R. and N.G.S.; investigation and writing, F.d.A.V.A., P.C.G.N., L.V.A., N.G.S. and K.R.; supervision, K.R. and N.G.S. All authors have read and agreed to the published version of the manuscript.

Funding: This research was funded by Oswaldo Cruz Institute (Fiocruz) and Roberto Alcantara Gomes Biology Institute (UERJ).

Conflicts of Interest: The authors declare no conflict of interest.

References

1.   Young, P.R. Arboviruses: A Family on the Move. *Adv. Exp. Med. Biol.* **2018**, *1062*, 1–10. [CrossRef]
2.   Guzman, M.G.; Harris, E. Dengue. *Lancet* **2015**, *385*, 453–465. [CrossRef] [PubMed]
3.   Vairo, F.; Haider, N.; Kock, R.; Ntoumi, F.; Ippolito, G.; Zumla, A. Chikungunya: Epidemiology, Pathogenesis, Clinical Features, Management, and Prevention. *Infect. Dis. Clin. N. Am.* **2019**, *33*, 1003–1025. [CrossRef]
4.   Uno, N.; Ross, T.M. Dengue Virus and the Host Innate Immune Response. *Emerg. Microbes Infect.* **2018**, *7*, 167. [CrossRef]
5.   Fernandes-Santos, C.; de Azeredo, E.L. Innate Immune Response to Dengue Virus: Toll-like Receptors and Antiviral Response. *Viruses* **2022**, *14*, 992. [CrossRef]
6.   Tanabe, I.S.B.; Tanabe, E.L.L.; Santos, E.C.; Martins, W.V.; Araújo, I.M.T.C.; Cavalcante, M.C.A.; Lima, A.R.V.; Câmara, N.O.S.; Anderson, L.; Yunusov, D.; et al. Cellular and Molecular Immune Response to Chikungunya Virus Infection. *Front. Cell. Infect. Microbiol.* **2018**, *8*, 345. [CrossRef] [PubMed]
7.   Gubler, D.J. Dengue and Dengue Hemorrhagic Fever. *Clin. Microbiol. Rev.* **1998**, *11*, 480–496. [CrossRef] [PubMed]
8.   Gubler, D.J.; Clark, G.G. Dengue/Dengue Hemorrhagic Fever: The Emergence of a Global Health Problem. *Emerg. Infect. Dis.* **1995**, *1*, 55–57. [CrossRef] [PubMed]
9.   Hotta, S. Experimental Studies on Dengue. I. Isolation, Identification and Modification of the Virus. *J. Infect. Dis.* **1952**, *90*, 1–9. [CrossRef]
10.   Silva, N.M.; Santos, N.C.; Martins, I.C. Dengue and Zika Viruses: Epidemiological History, Potential Therapies, and Promising Vaccines. *Trop. Med. Infect. Dis.* **2020**, *5*, 150. [CrossRef]
11.   Valle, D.; Pimenta, D.N.; Da Cunha, R.V. *Dengue: Teorias E Práticas*; Editora Fiocruz: Rio de Janeiro, Brazil, 2015; ISBN 9788575415528.
12.   World Health Organization. *Global Strategy for Dengue Prevention and Control, 2012–2020*; World Health Organization: Geneva, Switzerland, 2012.
13.   Jing, Q.; Wang, M. Dengue Epidemiology. *Glob. Health J.* **2019**, *3*, 37–45. [CrossRef]
14.   World Health Organization. *Dengue: Guidelines for Diagnosis, Treatment, Prevention and Control*; World Health Organization: Geneva, Switzerland, 2009; ISBN 9789241547871.
15.   Chambers, T.J.; Hahn, C.S.; Galler, R.; Rice, C.M. Flavivirus genome organization, expression, and replication. *Annu. Rev. Microbiol.* **1990**, *44*, 649–688. [CrossRef] [PubMed]
16.   Zhang, Y.; Corver, J.; Chipman, P.R.; Zhang, W.; Pletnev, S.V.; Sedlak, D.; Baker, T.S.; Strauss, J.H.; Kuhn, R.J.; Rossman, M.G. Structures of Immature Flavivirus Particles. *EMBO J.* **2003**, *22*, 2604–2613. [CrossRef] [PubMed]
17.   Mukhopadhyay, S.; Kuhn, R.J.; Rossmann, M.G. A Structural Perspective of the Flavivirus Life Cycle. *Nat. Rev. Microbiol.* **2005**, *3*, 13–22. [CrossRef] [PubMed]
18.   Lindenbach, B.D.; Rice, C.M. Molecular Biology of Flaviviruses. *Adv. Virus Res.* **2003**, *59*, 23–61. [CrossRef]
19.   Lindenbach, B.D.; Thiel, H.J.; Rice, C.M. Flaviviridae: The Viruses and Their Replication. In *Fields Virology*; Knipe, D.M., Howley, P.M., Eds.; Lippincott Williams and Wilkins: Philadelphia, PA, USA, 2007; pp. 1102–1152.
20.   Sharma, A.; Gupta, S.P. Fundamentals of Viruses and Their Proteases. In *Viral Proteases and Their Inhibitors*; Academic Press: Cambridge, MA, USA, 2017; pp. 1–24. [CrossRef]
21.   Nowee, G.; Bakker, J.W.; Geertsema, C.; Ros, V.I.D.; Göertz, G.P.; Fros, J.J.; Pijlman, G.P. A Tale of 20 Alphaviruses; Inter-Species Diversity and Conserved Interactions between Viral Non-Structural Protein 3 and Stress Granule Proteins. *Front. Cell Dev. Biol.* **2021**, *9*, 625711. [CrossRef] [PubMed]
22.   Kim, D.Y.; Reynaud, J.M.; Rasalouskaya, A.; Akhrymuk, I.; Mobley, J.A.; Frolov, I.; Frolova, E.I. New World and Old World Alphaviruses Have Evolved to Exploit Different Components of Stress Granules, FXR and G3BP Proteins, for Assembly of Viral Replication Complexes. *PLoS Pathog.* **2016**, *12*, e1005810. [CrossRef] [PubMed]
23.   Ojeda Rodriguez, J.A.; Haftel, A.; Walker, J.R., III. *Chikungunya Fever*; StatPearls: Treasure Island, FL, USA, 2023. Available online: https://www.ncbi.nlm.nih.gov/books/NBK534224/ (accessed on 26 June 2023).
24.   Powers, A.M.; Brault, A.C.; Tesh, R.B.; Weaver, S.C. Re-Emergence of Chikungunya and O'nyong-Nyong Viruses: Evidence for Distinct Geographical Lineages and Distant Evolutionary Relationships. *Microbiology* **2000**, *81*, 471–479. [CrossRef]
25.   Khan, A.H.; Morita, K.; Parquet, M.d.C.; Hasebe, F.; Mathenge, E.G.M.; Igarashi, A. Complete Nucleotide Sequence of Chikungunya Virus and Evidence for an Internal Polyadenylation Site. *J. Gen. Virol.* **2002**, *83*, 3075–3084. [CrossRef] [PubMed]
26.   Schuffenecker, I.; Iteman, I.; Michault, A.; Murri, S.; Frangeul, L.; Vaney, M.-C.; Lavenir, R.; Pardigon, N.; Reynes, J.-M.; Pettinelli, F.; et al. Genome Microevolution of Chikungunya Viruses Causing the Indian Ocean Outbreak. *PLoS Med.* **2006**, *3*, e263. [CrossRef]
27.   Strauss, J.H.; Strauss, E.G. The Alphaviruses: Gene Expression, Replication, and Evolution. *Microbiol. Rev.* **1994**, *58*, 491–562. [CrossRef] [PubMed]
28.   Jin, J.; Simmons, G. Antiviral Functions of Monoclonal Antibodies against Chikungunya Virus. *Viruses* **2019**, *11*, 305. [CrossRef] [PubMed]
29.   Guzman, M.G.; Gubler, D.J.; Izquierdo, A.; Martinez, E.; Halstead, S.B. Dengue Infection. *Nat. Rev. Dis. Primers* **2016**, *2*, 16055. [CrossRef] [PubMed]

30. Kraemer, M.U.G.; Reiner, R.C.; Brady, O.J.; Messina, J.P.; Gilbert, M.; Pigott, D.M.; Yi, D.; Johnson, K.; Earl, L.; Marczak, L.B.; et al. Past and Future Spread of the Arbovirus Vectors Aedes Aegypti and Aedes Albopictus. *Nat. Microbiol.* **2019**, *4*, 854–863. [CrossRef] [PubMed]

31. Tambyah, P.A.; Koay, E.S.C.; Poon, M.L.M.; Lin, R.V.T.P.; Ong, B.K.C. Dengue Hemorrhagic Fever Transmitted by Blood Transfusion. *N. Engl. J. Med.* **2008**, *359*, 1526–1527. [CrossRef] [PubMed]

32. Tan, F.L.-S.; Loh, D.L.S.K.; Prabhakaran, K. Dengue Haemorrhagic Fever after Living Donor Renal Transplantation. *Nephrol. Dial. Transplant.* **2005**, *20*, 447–448. [CrossRef] [PubMed]

33. Chye, J.K.; Lim, C.T.; Ng, K.B.; Lim, J.M.H.; George, R.; Lam, S.K. Vertical Transmission of Dengue. *Clin. Infect. Dis.* **1997**, *25*, 1374–1377. [CrossRef] [PubMed]

34. World Health Organization. *Dengue Haemorrhagic Fever: Diagnosis, Treatment, Prevention, and Control*; World Health Organization: Geneva, Switzerland, 1997; ISBN 9789241545006.

35. Ross, T.M. Dengue Virus. *Clin. Lab. Med.* **2010**, *30*, 149–160. [CrossRef]

36. Guzmán, M.G.; Kourí, G. Dengue Haemorrhagic Fever Integral Hypothesis: Confirming Observations, 1987–2007. *Trans. R. Soc. Trop. Med. Hyg.* **2008**, *102*, 522–523. [CrossRef]

37. Horstick, O.; Jaenisch, T.; Martinez, E.; Kroeger, A.; Lum, L.; Farrar, J.; Ranzinger, S.R. Comparing the Usefulness of the 1997 and 2009 WHO Dengue Case Classification: A Systematic Literature Review. *Am. J. Trop. Med. Hyg.* **2014**, *91*, 621–634. [CrossRef]

38. Dhanoa, A.; Hassan, S.S.; Ngim, C.F.; Lau, C.F.; Chan, T.S.; Adnan, N.A.A.; Eng, W.W.H.; Gan, H.M.; Rajasekaram, G. Impact of Dengue Virus (DENV) Co-Infection on Clinical Manifestations, Disease Severity and Laboratory Parameters. *BMC Infect. Dis.* **2016**, *16*, 406. [CrossRef] [PubMed]

39. Verdeal, J.C.R.; Costa Filho, R.; Vanzillotta, C.; de Macedo, G.L.; Bozza, F.A.; Toscano, L.; Prata, A.; Tanner, A.C.; Machado, F.R. Recomendações Para O Manejo de Pacientes Com Formas Graves de Dengue. *Rev. Bras. Ter. Intensiv.* **2011**, *23*, 125–133. [CrossRef]

40. Leowattana, W.; Leowattana, T. Dengue Hemorrhagic Fever and the Liver. *World J. Hepatol.* **2021**, *13*, 1968–1976. [CrossRef] [PubMed]

41. Huerre, M.R.; Trong Lan, N.; Marianneau, P.; Bac Hue, N.; Khun, H.; Thanh Hung, N.; Thi Khen, N.; Drouet, M.T.; Que Huong, V.T.; Quang Ha, D.; et al. Liver Histopathology and Biological Correlates in Five Cases of Fatal Dengue Fever in Vietnamese Children. *Virchow's Arch.* **2001**, *438*, 107–115. [CrossRef] [PubMed]

42. Mohan, B.; Patwari, A.K.; Anand, V.K. Brief Report. Hepatic Dysfunction in Childhood Dengue Infection. *J. Trop. Pediatr.* **2000**, *46*, 40–43. [CrossRef]

43. Burke, T. Dengue Haemorrhagic Fever: A Pathological Study. *Trans. R. Soc. Trop. Med. Hyg.* **1968**, *62*, 682–692. [CrossRef]

44. Póvoa, T.F.; Alves, A.M.B.; Oliveira, C.A.B.; Nuovo, G.J.; Chagas, V.L.A.; Paes, M.V. The Pathology of Severe Dengue in Multiple Organs of Human Fatal Cases: Histopathology, Ultrastructure and Virus Replication. *PLoS ONE* **2014**, *9*, e83386. [CrossRef]

45. Trung, D.T.; Thao, L.T.T.; Vinh, N.N.; Hien, T.T.; Simmons, C.; Hien, P.T.D.; Wills, B.; Chinh, N.T.; Hung, N.T. Liver Involvement Associated with Dengue Infection in Adults in Vietnam. *Am. J. Trop. Med. Hyg.* **2010**, *83*, 774–780. [CrossRef]

46. Jessie, K.; Fong, M.; Devi, S.; Lam, S.; Wong, K. Thong Localization of Dengue Virus in Naturally Infected Human Tissues, by Immunohistochemistry and in Situ Hybridization. *J. Infect. Dis.* **2004**, *189*, 1411–1418. [CrossRef]

47. Basílio-de-Oliveira, C.A.; Aguiar, G.R.; Baldanza, M.S.; Barth, O.M.; Eyer-Silva, W.A.; Paes, M.V. Pathologic Study of a Fatal Case of Dengue-3 Virus Infection in Rio de Janeiro, Brazil. *Braz. J. Infect. Dis.* **2005**, *9*, 341–347. [CrossRef]

48. Salgado, D.M.; Eltit, J.M.; Mansfield, K.; Panqueba, C.; Castro, D.; Vega, M.R.; Xhaja, K.; Schmidt, D.; Martin, K.J.; Allen, P.D.; et al. Heart and Skeletal Muscle Are Targets of Dengue Virus Infection. *Pediatr. Infect. Dis. J.* **2010**, *29*, 238–242. [CrossRef] [PubMed]

49. Balsitis, S.J.; Flores, D.; Coloma, J.; Beatty, P.R.; Alava, A.; McKerrow, J.H.; Castro, G.; Harris, E. Tropism of Dengue Virus in Mice and Humans Defined by Viral Nonstructural Protein 3-Specific Immunostaining. *Am. J. Trop. Med. Hyg.* **2009**, *80*, 416–424. [CrossRef]

50. Oliveira, L.d.L.S.; Alves, F.d.A.V.; Rabelo, K.; Moragas, L.J.; Mohana-Borges, R.; de Carvalho, J.J.; Basílio-de-Oliveira, C.; Basílio-de-Oliveira, R.P.; Rosman, F.C.; Salomão, N.G.; et al. Immunopathology of Renal Tissue in Fatal Cases of Dengue in Children. *Pathogens* **2022**, *11*, 1543. [CrossRef] [PubMed]

51. Alves, F.d.A.V.; Oliveira, L.d.L.S.; Salomão, N.G.; Provance, D.W.; Basílio-de-Oliveira, C.A.; Basílio-de-Oliveira, R.P.; Moragas, L.J.; de Carvalho, J.J.; Mohana-Borges, R.; Rabelo, K.; et al. Cytokines and Inflammatory Mediators: Markers Involved in Interstitial Damage to the Pancreas in Two Dengue Fever Cases Associated with Acute Pancreatitis. *PLoS ONE* **2022**, *17*, e0262785. [CrossRef] [PubMed]

52. Nunes, P.; Nogueira, R.; Coelho, J.; Rodrigues, F.; Salomão, N.; José, C.; de Carvalho, J.; Rabelo, K.; de Azeredo, E.; Basílio-de-Oliveira, R.; et al. A Stillborn Multiple Organs' Investigation from a Maternal DENV-4 Infection: Histopathological and Inflammatory Mediators Characterization. *Viruses* **2019**, *11*, 319. [CrossRef] [PubMed]

53. Powers, A.M.; Logue, C.H. Changing Patterns of Chikungunya Virus: Re-Emergence of a Zoonotic Arbovirus. *J. Gen. Virol.* **2007**, *88*, 2363–2377. [CrossRef]

54. Martins, E.B.; Silva, M.F.B.; Tassinari, W.S.; de Bruycker-Nogueira, F.; Moraes, I.C.V.; Rodrigues, C.D.S.; Santos, C.C.; Sampaio, S.A.; Pina-Costa, A.; Fabri, A.A.; et al. Detection of Chikungunya Virus in Bodily Fluids: The INOVACHIK Cohort Study. *PLoS Negl. Trop. Dis.* **2022**, *16*, e0010242. [CrossRef] [PubMed]

55. Schwartz, O.; Albert, M.L. Biology and Pathogenesis of Chikungunya Virus. *Nat. Rev. Microbiol.* **2010**, *8*, 491–500. [CrossRef]

56. Robinson, M.C. An epidemic of virus disease in Southern Province, Tanganyika Territory, in 1952–1953 I. Clinical features. *Trans. R. Soc. Trop. Med. Hyg.* **1955**, *49*, 28–32. [CrossRef] [PubMed]
57. Mavalankar, D.; Shastri, P.; Bandyopadhyay, T.; Parmar, J.; Ramani, K.V. Increased Mortality Rate Associated with Chikungunya Epidemic, Ahmedabad, India. *Emerg. Infect. Dis.* **2008**, *14*, 412–415. [CrossRef]
58. Simon, F.; Savini, H.; Parola, P. Chikungunya: A Paradigm of Emergence and Globalization of Vector-Borne Diseases. *Med. Clin. N. Am.* **2008**, *92*, 1323–1343. [CrossRef] [PubMed]
59. Mohan, A.; Kiran, D.H.N.; Manohar, C.; Kumar, P. Epidemiology, Clinical Manifestations, and Diagnosis of Chikungunya Fever: Lessons Learned from the Re-Emerging Epidemic. *Indian J. Dermatol.* **2010**, *55*, 54. [CrossRef]
60. Kumar, R.; Sharma, M.; Jain, S.; Yadav, S.; Singhal, A.K. Cutaneous Manifestations of Chikungunya Fever: Observations from an Outbreak at a Tertiary Care Hospital in Southeast Rajasthan, India. *Indian Dermatol. Online J.* **2017**, *8*, 336. [CrossRef]
61. Bandyopadhyay, D.; Ghosh, S. Mucocutaneous Manifestations of Chikungunya Fever. *Indian J. Dermatol.* **2010**, *55*, 64. [CrossRef] [PubMed]
62. Chandrashekar, L.; Singh, N.; Konda, D.; Thappa, D.; Srinivas, B.; Dhodapkar, R. Vesiculobullous Viral Exanthem due to Chikungunya in an Infant. *Indian Dermatol. Online J.* **2014**, *5*, 119. [CrossRef] [PubMed]
63. Villamil-Gómez, W.E.; Alba-Silvera, L.; Menco-Ramos, A.; Gonzalez-Vergara, A.; Molinares-Palacios, T.; Barrios-Corrales, M.; Rodriguez-Morales, A.J. Congenital Chikungunya Virus Infection in Sincelejo, Colombia: A Case Series. *J. Trop. Pediatr.* **2015**, *61*, 386–392. [CrossRef]
64. Duarte, M.d.C.M.B.; de Oliveira Neto, A.F.; Bezerra, P.G.d.M.; Cavalcanti, L.A.; Silva, V.M.d.B.; de Abreu, S.G.A.A.; Leite, S.F.B.; Cavalcanti, N.V. Chikungunya Infection in Infants. *Rev. Bras. Saúde Matern. Infant.* **2016**, *16*, S63–S71. [CrossRef]
65. Farias, L.A.B.G.; Pires Neto, R.D.J.; Leite, R.D. Vesiculobullous Exanthema in a 3-Month-Old Child with Probable Acute Chikungunya Infection. *J. Health Biol. Sci.* **2019**, *7*, 429. [CrossRef]
66. Simon, F.; Javelle, E.; Cabie, A.; Bouquillard, E.; Troisgros, O.; Gentile, G.; Leparc-Goffart, I.; Hoen, B.; Gandjbakhch, F.; Rene-Corail, P.; et al. French Guidelines for the Management of Chikungunya (Acute and Persistent Presentations). *Médecine Mal. Infect.* **2015**, *45*, 243–263. [CrossRef]
67. de Lima, S.T.S.; de Souza, W.M.; Cavalcante, J.W.; da Silva Candido, D.; Fumagalli, M.J.; Carrera, J.-P.; Simões Mello, L.M.; De Carvalho Araújo, F.M.; Cavalcante Ramalho, I.L.; de Almeida Barreto, F.K.; et al. Fatal Outcome of Chikungunya Virus Infection in Brazil. *Clin. Infect. Dis.* **2020**, *73*, e2436–e2443. [CrossRef]
68. Economopoulou, A.; Dominguez, M.; Helynck, B.; Sissoko, D.; Wichmann, O.; Quenel, P.; Germonneau, P.; Quatresous, I. Atypical Chikungunya Virus Infections: Clinical Manifestations, Mortality and Risk Factors for Severe Disease during the 2005–2006 Outbreak on Réunion. *Epidemiol. Infect.* **2008**, *137*, 534–541. [CrossRef] [PubMed]
69. Rajapakse, S.; Rodrigo, C.; Rajapakse, A. Atypical Manifestations of Chikungunya Infection. *Trans. R. Soc. Trop. Med. Hyg.* **2010**, *104*, 89–96. [CrossRef]
70. Deeba, I.M.; Hasan, M.; Mosabbir, A.A.; Banna, H.; Islam, M.N.; Raheem, E.; Hossain, M.M. Manifestations of Atypical Symptoms of Chikungunya during the Dhaka Outbreak (2017) in Bangladesh. *Am. J. Trop. Med. Hyg.* **2019**, *100*, 1545–1548. [CrossRef] [PubMed]
71. Renault, P.; de Valk, H.; Balleydier, E.; Solet, J.-L.; Ilef, D.; Rachou, E.; Larrieu, S.; Ledrans, M.; Filleul, L.; Quatresous, I.; et al. A Major Epidemic of Chikungunya Virus Infection on Réunion Island, France, 2005–2006. *Am. J. Trop. Med. Hyg.* **2007**, *77*, 727–731. [CrossRef]
72. Da Cunha, R.V.; Trinta, K.S. Chikungunya Virus: Clinical Aspects and Treatment—A Review. *Memórias Inst. Oswaldo Cruz* **2017**, *112*, 523–531. [CrossRef] [PubMed]
73. Bonn, D. How Did Chikungunya Reach the Indian Ocean? *Lancet Infect. Dis.* **2006**, *6*, 543. [CrossRef]
74. Tsetsarkin, K.A.; Vanlandingham, D.L.; McGee, C.E.; Higgs, S. A Single Mutation in Chikungunya Virus Affects Vector Specificity and Epidemic Potential. *PLoS Pathog.* **2007**, *3*, e201. [CrossRef] [PubMed]
75. Martin, B.M.; Evans, A.A.; de Carvalho, D.S.; Shimakura, S.E. Clinical Outcomes of Dengue Virus Infection in Pregnant and Non-Pregnant Women of Reproductive Age: A Retrospective Cohort Study from 2016 to 2019 in Paraná, Brazil. *BMC Infect. Dis.* **2022**, *22*, 5. [CrossRef]
76. Escobar, M.; Nieto, A.J.; Loaiza-Osorio, S.; Barona, J.S.; Rosso, F. Pregnant Women Hospitalized with Chikungunya Virus Infection, Colombia, 2015. *Emerg. Infect. Dis.* **2017**, *23*, 1777–1783. [CrossRef]
77. Salomão, N.G.; Rabelo, K.; Avvad-Portari, E.; Basílio-de-Oliveira, C.; Basílio-de-Oliveira, R.P.; Ferreira, F.; Ferreira, L.; de Souza, T.M.; Nunes, P.; Lima, M.F.; et al. Histopathological and Immunological Characteristics of Placentas Infected with Chikungunya Virus. *Front. Microbiol.* **2022**, *13*, 1055536. [CrossRef]
78. Salomão, N.G.; Araújo, L.R.; Rabelo, K.; Avvad-Portari, E.; de Souza, L.J.; Maria, R.; Valle, N.; Romanholo, F.; Basílio-de-Oliveira, C.A.; Basílio-de-Oliveira, R.P.; et al. Placental Alterations in a Chikungunya-Virus-Infected Pregnant Woman: A Case Report. *Microorganisms* **2022**, *10*, 872. [CrossRef]
79. Lyra, P.P.; Campos, G.S.; Bandeira, I.D.; Sardi, S.I.; Costa, L.F.d.M.; Santos, F.R.; Ribeiro, C.A.S.; Jardim, A.M.B.; Santiago, A.C.T.; Oliveira, P.M.R.; et al. Congenital Chikungunya Virus Infection after an Outbreak in Salvador, Bahia, Brazil. *AJP Rep.* **2016**, *6*, e299–e300. [CrossRef] [PubMed]

80. Arragain, L.; Dupont-Rouzeyrol, M.; O'Connor, O.; Sigur, N.; Grangeon, J.-P.; Huguon, E.; Dechanet, C.; Cazorla, C.; Gourinat, A.-C.; Descloux, E. Vertical Transmission of Dengue Virus in the Peripartum Period and Viral Kinetics in Newborns and Breast Milk: New Data. *J. Pediatr. Infect. Dis. Soc.* **2016**, *6*, 324–331. [CrossRef]
81. Vats, A.; Ho, T.-C.; Puc, I.; Chen, Y.-J.; Chang, C.-H.; Chien, Y.-W.; Perng, G.-C. Evidence That Hematopoietic Stem Cells in Human Umbilical Cord Blood Is Infectable by Dengue Virus: Proposing a Vertical Transmission Candidate. *Heliyon* **2021**, *7*, e06785. [CrossRef] [PubMed]
82. Watson, E.D.; Cross, J.C. Development of Structures and Transport Functions in the Mouse Placenta. *Physiology* **2005**, *20*, 180–193. [CrossRef]
83. Gude, N.M.; Roberts, C.T.; Kalionis, B.; King, R.G. Growth and Function of the Normal Human Placenta. *Thromb. Res.* **2004**, *114*, 397–407. [CrossRef] [PubMed]
84. Myatt, L. Placental Adaptive Responses and Fetal Programming. *J. Physiol.* **2006**, *572*, 25–30. [CrossRef] [PubMed]
85. Ander, S.E.; Diamond, M.S.; Coyne, C.B. Immune Responses at the Maternal-Fetal Interface. *Sci. Immunol.* **2019**, *4*, eaat6114. [CrossRef] [PubMed]
86. Ji, L.; Brkić, J.; Liu, M.; Fu, G.; Peng, C.; Wang, Y.-L. Placental Trophoblast Cell Differentiation: Physiological Regulation and Pathological Relevance to Preeclampsia. *Mol. Asp. Med.* **2013**, *34*, 981–1023. [CrossRef] [PubMed]
87. Zaga-Clavellina, V.; Diaz, L.; Olmos-Ortiz, A.; Godínez-Rubí, M.; Rojas-Mayorquín, A.E.; Ortuño-Sahagún, D. Central Role of the Placenta during Viral Infection: Immuno-Competences and MiRNA Defensive Responses. *Biochim. Biophys. Acta Mol. Basis Dis.* **2021**, *1867*, 166182. [CrossRef]
88. Olmos-Ortiz, A.; Flores-Espinosa, P.; Mancilla-Herrera, I.; Vega-Sánchez, R.; Díaz, L.; Zaga-Clavellina, V. Innate Immune Cells and Toll-like Receptor–Dependent Responses at the Maternal–Fetal Interface. *Int. J. Mol. Sci.* **2019**, *20*, 3654. [CrossRef]
89. Pazos, M.; Sperling, R.S.; Moran, T.M.; Kraus, T.A. The Influence of Pregnancy on Systemic Immunity. *Immunol. Res.* **2012**, *54*, 254–261. [CrossRef] [PubMed]
90. Cornish, E.F.; Filipovic, I.; Åsenius, F.; Williams, D.J.; McDonnell, T. Innate Immune Responses to Acute Viral Infection during Pregnancy. *Front. Immunol.* **2020**, *11*, 572567. [CrossRef] [PubMed]
91. Rowe, J.H.; Ertelt, J.M.; Xin, L.; Way, S.S. Pregnancy Imprints Regulatory Memory That Sustains Anergy to Fetal Antigen. *Nature* **2012**, *490*, 102–106. [CrossRef]
92. Jabrane-Ferrat, N.; Siewiera, J. The up Side of Decidual Natural Killer Cells: New Developments in Immunology of Pregnancy. *Immunology* **2014**, *141*, 490–497. [CrossRef] [PubMed]
93. Hanna, J.; Goldman-Wohl, D.; Hamani, Y.; Avraham, I.; Greenfield, C.; Natanson-Yaron, S.; Prus, D.; Cohen-Daniel, L.; Arnon, T.I.; Manaster, I.; et al. Decidual NK Cells Regulate Key Developmental Processes at the Human Fetal-Maternal Interface. *Nat. Med.* **2006**, *12*, 1065–1074. [CrossRef]
94. Manaster, I.; Mandelboim, O. The Unique Properties of Uterine NK Cells. *Am. J. Reprod. Immunol.* **2010**, *63*, 434–444. [CrossRef]
95. Barrientos, G.; Tirado-González, I.; Freitag, N.; Kobelt, P.; Moschansky, P.; Klapp, B.F.; Thijssen, V.L.J.L.; Blois, S.M. CXCR4+ Dendritic Cells Promote Angiogenesis during Embryo Implantation in Mice. *Angiogenesis* **2012**, *16*, 417–427. [CrossRef] [PubMed]
96. Liu, S.; Diao, L.; Huang, C.; Li, Y.; Zeng, Y.; Kwak-Kim, J.Y.H. The Role of Decidual Immune Cells on Human Pregnancy. *J. Reprod. Immunol.* **2017**, *124*, 44–53. [CrossRef] [PubMed]
97. Bulmer, J.N.; Williams, P.J.; Lash, G.E. Immune Cells in the Placental Bed. *Int. J. Dev. Biol.* **2010**, *54*, 281–294. [CrossRef]
98. Vishnyakova, P.; Elchaninov, A.; Fatkhudinov, T.; Sukhikh, G. Role of the Monocyte–Macrophage System in Normal Pregnancy and Preeclampsia. *Int. J. Mol. Sci.* **2019**, *20*, 3695. [CrossRef] [PubMed]
99. Arruda, L.V.; Salomão, N.G.; Alves, F.d.A.V.; Rabelo, K. The Innate Defense in the Zika-Infected Placenta. *Pathogens* **2022**, *11*, 1410. [CrossRef] [PubMed]
100. Freitag, N.; Zwier, M.V.; Barrientos, G.; Tirado-Gonzalez, I.; Conrad, M.L.; Rose, M.; Scherjon, S.A.; Plösch, T.; Blois, S.M. Influence of Relative NK–DC Abundance on Placentation and Its Relation to Epigenetic Programming in the Offspring. *Cell Death Dis.* **2014**, *5*, e1392. [CrossRef]
101. Yong, H.E.J.; Chan, S.-Y.; Chakraborty, A.; Rajaraman, G.; Ricardo, S.D.; Benharouga, M.; Alfaidy, N.; Staud, F.; Wallace, E.M. Significance of the Placental Barrier in Antenatal Viral Infections. *Biochim. Biophys. Acta Mol. Basis Dis.* **2021**, *1867*, 166244. [CrossRef] [PubMed]
102. Taguchi, T.; Mukai, K. Innate Immunity Signalling and Membrane Trafficking. *Curr. Opin. Cell Biol.* **2019**, *59*, 1–7. [CrossRef] [PubMed]
103. Basurko, C.; Carles, G.; Youssef, M.; Guindi, W.E.L. Maternal and Foetal Consequences of Dengue Fever during Pregnancy. *Eur. J. Obstet. Gynecol. Reprod. Biol.* **2009**, *147*, 29–32. [CrossRef]
104. do Nascimento, L.B.; Siqueira, C.M.; Coelho, G.E.; Siqueira, J.B. Dengue in Pregnant Women: Characterization of Cases in Brazil, 2007–2015. *Epidemiol. Serviços Saúde* **2017**, *26*, 433–442. [CrossRef]
105. Paixão, E.S.; Teixeira, M.G.; Costa, M.d.C.N.; Rodrigues, L.C. Dengue during Pregnancy and Adverse Fetal Outcomes: A Systematic Review and Meta-Analysis. *Lancet Infect. Dis.* **2016**, *16*, 857–865. [CrossRef]
106. Paixão, E.S.; Costa, M.d.C.N.; Rodrigues, L.C.; Rasella, D.; Cardim, L.L.; Brasileiro, A.C.; Teixeira, M.G.L.C. Trends and Factors Associated with Dengue Mortality and Fatality in Brazil. *Rev. Soc. Bras. Med. Trop.* **2015**, *48*, 399–405. [CrossRef]
107. Paixao, E.S.; Harron, K.; Campbell, O.; Teixeira, M.G.; Costa, M.d.C.N.; Barreto, M.L.; Rodrigues, L.C. Dengue in Pregnancy and Maternal Mortality: A Cohort Analysis Using Routine Data. *Sci. Rep.* **2018**, *8*, 9938. [CrossRef]

108. Fatimil, L.E.; Mollah, A.H.; Ahmed, S.; Rahman, M. Vertical Transmission of Dengue: First Case Report from Bangladesh. *Southeast Asian J. Trop. Med. Public Health* **2003**, *34*, 800–803.
109. Carles, G.; Talarmin, A.; Peneau, C.; Bertsch, M. Dengue Fever and Pregnancy. A Study of 38 Cases in French Guiana. *J. Gynecol. Obstet. Biol. Reprod.* **2000**, *29*, 758–762.
110. Tan, P.C.; Rajasingam, G.; Devi, S.; Omar, S.Z. Dengue Infection in Pregnancy. *Obstet. Gynecol.* **2008**, *111*, 1111–1117. [CrossRef]
111. Phongsamart, W.; Yoksan, S.; Vanaprapa, N.; Chokephaibulkit, K. Dengue Virus Infection in Late Pregnancy and Transmission to the Infants. *Pediatr. Infect. Dis. J.* **2008**, *27*, 500–504. [CrossRef]
112. Sirinavin, S.; Nuntnarumit, P.; Supapannachart, S.; Boonkasidecha, S.; Techasaensiri, C.; Yoksarn, S. Vertical Dengue Infection. *Pediatr. Infect Dis. J.* **2004**, *23*, 1042–1047. [CrossRef] [PubMed]
113. Kariyawasam, S.; Senanayake, H. Dengue Infections during Pregnancy: Case Series from a Tertiary Care Hospital in Sri Lanka. *J. Infect. Dev. Ctries.* **2010**, *4*, 767–775. [CrossRef]
114. Ribeiro, C.F.; Lopes, V.G.S.; Brasil, P.; Coelho, J.; Muniz, A.G.; Nogueira, R.M.R. Perinatal Transmission of Dengue: A Report of 7 Cases. *J. Pediatr.* **2013**, *163*, 1514–1516. [CrossRef] [PubMed]
115. Feitoza, H.A.C.; Koifman, S.; Koifman, R.J.; Saraceni, V. Os Efeitos Maternos, Fetais e Infantis Decorrentes da Infecção por Dengue Durante a Gestação em Rio Branco, Acre, Brasil, 2007–2012. *Cad. Saúde Pública* **2017**, *33*, 1–11. [CrossRef] [PubMed]
116. Brar, R.; Sikka, P.; Suri, V.; Singh, M.P.; Suri, V.; Mohindra, R.; Biswal, M. Maternal and Fetal Outcomes of Dengue Fever in Pregnancy: A Large Prospective and Descriptive Observational Study. *Arch. Gynecol. Obstet.* **2021**, *304*, 91–100. [CrossRef]
117. Machain-Williams, C.; Raga, E.; Baak-Baak, C.M.; Kiem, S.; Blitvich, B.J.; Ramos, C. Maternal, Fetal, and Neonatal Outcomes in Pregnant Dengue Patients in Mexico. *BioMed Res. Int.* **2018**, *2018*, e9643083. [CrossRef] [PubMed]
118. Tien Dat, T.; Kotani, T.; Yamamoto, E.; Shibata, K.; Moriyama, Y.; Tsuda, H.; Yamashita, M.; Kajiyama, H.; Duc Thien Minh, D.; Quang Thanh, L.; et al. Dengue Fever during Pregnancy. *Nagoya J. Med. Sci.* **2018**, *80*, 241–247. [CrossRef]
119. Figueiredo, L.T.M.; Carlucci, R.H.; Duarte, G. A Prospective Study with Children Whose Mothers Had Dengue during Pregnancy. *Rev. Inst. Med. Trop. São Paulo* **1994**, *36*, 417–421. [CrossRef] [PubMed]
120. Kliks, S.C.; Nimmanitya, S.; Burke, D.S.; Nisalak, A. Evidence That Maternal Dengue Antibodies Are Important in the Development of Dengue Hemorrhagic Fever in Infants. *Am. J. Trop. Med. Hyg.* **1988**, *38*, 411–419. [CrossRef] [PubMed]
121. Maroun, S.L.C.; Marliere, R.C.C.; Barcellus, R.C.; Barbosa, C.N.; Ramos, J.R.M.; Moreira, M.E.L. Case Report: Vertical Dengue Infection. *J. Pediatr.* **2008**, *84*, 556–559. [CrossRef]
122. Da Mota, A.K.M.; Miranda Filho, A.L.; Saraceni, V.; Koifman, S. Mortalidade Materna E Incidência de Dengue Na Região Sudeste Do Brasil: Estudo Ecológico No Período 2001-2005. *Cad. Saúde Pública* **2012**, *28*, 1057–1066. [CrossRef]
123. Bingham, J.; Chauhan, S.P.; Hayes, E.; Gherman, R.; Lewis, D. Recurrent Shoulder Dystocia: A Review. *Obstet. Gynecol. Surv.* **2010**, *65*, 183–188. [CrossRef] [PubMed]
124. Basurko, C.; Matheus, S.; Hildéral, H.; Everhard, S.; Restrepo, M.; Cuadro-Alvarez, E.; Lambert, V.; Boukhari, R.; Duvernois, J.-P.; Favre, A.; et al. Estimating the Risk of Vertical Transmission of Dengue: A Prospective Study. *Am. J. Trop. Med. Hyg.* **2018**, *98*, 1826–1832. [CrossRef]
125. Coyne, C.B. The tree(S) of life: The human placenta and my journey to learn more about it. *PLoS Pathog.* **2016**, *12*, e1005515. [CrossRef]
126. Delorme-Axford, E.; Sadovsky, Y.; Coyne, C.B. The Placenta as a Barrier to Viral Infections. *Annu. Rev. Virol.* **2014**, *1*, 133–146. [CrossRef]
127. Watanabe, S.; Vasudevan, S.G. Clinical and Experimental Evidence for Transplacental Vertical Transmission of Flaviviruses. *Antivir. Res.* **2023**, *210*, 105512. [CrossRef] [PubMed]
128. McClure, E.M.; Goldenberg, R.L. Infection and Stillbirth. *Semin. Fetal Neonatal Med.* **2009**, *14*, 182–189. [CrossRef] [PubMed]
129. Ribeiro, C.P.; Gloria, V.; Brasil, P.; Rodrigues, A.; Rohloff, R.D.; Maria, R. Dengue Infection in Pregnancy and Its Impact on the Placenta. *Int. J. Infect. Dis* **2017**, *55*, 109–112. [CrossRef]
130. Basurko, C.; Hcini, N.; Demar, M.; Abboud, P.; Nacher, M.; Carles, G.; Lambert, V.; Matheus, S. Symptomatic Chikungunya Virus Infection and Pregnancy Outcomes: A Nested Case-Control Study in French Guiana. *Viruses* **2022**, *14*, 2705. [CrossRef]
131. Foeller, M.E.; Nosrat, C.; Krystosik, A.; Noel, T.; Gérardin, P.; Cudjoe, N.; Mapp-Alexander, V.; Mitchell, G.; Macpherson, C.; Waechter, R.; et al. Chikungunya Infection in Pregnancy—Reassuring Maternal and Perinatal Outcomes: A Retrospective Observational Study. *BJOG Int. J. Obstet. Gynaecol.* **2021**, *128*, 1077–1086. [CrossRef] [PubMed]
132. Gupta, N.; Gupta, S. Short-Term Pregnancy Outcomes in Patients Chikungunya Infection: An Observational Study. *J. Fam. Med. Prim. Care* **2019**, *8*, 985. [CrossRef]
133. Ali, A.A.; Abdallah, T.M.; Alshareef, S.A.; Al-Nafeesah, A.; Adam, I. Maternal and Perinatal Outcomes during a Chikungunya Outbreak in Kassala, Eastern Sudan. *Arch. Gynecol. Obstet.* **2021**, *305*, 855–858. [CrossRef]
134. Robillard, P.-Y.; Boumahni, B.; Gérardin, P.; Michault, A.; Fourmaintraux, A.; Schuffenecker, I.; Carbonnier, M.; Djémili, S.; Choker, G.; Roge-Wolter, M.; et al. Transmission Verticale Materno-Fœtale Du Virus Chikungunya. *Presse Médicale* **2006**, *35*, 785–788. [CrossRef]
135. Gérardin, P.; Barau, G.; Michault, A.; Bintner, M.; Randrianaivo, H.; Choker, G.; Lenglet, Y.; Touret, Y.; Bouveret, A.; Grivard, P.; et al. Multidisciplinary Prospective Study of Mother-To-Child Chikungunya Virus Infections on the Island of La Réunion. *PLoS Med.* **2008**, *5*, e60. [CrossRef]

136. Gérardin, P.; Couderc, T.; Bintner, M.; Tournebize, P.; Renouil, M.; Lémant, J.; Boisson, V.; Borgherini, G.; Staikowsky, F.; Schramm, F.; et al. Chikungunya Virus–Associated Encephalitis. *Neurology* **2015**, *86*, 94–102. [CrossRef]
137. Ramful, D.; Carbonnier, M.; Pasquet, M.; Bouhmani, B.; Ghazouani, J.; Noormahomed, T.; Beullier, G.; Attali, T.; Samperiz, S.; Fourmaintraux, A.; et al. Mother-To-Child Transmission of Chikungunya Virus Infection. *Pediatr. Infect. Dis. J.* **2007**, *26*, 811–815. [CrossRef] [PubMed]
138. Salomão, N.; Brendolin, M.; Rabelo, K.; Wakimoto, M.; de Filippis, A.M.; dos Santos, F.; Moreira, M.E.; Basílio-de-Oliveira, C.A.; Avvad-Portari, E.; Paes, M.; et al. Spontaneous Abortion and Chikungunya Infection: Pathological Findings. *Viruses* **2021**, *13*, 554. [CrossRef] [PubMed]
139. Touret, Y.; Randrianaivo, H.; Michault, A.; Schuffenecker, I.; Kauffmann, E.; Lenglet, Y.; Barau, G.; Fourmaintraux, A. Transmission Materno-Fœtale Précoce Du Virus Chikungunya. *Presse Médicale* **2006**, *35*, 1656–1658. [CrossRef] [PubMed]
140. Prata-Barbosa, A.; Cleto-Yamane, T.L.; Robaina, J.R.; Guastavino, A.; Clara, M.; Brindeiro, R.; Medronho, R.; José, A. Co-Infection with Zika and Chikungunya Viruses Associated with Fetal Death—A Case Report. *Int. J. Infect. Dis.* **2018**, *72*, 25–27. [CrossRef]
141. Grivard, P.; Le Roux, K.; Laurent, P.; Fianu, A.; Perrau, J.; Gigan, J.; Hoarau, G.; Grondin, N.; Staikowsky, F.; Favier, F.; et al. Molecular and Serological Diagnosis of Chikungunya Virus Infection. *Pathol. Biol.* **2007**, *55*, 490–494. [CrossRef] [PubMed]
142. Shenoy, S.; Pradeep, G.C.M. Neurodevelopmental Outcome of Neonates with Vertically Transmitted Chikungunya Fever with Encephalopathy. Available online: https://www.indianpediatrics.net/mar2012/mar-238-240.htm (accessed on 27 June 2023).
143. Campos, G.S.; Bandeira, A.; França, V.; Dias, J.P.; Carvalho, R.H.; Sardi, S.I. First Detection of Chikungunya Virus in Breast Milk. *Pediatr. Infect. Dis. J.* **2017**, *36*, 1015–1017. [CrossRef]
144. Van Enter, B.J.D.; Huibers, M.H.W.; van Rooij, L.; Steingrover, R.; van Hensbroek, M.B.; Voigt, R.R.; Hol, J. Perinatal Outcomes in Vertically Infected Neonates during a Chikungunya Outbreak on the Island of Curaçao. *Am. J. Trop. Med. Hyg.* **2018**, *99*, 1415–1418. [CrossRef]
145. Lenglet, Y.; Barau, G.; Robillard, P.-Y.; Randrianaivo, H.; Michault, A.; Bouveret, A.; Gérardin, P.; Boumahni, B.; Touret, Y.; Kauffmann, E.; et al. Infection à Chikungunya Chez La Femme Enceinte et Risque de Transmission Materno-Fœtale. *J. Gynécologie Obs. Biol. Reprod.* **2006**, *35*, 578–583. [CrossRef]
146. Mangalgi, S.; Shenoy, S.; Maralusiddappa, P.; Aprameya, I. Neonatal Chikungunya: A Case Series. *J. Pediatr. Sci.* **2011**, *3*, 103–105. [CrossRef]
147. Ramos, R.; Viana, R.R.; Brainer-Lima, A.M.; FloreÂncio, T.; Carvalho, M.D.; van der Linden, V.; Amorim, A.; Rocha, M.A.; Medeiros, F. Perinatal Chikungunya Virus–Associated Encephalitis Leading to Postnatal-Onset Microcephaly and Optic Atrophy. *Pediatr. Infect. Dis. J.* **2018**, *37*, 94–95. [CrossRef]
148. Bandeira, A.C.; Campos, G.S.; Sardi, S.I.; Rocha, V.F.D.; Rocha, G.C.M. Neonatal Encephalitis due to Chikungunya Vertical Transmission: First Report in Brazil. *IDCases* **2016**, *5*, 57–59. [CrossRef]
149. Gérardin, P.; Sampériz, S.; Ramful, D.; Boumahni, B.; Bintner, M.; Alessandri, J.-L.; Carbonnier, M.; Tiran-Rajaoefera, I.; Beullier, G.; Boya, I.; et al. Neurocognitive Outcome of Children Exposed to Perinatal Mother-To-Child Chikungunya Virus Infection: The CHIMERE Cohort Study on Reunion Island. *PLoS Negl. Trop. Dis.* **2014**, *8*, e2996. [CrossRef] [PubMed]
150. Yin, X.; Zhong, X.; Pan, S. Vertical transmission of dengue infection: The first putative case reported in China. *Rev. Inst. Med. Trop. São Paulo* **2016**, *58*, 90. [CrossRef] [PubMed]
151. Mor, G.; Cardenas, I. The Immune System in Pregnancy: A Unique Complexity. *Am. J. Reprod. Immunol.* **2010**, *63*, 425–433. [CrossRef] [PubMed]
152. Barrientos, G.; Tirado-González, I.; Klapp, B.F.; Karimi, K.; Arck, P.C.; Garcia, M.U.; Blois, S.M. The Impact of Dendritic Cells on Angiogenic Responses at the Fetal–Maternal Interface. *J. Reprod. Immunol.* **2009**, *83*, 85–94. [CrossRef]
153. Tagliani, E.; Erlebacher, A. Dendritic Cell Function at the Maternal–Fetal Interface. *Expert Rev. Clin. Immunol.* **2011**, *7*, 593–602. [CrossRef]
154. Moizéis, R.N.C.; Fernandes, T.A.A.d.M.; Guedes, P.M.d.M.; Pereira, H.W.B.; Lanza, D.C.F.; de Azevedo, J.W.V.; Galvão, J.M.d.A.; Fernandes, J.V. Chikungunya Fever: A Threat to Global Public Health. *Pathog. Glob. Health* **2018**, *112*, 182–194. [CrossRef]
155. Alves, A.M.B.; del Angel, R.M. Dengue Virus and Other Flaviviruses (Zika): Biology, Pathogenesis, Epidemiology, and Vaccine Development. In *Human Virology in Latin America*; Ludert, J., Pujol, F., Arbiza, J., Eds.; Springer International Publishing: Berlin/Heidelberg, Germany, 2017; pp. 141–167.
156. Tabata, T.; Petitt, M.; Puerta-Guardo, H.; Michlmayr, D.; Harris, E.; Pereira, L. Zika Virus Replicates in Proliferating Cells in Explants from First-Trimester Human Placentas, Potential Sites for Dissemination of Infection. *J. Infect. Dis.* **2017**, *217*, 1202–1213. [CrossRef]
157. Tabata, T.; Petitt, M.; Puerta-Guardo, H.; Michlmayr, D.; Wang, C.; Fang-Hoover, J.; Harris, E.; Pereira, L. Zika Virus Targets Different Primary Human Placental Cells, Suggesting Two Routes for Vertical Transmission. *Cell Host Microbe* **2016**, *20*, 155–166. [CrossRef]
158. Owen, J.L.; Mohamadzadeh, M. Macrophages and Chemokines as Mediators of Angiogenesis. *Front. Physiol.* **2013**, *4*, 159. [CrossRef]
159. Abrahams, V.M.; Kim, Y.M.; Straszewski, S.L.; Romero, R.; Mor, G. Macrophages and Apoptotic Cell Clearance during Pregnancy. *Am. J. Reprod. Immunol.* **2004**, *51*, 275–282. [CrossRef]
160. Zulu, M.Z.; Martinez, F.O.; Gordon, S.; Gray, C.M. The Elusive Role of Placental Macrophages: The Hofbauer Cell. *J. Innate Immun.* **2019**, *11*, 447–456. [CrossRef]

161. Watanabe, S.; Chan, K.W.K.; Tan, N.W.W.; Mahid, M.B.A.; Chowdhury, A.; Chang, K.T.E.; Vasudevan, S.G. Experimental Evidence for a High Rate of Maternal-Fetal Transmission of Dengue Virus in the Presence of Antibodies in Immunocompromised Mice. *eBioMedicine* **2022**, *77*, 103930. [CrossRef] [PubMed]
162. Moffett-King, A. Natural Killer Cells and Pregnancy. *Nat. Rev. Immunol.* **2002**, *2*, 656–663. [CrossRef]
163. Keskin, D.B.; Allan, D.S.J.; Rybalov, B.; Andzelm, M.M.; Stern, J.N.H.; Kopcow, H.D.; Koopman, L.A.; Strominger, J.L. TGFbeta Promotes Conversion of CD16+ Peripheral Blood NK Cells into CD16- NK Cells with Similarities to Decidual NK Cells. *Proc. Natl. Acad. Sci. USA* **2007**, *104*, 3378–3383. [CrossRef] [PubMed]
164. Laškarin, G.; Tokmadžić, V.S.; Štrbo, N.; Bogović, T.; Szekeres-Bartho, J.; Randić, L.; Podack, E.R.; Rukavina, D. Progesterone Induced Blocking Factor (PIBF) Mediates Progesterone Induced Suppression of Decidual Lymphocyte Cytotoxicity. *Am. J. Reprod. Immunol.* **2002**, *48*, 201–209. [CrossRef] [PubMed]
165. Tong, M.; Abrahams, V.M. Immunology of the Placenta. *Obstet. Gynecol. Clin. N. Am.* **2020**, *47*, 49–63. [CrossRef]
166. Vieira, R.d.M.; Meagher, A.; Crespo, Â.C.; Kshirsagar, S.; Iyer, V.; Norwitz, E.R.; Strominger, J.L.; Tilburgs, T. Human Term Pregnancy Decidual NK Cells Generate Distinct Cytotoxic Responses. *J. Immunol.* **2020**, *204*, 3149–3159. [CrossRef] [PubMed]
167. Crespo, Â.C.; Strominger, J.L.; Tilburgs, T. Expression of KIR2DS1 by Decidual Natural Killer Cells Increases Their Ability to Control Placental HCMV Infection. *Proc. Natl. Acad. Sci. USA* **2016**, *113*, 15072–15077. [CrossRef]
168. Parker, E.L.; Silverstein, R.B.; Verma, S.; Mysorekar, I.U. Viral-Immune Cell Interactions at the Maternal-Fetal Interface in Human Pregnancy. *Front. Immunol.* **2020**, *11*, 522047. [CrossRef]
169. Bortolotti, D.; Gentili, V.; Rotola, A.; Cultrera, R.; Marci, R.; Di Luca, D.; Rizzo, R. HHV-6A Infection of Endometrial Epithelial Cells Affects Immune Profile and Trophoblast Invasion. *Am. J. Reprod. Immunol.* **2019**, *82*, e13174. [CrossRef]
170. Crespo, Â.C.; Mulik, S.; Dotiwala, F.; Ansara, J.A.; Sen Santara, S.; Ingersoll, K.; Ovies, C.; Junqueira, C.; Tilburgs, T.; Strominger, J.L.; et al. Decidual NK Cells Transfer Granulysin to Selectively Kill Bacteria in Trophoblasts. *Cell* **2020**, *182*, 1125–1139.e18. [CrossRef]
171. Jabrane-Ferrat, N. Features of Human Decidual NK Cells in Healthy Pregnancy and during Viral Infection. *Front. Immunol.* **2019**, *10*, 1397. [CrossRef]
172. Bouteiller, P.L.; Siewiera, J.; Casart, Y.C.; Aguerre-Girr, M.; El Costa, H.; Berrebi, A.; Tabiasco, J.; Jabrane-Ferrat, N. The Human Decidual NK-Cell Response to Virus Infection: What Can We Learn from Circulating NK Lymphocytes? *J. Reprod. Immunol.* **2011**, *88*, 170–175. [CrossRef] [PubMed]
173. Nair, S.; Poddar, S.; Shimak, R.M.; Diamond, M.S. Interferon Regulatory Factor 1 Protects against Chikungunya Virus-Induced Immunopathology by Restricting Infection in Muscle Cells. *J. Virol.* **2017**, *91*, e01419-17. [CrossRef] [PubMed]
174. Cook, L.E.; Locke, M.C.; Young, A.R.; Monte, K.; Hedberg, M.L.; Shimak, R.M.; Sheehan, K.C.F.; Veis, D.J.; Diamond, M.S.; Lenschow, D.J. Distinct Roles of Interferon Alpha and Beta in Controlling Chikungunya Virus Replication and Modulating Neutrophil-Mediated Inflammation. *J. Virol.* **2019**, *94*, e00841-19. [CrossRef] [PubMed]
175. McNab, F.; Mayer-Barber, K.; Sher, A.; Wack, A.; O'Garra, A. Type I Interferons in Infectious Disease. *Nat. Rev. Immunol.* **2015**, *15*, 87–103. [CrossRef] [PubMed]
176. Koyama, S.; Ishii, K.J.; Coban, C.; Akira, S. Innate Immune Response to Viral Infection. *Cytokine* **2008**, *43*, 336–341. [CrossRef]
177. Locke, M.C.; Fox, L.; Dunlap, B.F.; Young, A.R.; Monte, K.; Lenschow, D.J. Interferon Alpha, but Not Interferon Beta, Acts Early to Control Chronic Chikungunya Virus Pathogenesis. *J. Virol.* **2022**, *96*, e0114321. [CrossRef]
178. Ivashkiv, L.B.; Donlin, L.T. Regulation of Type I Interferon Responses. *Nat. Rev. Immunol.* **2013**, *14*, 36–49. [CrossRef]
179. Sadler, A.J.; Williams, B.R.G. Interferon-Inducible Antiviral Effectors. *Nat. Rev. Immunol.* **2008**, *8*, 559–568. [CrossRef] [PubMed]
180. Schoggins, J.W.; Rice, C.M. Interferon-Stimulated Genes and Their Antiviral Effector Functions. *Curr. Opin. Virol.* **2011**, *1*, 519–525. [CrossRef]
181. Lavoie, T.B.; Kalie, E.; Crisafulli-Cabatu, S.; Abramovich, R.; DiGioia, G.; Moolchan, K.; Pestka, S.; Schreiber, G. Binding and Activity of All Human Alpha Interferon Subtypes. *Cytokine* **2011**, *56*, 282–289. [CrossRef] [PubMed]
182. Thomas, C.; Moraga, I.; Levin, D.; Krutzik, P.O.; Podoplelova, Y.; Trejo, A.; Lee, C.; Yarden, G.; Vleck, S.E.; Glenn, J.S.; et al. Structural Linkage between Ligand Discrimination and Receptor Activation by Type I Interferons. *Cell* **2011**, *146*, 621–632. [CrossRef] [PubMed]
183. Castillo Ramirez, J.A.; Urcuqui-Inchima, S. Dengue Virus Control of Type I IFN Responses: A History of Manipulation and Control. *J. Interferon Cytokine Res.* **2015**, *35*, 421–430. [CrossRef]
184. Elong Ngono, A.; Shresta, S. Immune Response to Dengue and Zika. *Annu. Rev. Immunol.* **2018**, *36*, 279–308. [CrossRef] [PubMed]
185. Morrison, J.; Aguirre, S.; Fernandez-Sesma, A. Innate Immunity Evasion by Dengue Virus. *Viruses* **2012**, *4*, 397–413. [CrossRef]
186. Luo, H.; Winkelmann, E.R.; Fernandez-Salas, I.; Li, L.; Mayer, S.V.; Danis-Lozano, R.; Sanchez-Casas, R.M.; Vasilakis, N.; Tesh, R.; Barrett, A.D.; et al. Zika, Dengue and Yellow Fever Viruses Induce Differential Anti-Viral Immune Responses in Human Monocytic and First Trimester Trophoblast Cells. *Antivir. Res.* **2018**, *151*, 55–62. [CrossRef]
187. Viettri, M.; Caraballo, G.; Salazar, E.; Espejel-Nuñez, A.; Betanzos, A.; Ortiz-Navarrete, V.; Estrada-Gutierrez, G.; Nava, P.; Ludert, J.E. Comparative Infections of Zika, Dengue, and Yellow Fever Viruses in Human Cytotrophoblast-Derived Cells Suggest a Gating Role for the Cytotrophoblast in Zika Virus Placental Invasion. *Microbiol. Spectr.* **2023**, *11*, e0063023. [CrossRef]
188. Bayer, A.; Lennemann, N.J.; Ouyang, Y.; Bramley, J.C.; Morosky, S.; Marques, E.; Cherry, S.; Sadovsky, Y.; Coyne, C.B. Type III Interferons Produced by Human Placental Trophoblasts Confer Protection against Zika Virus Infection. *Cell Host Microbe* **2016**, *19*, 705–712. [CrossRef] [PubMed]

189. Randall, R.E.; Goodbourn, S. Interferons and Viruses: An Interplay between Induction, Signalling, Antiviral Responses and Virus Countermeasures. *J. Gen. Virol.* **2008**, *89*, 1–47. [CrossRef] [PubMed]
190. Chen, H.; Min, N.; Ma, L.; Mok, C.-K.; Chu, J.J.H. Adenovirus Vectored IFN-α Protects Mice from Lethal Challenge of Chikungunya Virus Infection. *PLoS Negl. Trop. Dis.* **2020**, *14*, e0008910. [CrossRef] [PubMed]
191. Venugopalan, A.; Ghorpade, R.P.; Chopra, A. Cytokines in Acute Chikungunya. *PLoS ONE* **2014**, *9*, e111305. [CrossRef]
192. Neupane, B.; Acharya, D.; Nazneen, F.; Gonzalez-Fernandez, G.; Flynt, A.S.; Bai, F. Interleukin-17A Facilitates Chikungunya Virus Infection by Inhibiting IFN-α2 Expression. *Front. Immunol.* **2020**, *11*, 588382. [CrossRef] [PubMed]
193. Rizzuto, G.; Tagliani, E.; Manandhar, P.; Erlebacher, A.; Bakardjiev, A.I. Limited Colonization Undermined by Inadequate Early Immune Responses Defines the Dynamics of Decidual Listeriosis. *Infect. Immun.* **2017**, *85*, e00153-17. [CrossRef] [PubMed]
194. Racicot, K.; Aldo, P.; El-Guindy, A.; Kwon, J.-Y.; Romero, R.; Mor, G. Cutting Edge: Fetal/Placental Type I IFN Can Affect Maternal Survival and Fetal Viral Load during Viral Infection. *J. Immunol.* **2017**, *198*, 3029–3032. [CrossRef] [PubMed]
195. Haese, N.N.; Smith, H.; Onwuzu, K.; Kreklywich, C.N.; Smith, J.L.; Denton, M.; Kreklywich, N.; Streblow, A.D.; Frias, A.E.; Morgan, T.K.; et al. Differential Type 1 IFN Gene Expression in CD14+ Placenta Cells Elicited by Zika Virus Infection during Pregnancy. *Front. Virol.* **2021**, *1*, 783407. [CrossRef]
196. Quanquin, N.; Barres, L.G.; Aliyari, S.R.; Day, N.; Gerami, H.; Fisher, S.J.; Kakuru, A.; Kamya, M.R.; Havlir, D.V.; Feeney, M.E.; et al. Gravidity-Dependent Associations between Interferon Response and Birth Weight in Placental Malaria. *Malar. J.* **2020**, *19*, 280. [CrossRef] [PubMed]

Article

Multiple Porcine Innate Immune Signaling Pathways Are Involved in the Anti-PEDV Response

Youwen Zhang [1,2,3,4], Yulin Xu [1,2,3,4], Sen Jiang [1,2,3,4], Shaohua Sun [1,2,3,4], Jiajia Zhang [1,2,3,4], Jia Luo [1,2,3,4], Qi Cao [1,2,3,4], Wanglong Zheng [1,2,3,4], François Meurens [5,6], Nanhua Chen [1,2,3,4] and Jianzhong Zhu [1,2,3,4,*]

[1] College of Veterinary Medicine, Yangzhou University, Yangzhou 225009, China; dx120200161@stu.yzu.edu.cn (S.J.); hnchen@yzu.edu.cn (N.C.)
[2] Joint International Research Laboratory of Agriculture and Agri-Product Safety, Yangzhou 225009, China
[3] Comparative Medicine Research Institute, Yangzhou University, Yangzhou 225009, China
[4] Jiangsu Co-Innovation Center for Prevention and Control of Important Animal Infectious Diseases and Zoonoses, Yangzhou 225009, China
[5] Swine and Poultry Infectious Diseases Research Center, Faculty of Veterinary Medicine, University of Montreal, St. Hyacinthe, QC J2S 2M2, Canada; francois.meurens@umontreal.ca
[6] Department of Veterinary Microbiology and Immunology, Western College of Veterinary Medicine, University of Saskatchewan, Saskatoon, SK S7N 5E2, Canada
* Correspondence: jzzhu@yzu.edu.cn

Abstract: Porcine epidemic diarrhea virus (PEDV) has caused great damage to the global pig industry. Innate immunity plays a significant role in resisting viral infection; however, the exact role of innate immunity in the anti-PEDV response has not been fully elucidated. In this study, we observed that various porcine innate immune signaling adaptors are involved in anti-PEDV (AJ1102-like strain) activity in transfected Vero cells. Among these, TRIF and MAVS showed the strongest anti-PEDV activity. The endogenous TRIF, MAVS, and STING were selected for further examination of anti-PEDV activity. Agonist stimulation experiments showed that TRIF, MAVS, and STING signaling all have obvious anti-PEDV activity. The siRNA knockdown assay showed that TRIF, MAVS, and STING are also all involved in anti-PEDV response, and their remarkable effects on PEDV replication were confirmed in TRIF$^{-/-}$, MAVS$^{-/-}$ and STING$^{-/-}$ Vero cells via the CRISPR approach. For further verification, the anti-PEDV activity of TRIF, MAVS, and STING could be reproduced in porcine IPEC-DQ cells treated with siRNAs. In summary, this study reveals that multiple pattern-recognition receptor (PRR) signaling pathways of porcine innate immunity play an important role in the anti-PEDV infection, providing new and useful antiviral knowledge for prevention and control of PEDV spreading.

Keywords: PEDV; porcine innate immunity; pattern recognition receptor (PRR); signaling adaptors; knockout; siRNA

Citation: Zhang, Y.; Xu, Y.; Jiang, S.; Sun, S.; Zhang, J.; Luo, J.; Cao, Q.; Zheng, W.; Meurens, F.; Chen, N.; et al. Multiple Porcine Innate Immune Signaling Pathways Are Involved in the Anti-PEDV Response. *Viruses* **2023**, *15*, 1629. https://doi.org/10.3390/v15081629

Academic Editors: Caijun Sun and Feng Ma

Received: 5 July 2023
Revised: 22 July 2023
Accepted: 24 July 2023
Published: 26 July 2023

1. Introduction

Porcine epidemic diarrhea virus (PEDV) is an intestinal coronavirus that targets the small intestinal epithelial cells of pigs. It causes damage to the small intestinal epithelial tissue, inducing intestinal congestion, swelling, and watery diarrhea in pigs, which leads to anorexia and body wasting in fattening pigs and high mortality rate in suckling piglets [1]. In recent years, with the emergence of new PEDV variant strains, PED broke out again globally, especially in Asian countries, causing great economic loss [2,3].

As a porcine enteric virus, PEDV belongs to the α coronavirus class and is a positive-sense single-stranded RNA virus [4]. The total length of the virus genome is 28 kb and it contains open reading frame (ORF)1a, ORF1b, spike protein (S), accessory protein ORF3, envelope protein (E), membrane protein (M) and nucleocapsid protein (N) genes [4]. The S protein encoded by the S gene consists of two domains, S1 and S2, with the interaction between S1 and host cell receptors as the first step of infection and the main determinant of viral anisotropy [5]. Additionally, PEDV encodes two large polyproteins, pp1a and pp1ab. Via the protease activity of non-structural proteins (Nsp) 3 and Nsp5, these two large polyproteins are further processed into 16 non-structural proteins, Nsp1 to Nsp16 [6]. In addition to its genetic diversity, PEDV has also evolved a variety of strategies to antagonize host innate antiviral defense for successful infection [7].

The innate immune system is known to be the first line of defense against infection caused by pathogens [8]. Pathogenic microorganisms produce conserved molecules called pathogen-associated molecular patterns (PAMPs) when they invade the organism or replicate in the host organism [8]. Host cells express receptors for these molecules, named pattern recognition receptors (PRRs), which are composed of toll-like receptors (TLRs), RIG-I-like receptors (RLRs), NOD-like receptors (NLRs), C-type lectin receptors (CLRs), and cytosolic DNA receptors (CDRs) [9]. These receptors recognize PAMPs and activate intracellular cascade signaling pathways comprising adaptors, kinases, and transcription factors, ultimately initiating intracellular gene transcription or protease cleavage of cytokine precursors [9]. Thus, the produced antiviral interferons (IFNs), inflammatory factors, and chemokines play an anti-infection role while profoundly influencing subsequent adaptive immune responses [9,10].

The mutual evolution of virus and innate immune response of the host leads to significant viral diversification and enhanced host antiviral response [11]. In the competition between viruses and host cells, many viruses, including coronaviruses, have evolved various strategies to evade or disrupt antiviral immunity, such as type I and type III interferons (IFNs) [12]. There is increasing evidence that several PEDV proteins such as nsp1, nsp3, nsp5, nsp8, nsp14, nsp15, nsp16, E, M, and N can resist host IFN signaling [13]. In addition, PEDV has been reported to inhibit IFN induction via different mechanisms [13,14]. As for innate immunity for recognition of and defense against PEDV, it has not been totally understood until now. In this study, we demonstrated that PEDV can activate multiple PRR-mediated signaling pathways to induce anti-PEDV activity. This study provides new insight into host innate immune responses against PEDV infection.

2. Materials and Methods

2.1. Cells, Virus and Reagents

PEDV-permissive Vero cells and human embryonic kidney (HEK) 293T cells were cultured in Dulbecco's Modified Eagle's Medium (DMEM; Thermo Fisher Scientific, Waltham, MA, USA) supplemented with 10% fetal bovine serum (FBS, Thermo Fisher Scientific) and 100 U/mL of penicillin plus 100 g/mL streptomycin. IPEC-DQ cells were cultured in Roswell Park Memorial Institute (RPMI) medium (Hyclone Laboratories, Logan, UT, USA) containing 10% FBS with penicillin/streptomycin. All cells were maintained at 37 °C with 5% CO_2 in a humidified incubator. The PEDV strain used in this study was a mutant strain of PEDV AJ1102 (GenBank: JX188454.1). Poly(I:C)-LMW, poly(I:C)-HMW, 2′3′-cGAMP and poly(dA:dT) were all obtained from InvivoGen (Hongkong, China). Mouse anti-glyceraldehyde-3-phosphate dehydrogenase (GAPDH) mAb, mouse anti-green fluorescence protein (GFP) mAb, and mouse anti-β-Actin mAb were all brought from TransGen (Beijing, China). Mouse anti-MAVS mAb was from Santa Cruz Biotechnology (Dallas, TX, USA), rabbit anti-TRIF pAb was from Biodargon (Beijing, China), and rabbit anti-STING pAb was bought from Proteintech (Wuhan, China).

2.2. Preparation of Monoclonal Antibody against N Protein of PEDV

The PEDV N gene amplified by PCR from PEDV cDNA using the primers shown in Table 1 was inserted into the *EcoR* V and *Sal* I sites of pENTR4-2HA vector. The resultant recombinant pENTR4 vector was subjected for LR recombination with pDEST527 (Thermo Fisher Scientific) to construct the pDEST527-N prokaryotic expression vector. After induction at 25 °C for 12 h with 1 mM IPTG, the soluble N protein was expressed in transformed *E. coli* (DE3/BL21). The expressed N protein was purified via nickel column according to the instructions of the His-tag Protein Purification Kit (Beyotime, China). Female 6-week-old BALB/c mice were immunized four times with 30 µg purified N protein plus 1:1 immune adjuvant (QuickAntique-Mouse5w, Biodragon, China) each, via multiple subcutaneous injections in the back. The monoclonal antibody (mAb) of N protein was obtained via regular hybridoma technology from the immunized mice. The reaction specificity of N mAb is shown as in Figure S1.

Table 1. PCR primers used for PEDV N gene amplification.

Prime Names	Sequences (5′-3′)
PEDV-N-F	GGGCC<u>GTCGAC</u>ATGGCTTCTGTCAGTTTTCAGGATCG
PEDV-N-R	GGGCC<u>GATATC</u>ATTTCCTGTATCGAAGATCTCGTTGATAATTTCAAC

Note: the restriction enzyme sites are underlined.

2.3. Quantitative Reverse Transcription Polymerase Chain Reaction (RT-qPCR)

The total RNA of the Vero cells after PEDV infection or agonist stimulation was extracted with TRIzol reagent (Thermo Fisher Scientific). The extracted RNA was reverse-transcribed into cDNA with the HiScript® 1st Strand cDNA Synthesis Kit (Vazyme, China). qPCR was performed in 20 µL reactions with the Cham Q Universal SYBR qPCR Master Mix (Vazyme, China) using the StepOne Plus real-time PCR system (Applied Biosystems, Foster City, CA, USA). The qPCR primers were as shown in Table 2. Results of the relative mRNA expression were calculated via the $2^{-\Delta\Delta Ct}$ method, with β-Actin used as the internal reference control.

Table 2. The qPCR primers used for the detection of mRNA expression.

Prime Names	Sequences (5′-3′)
PEDV-N-f	CAAGAACAGAAACCAGTCAAATGACC
PEDV-N-R	AGAGTGGAGGAGAATTCCCAAGG
M-IFN-α-F	ATCTGCTCTCTGGGCTGTGATCT
M-IFN-α-R	TTCAGACAGGAGAAAGGAGAGATTCT
M-IFN-β-F	AAATTGCTCTCCTGTTGTGCTTCT
M-IFN-β-R	AAGCCTTCCATTCAATTGCCA
M-ISG15-F	CTCTGAGCATCCTGGTGAGGAA
M-ISG15-R	CGAAGGTCAGCCAGAACAGGT
M-β-Actin-F	AGAAGATGACCCAGATCATGTTTG
M-β-Actin-R	ATCCATCACGATGCCAGTGGTA

Note: M denotes monkey.

2.4. Western Blot Analysis

The treated cells were collected and lyzed in a radio-immunoprecipitation assay (RIPA) buffer (50 mM Tris pH 7.2, 150 mM NaCl, 1% sodium deoxycholate, 1% Triton X-100). The extracted whole-cell protein was separated via sodium dodecyl sulfate-polyacrylamide gel electrophoresis (SDS-PAGE), and then the protein in gel was transferred to a polyvinylidene fluoride (PVDF) membrane. The membrane was blocked with Tris-buffered saline with Tween (TBST) solution containing 5% skim milk at room temperature for 1 h and incubated overnight at 4 °C with mouse PEDV N mAb (1:1000) as specific primary antibody or other primary antibodies. Next, the membrane was washed with TBST 3 times each for 3–5 min, and then incubated with HRP-labeled goat anti-mouse or rabbit IgG (1:10,000, Transgen Biotech, Beijing, China) as the secondary antibody. The protein signals were visualized using an imaging system (Tanon, China) with the ECL chemiluminescence detection system (Tanon, China) according to the manufacturer's instructions.

2.5. Virus TCID50 Titration

Vero cells cultured in 96-well plates were infected with PEDV in 10-fold continuous dilutions at 37 °C for 2 h, and the supernatants were replaced with fresh DMEM containing

2% FBS. After 4 to 5 days of infection, Vero cells showed significant cytopathic effect (CPE), manifested as cell shedding. The virus titer was calculated according to the Reed–Muench method and expressed as 50% tissue culture infection dose (TCID50).

2.6. Plaque Assay

A suitable number of Vero cells were seeded into the different cell plates (1.5×10^6 cells/well in 6-well plate and $3–4 \times 10^5$ cells/well in 24-well plate). The PEDV sample to be tested was diluted 10^{-1}–10^{-4} times to infect cells in the plate wells. After 2 h of viral infection, the monolayer cells were immediately and slowly covered with 1:1 mixed 1.6–2% low-melting agarose solution with DMEM medium containing 4% FBS. About 3–5 days after infection, cells were fixed and stained with crystal violet dye containing 4% polyformaldehyde. After staining, the cells were rinsed gently and directly with tap water until the clear plaques appeared, then photographed.

2.7. RNA Interference by siRNA

All siRNAs targeting monkey and porcine TRIF, MAVS, and STING were designed and synthesized by Thermo Fisher Scientific. The knockdown efficiency of individual siRNAs for monkey cells has already been tested in our previous work [15]. To determine the knockdown efficiency of designed porcine siRNA sequences (Table 3). IPEC-DQ cells were transfected with 50–150 nM individual siRNAs with Lipofectamine 2000 (Thermo Fisher Scientific), and the expression of corresponding proteins were measured via Western blotting using anti-TRIF antibody (1:1000), anti-MAVS antibody (1:1000), and anti-STING antibody (1:2000), respectively. For the validated siRNA duplexes, Vero cells or IPEC-DQ cells in 12-well plates with 70–80% monolayers were transfected with 100 nM siRNAs using Lipofectamine 2000 and used for subsequent experiments.

Table 3. The designed siRNA sequences for porcine TRIF, MAVS, and STING genes.

siRNA Names	Sequences (5′-3′)
P-STING-11-F/R	CCAGCCUGCAUCCAUCCAUUU AUGGAUGGAUGCAGGCUGGUU
P-STING-689-F/R	CCGACCGUGCUGGCAUCAAUU UUGAUGCCAGCACGGUCGGUU
P-STING-529-F/R	GCUCGGAUCCAAGCUUAUAUU UAUAAGCUUGGAUCCGAGCUU
P-MAVS2329-F/R	CCACCACAGAGAUCUUUAAUU UUAAAGAUCUCUGUGGUGGUU
P-MAVS3138-F/R	GGCUGCACUACUGUAUUAUUU AUAAUACAGUAGUGCAGCCUU
P-TRIF5800-F/R	GCCUGUCCUUUACCCUUUAUU UAAAGGGUAAACGACACCCUU
P-TRIF11858-F/R	GGGUUCAUCACAUUAAUAAUU UUAUUAAUGUGAUGAACCCUU
siNC-F/R	UUCUCCGAACGUGUCACGUUU ACGUGACACGUUCGGAGAAUU

Note: The P denotes porcine.

2.8. Preparation of TRIF$^{-/-}$, MAVS$^{-/-}$ and STING$^{-/-}$ Vero Cells by CRISPR/Cas9 Approach

The clustered, regularly interspaced short palindromic repeat guide RNAs (CRISPR gRNAs) were designed based on the first exons of monkey STING (GenBank: CM014354.1), TRIF (GenBank: CM014354.1), and MAVS (GenBank: CM014354.1). For each gene, 2–3 gRNAs were selected (Table 4), and the annealed gRNA encoding DNA pairs were ligated with the *Bbs*I digested vector pSpCas9(BB)-2A-GFP (pX458, Addgene, Watertown, NY, USA). Subsequently, each gRNA expressing pX458 was transfected into Vero cells using Lipofectamine 2000, and the GFP-positive cells were sorted from transfected cells using a BD FACSAria III Sorter. The individual Vero cell clones obtained via limited dilution from the sorted GFP expressing cells were screened via PCR using the designed primers (Table 4). The PCR products were cloned into T vector, using the pClone007 Versatile Simple Vector Kit (TsingKe Biological Technology, Beijing, China). The inserted fragments were multiply sequenced, and the sequences were analyzed for base insertion and deletion (indel) mutations, based on which two TRIF$^{-/-}$, three MAVS$^{-/-}$ and four STING$^{-/-}$ Vero cell clones were obtained (Figure S2).

Table 4. The CRISPR gRNA encoding DNA sequences and PCR primers for monkey TRIF, MAVS, and STING genes.

Prime Names	Sequences (5′-3′)
M-STING gRNA1-F/R	CACCGTGGATGGATGCAGACTGGAG AAACCTCCAGTCTGCATCCATCCAC
M-STING gRNA2-F/R	CACCGCCATCCATCCCGTGTCCCAG AAACCTGGGACACGGGATGGATGGC
M-STING gRNA3-F/R	CACCGCTGGGACAGCTGTTAAATG AAACCATTTAACAGCTGTCCCAGC
M-TRIF gRNA1-F/R	CACCGTAGGCCACGTCCCGCAGCG AAACCGCTGCGGGACGTGGCCTAC
M-TRIF gRNA2-F/R	CACCGATGAGGCCCGAAACCGGTGT AAACACACCGGTTTCGGGCCTCATC
M-MAVS gRNA1-F/R	CACCGTCTTCAGTACCCTTCAGCGG AAACCCGCTGAAGGGTACTGAAGAC
M-MAVS gRNA2-F/R	CACCGCTGGTAGCTCTGGTAGACAC AAACGTGTCTACCAGAGCTACCAGC
M-STING-F/R	TCGCAGAGACAGGAGCTTTG GGCTGCAGACCCCATTTAAC
M-TRIF-F/R	ACTGAAGGCTGATGCAGCG TTTCCAAGTTGCTGGCCAGG
M-MAVS-F/R	GCTCTTCTGGCTTTCTTGGCG GCTCAGCCTGGATCTACACCC

Note: M denotes monkey.

2.9. Statistical Analysis

The software program Image J 5.0 was used to measure the gray values of Western blot protein bands. The data were represented by mean $\pm$ SD and analyzed using GraphPad Prism 7.0. The *t* test was used to determine whether there was any significant difference between the results: $p < 0.05$ (*) and $p < 0.01$ (**). $p < 0.05$ indicated statistical significance.

3. Results

3.1. Cell Infection and Titer Determination of PEDV Strain

The PEDV strain used in the study was able to effectively infect Vero cells and cause significant cytopathic effect (CPE). Both TCID50 and plaque assays were performed in Vero cells to determine the virus titer. As shown in Figure 1A,B and Figure S3, the titers measured via TCID50 assay and plaque assay were $10^{4.75}/0.1$ mL and 4×10^5 PFU/0.1 mL at 72 h post-infection, respectively.

3.2. Detection of Activated IFN Levels and Transfected Adaptor Proteins in Vero Cells

Some studies have reported that an interferon (IFN) defect exists in Vero cells [16], whereas others have implied that Vero cells were able to produce IFN [17]. In this article, the anti-PEDV activity mediated by innate immune signaling adaptor proteins was studied mainly in Vero cells, and IFNs are an important effector affecting the antiviral results. To overcome this potential issue, we treated Vero cells with RLRs RIG-I/MDA5 agonist poly (I:C)-LMW and TLR3 agonist poly (I:C)-HMW for 12 h and 36 h, respectively, and then the collected cells were analyzed for IFN and IFN-stimulated gene (ISG) expressions via RT-qPCR. As shown in Figure 2A, the expression levels of IFN-α, IFN-β, and ISG15 genes were significantly upregulated upon both stimulations, indicating that the downstream IFN and ISG genes could be normally expressed in our Vero cells.

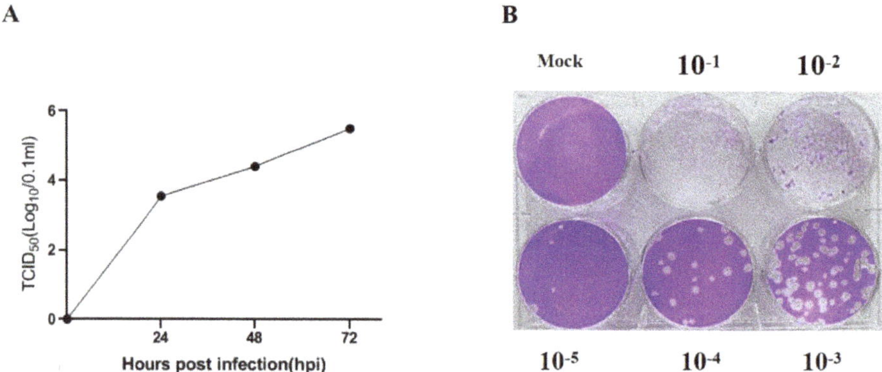

Figure 1. Titration of PEDV via TCID50 assay (**A**) and plaque assay (**B**), respectively.

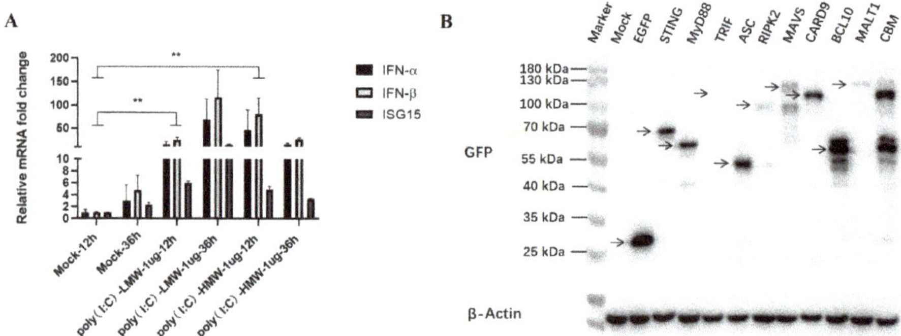

Figure 2. Detection of IFN levels and the adaptor expressions in Vero cells. (**A**) RT-qPCR detection of interferons (IFNs) and related genes in Vero cells. ** $p < 0.01$. (**B**) Expressions of 9 signaling adaptor proteins in transfected Vero cells, which are fused with GFP and marked with arrows.

We previously cloned and characterized the nine porcine innate immune signaling adaptors which represent and reflect all currently known innate immune signaling pathways [18,19]. The recombinant pEGFP plasmids encoding MAVS, TRIF, STING, MyD88, RIPK2, ASC, CARD9, BCL10, MALT1, and control vector pEGFP-N1 were transferred into Vero cells for 24 h, and the expressions of transfected adaptor proteins were determined via Western blotting (WB). As shown in Figure 2B, all the adaptor proteins could be correctly expressed in the transfected Vero cells, although the expression level of TRIF was poor.

3.3. The Anti-PEDV Activities of Ectopic Porcine Signaling Adaptors in Vero Cells

To explore the impacts of nine adaptor proteins on PEDV replication, the adaptor-transfected Vero cells were infected with PEDV, and the PEDV replication was examined by measuring the viral N gene transcription by RT-qPCR and protein expression via Western blotting, respectively. The RT-qPCR results showed that all the signaling adaptors decreased the expressions of PEDV N mRNA at various levels at 48 h and 72 h post-infection, with TRIF, MAVS, and STING being very significant in inhibition of PEDV at 72 h post-infection (Figure 3A). The Western blotting results showed that TRIF, MAVS, and STING were most effective in inhibition, while MyD88, RIPK2, ASC, and CARD9-BCL10-MALT1 (CBM) were less effective in inhibiting N protein expression at 72 h post-infection (Figure 3B). The collected cell supernatants were measured for PEDV titer via plaque assay, which further confirmed the above results (Figure 3C and Figure S3).

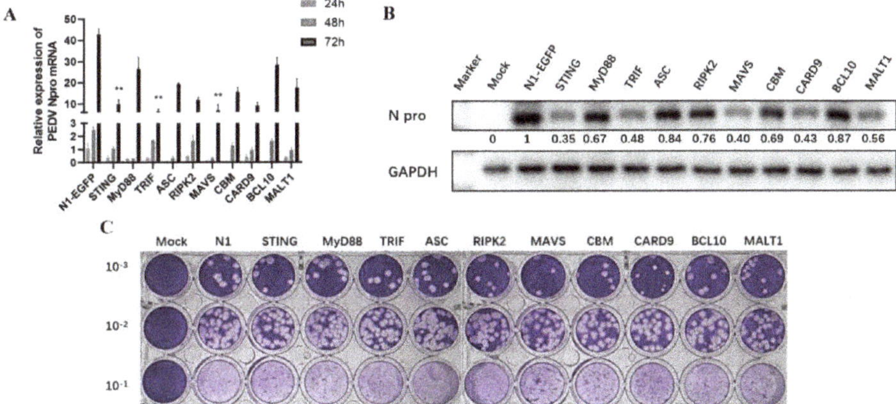

Figure 3. Detection of the anti-PEDV activity of exogenous adaptor proteins in Vero cells. (**A**) The effect of 9 adaptor proteins on PEDV N gene transcription at different time points was detected via RT-qPCR. ** $p < 0.01$. (**B**) The effect of 9 adaptor proteins on N protein expression of PEDV at 72 h post-infection was detected via WB. The gray values of each N protein bands after normalized by GAPDH are shown below the corresponding protein bands. (**C**) The effect of 9 adaptor proteins on virus amounts in cell supernatant after 72 h infection was detected via plaque assay.

3.4. The Antiviral Activities of Endogenous TRIF, MAVS, and STING Signaling against PEDV Replication

Based on the above screening results and considering the significance of the nucleic acid-sensing signal in the antiviral activity, TRIF, MAVS, and STING were chosen and their endogenous signaling was investigated for antiviral roles in PEDV infection. Several agonists were used to trigger endogenous innate signaling, including poly(I:C)-HMW for TLR3-TRIF signaling, poly(I:C)-LMW for RIG-I/MDA5-MAVS signaling, 2′3′-cGAMP for STING signaling, and poly(dA:dT) for cGAS-STING signaling. RT-qPCR (Figure 4A,B), WB detection (Figure 4C), and plaque assay (Figure 4D and Figure S3) all demonstrated that the poly(I:C)-HMW-induced TRIF signaling, poly(I:C)-LMV-induced MAVS signaling, 2′-3′-cGAMP, and poly(dA:dT)-activated STING signaling possess obvious anti-PEDV activity. The anti-PEDV effect in Vero cells was more obvious at 72 h post-infection than at 48 h post-infection (Figure 4A–C).

3.5. Knockdown of Endogenous TRIF, MAVS, and STING Enhances PEDV Replication

To further explore the roles of endogenous TRIF, MAVS, and STING in PEDV replication, the previous validated siRNAs were used to knock down TRIF, MAVS, and STING in Vero cells prior to PEDV infection [15]. In the TRIF, MAVS, and STING siRNA-treated Vero cells, PEDV N mRNA and N protein expressions were obviously increased at 72 h post-infection relative to control siRNA-treated cells (Figure 5A,B). Similarly, compared with control siRNA-treated cells, PEDV titers were obviously heightened in TRIF, MAVS, and STING siRNA-treated cells at 72 h post-infection (Figures 5C and S3). These results suggest that endogenous TRIF, MAVS, and STING are all part of the host defense mechanism and necessary for inhibition of PEDV replication.

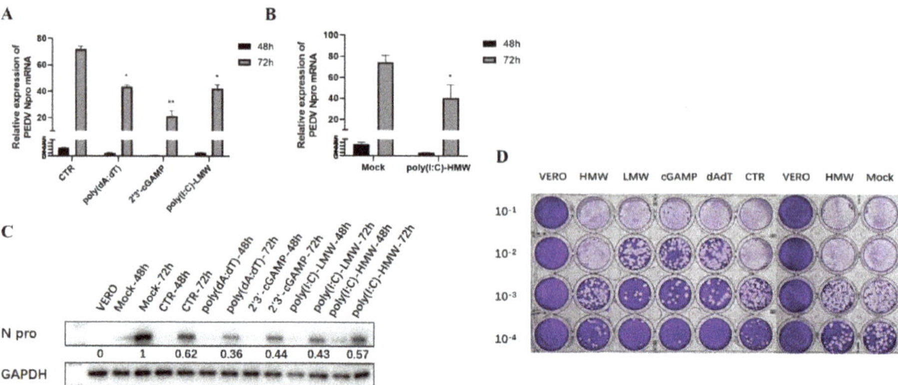

Figure 4. The effects of endogenous TRIF, MAVS, and STING signaling on PEDV replication in Vero cells. (**A**,**B**) The effect of different stimulants on PEDV replication at both 48 h and 72 h post-infection was detected via RT-qPCR. CTR denotes the transfection control and Mock the non-treatment control. * $p < 0.05$, ** $p < 0.01$. (**C**) The effect of stimulation on PEDV replication at both 48 h and 72 h post-infection was detected via WB. The gray values of each N protein bands after normalization by GAPDH are shown below the corresponding protein bands. (**D**) The effect of stimulants on the amount of virus in the supernatant after 72 h infection was detected via plaque assay.

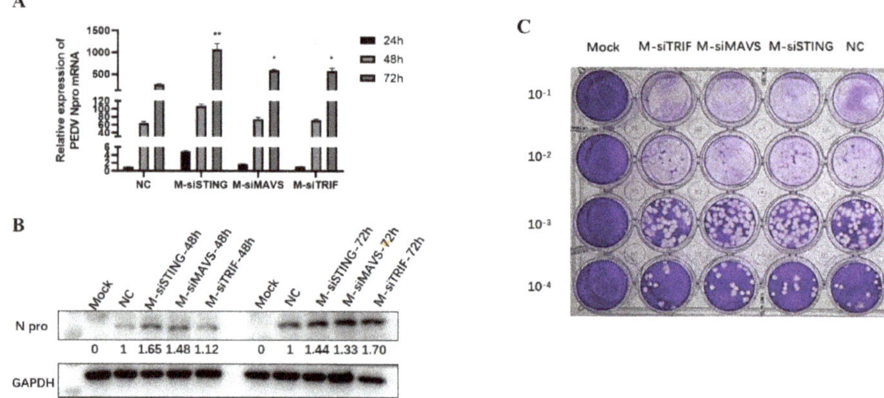

Figure 5. The effects of knockdown of endogenous TRIF, MAVS, and STING on PEDV replication in Vero cells. (**A**) The effect of signaling adaptor knockdown on PEDV N gene expression was detected via RT-qPCR. NC denotes the negative control siRNA. * $p < 0.05$, ** $p < 0.01$. (**B**) The effect of signaling adaptor knockdown on PEDV N protein expression. The gray values of each N protein bands after normalized by GAPDH are shown below the corresponding protein bands. (**C**) The effect of signaling adaptor knockdown on PEDV replication after 72 h infection in Vero cell was detected via plaque assay.

3.6. PEDV Replication Was Enhanced in TRIF$^{-/-}$, MAVS$^{-/-}$ and STING$^{-/-}$ Vero Cells

In order to further verify the anti-PEDV infection effects of STING, TRIF, and MAVS in Vero cells, we used CRISPR/Cas9 gene editing technology to construct STING, TRIF, and MAVS knockout Vero cell lines, respectively. Clones of TRIF$^{-/-}$, MAVS$^{-/-}$, and STING$^{-/-}$ Vero cells were obtained and used for PEDV infection together with normal Vero cells as a control. Compared with the control Vero cells (WT), the expression of N gene mRNA and N protein in TRIF$^{-/-}$, MAVS$^{-/-}$, and STING$^{-/-}$ Vero cells after PEDV infection and the virus content in culture supernatant were significantly increased (Figure 6A–C and Figure S3). Additionally, WB again confirmed that the adaptor protein expression in the corresponding knockout cells disappeared as expected (Figure 6B). These results further proved the important roles of TRIF, MAVS, and STING in host defense against PEDV infection, especially the important influence of TRIF and MAVS in PEDV infections.

3.7. Knockdown of Endogenous TRIF, MAVS, and STING in IPEC-DQ Cells Promoted PEDV Replication

PEDV is a diarrhea virus prevalent in pigs. So as to further detect the effects of endogenous TRIF, MAVS, and STING on the replication of PEDV, we used relevant IPEC-DQ cell line of piglet jejunal epithelium for PEDV infection. First, different individual siRNA sequences of porcine TRIF, MAVS, and STING were transfected into IPEC-DQ cells at a concentration of 50–150 nM, respectively, and cells were collected 24 h later for detection to determine siRNA knockdown efficiency. In accordance with the knockdown effect on protein level detected via WB (Figure 7A–C), STING 11 siRNA, MAVS 2329 siRNA, and TRIF 5800 siRNA were finally selected for subsequent experiments at a working concentration of 100 nM.

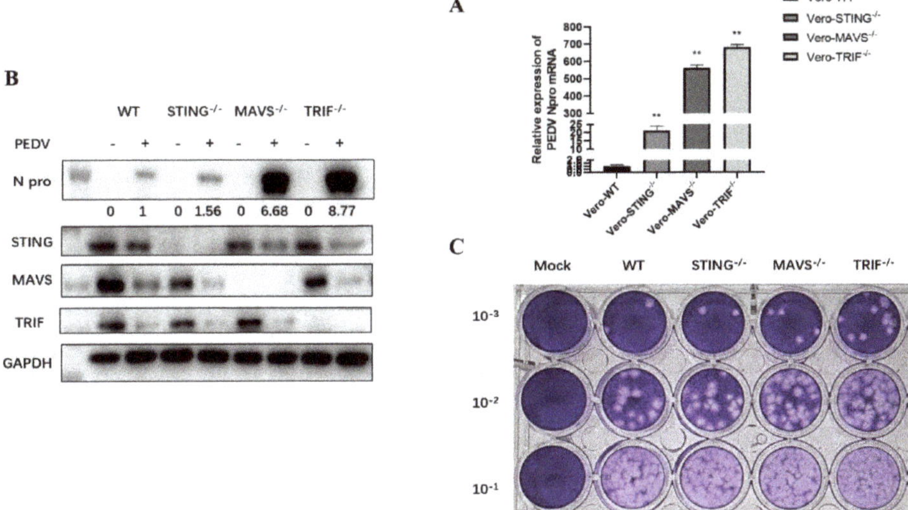

Figure 6. PEDV replication was increased in TRIF$^{-/-}$, MAVS$^{-/-}$, and STING$^{-/-}$ Vero cells. (**A**) RT-qPCR was used to detect PEDV N gene mRNA expression in knockout (KO) cells and wild-type (WT) normal Vero cells. ** $p < 0.01$. (**B**) The expression of PEDV N protein in knockout cells was detected via WB. The gray values of each N protein bands after normalization by GAPDH are shown below the corresponding protein bands. (**C**) The PEDV content in the knockout cell supernatants was detected via plaque assay.

The three siRNA duplexes and negative control siRNA duplex were transfected into IPEC-DQ cells using Lipofectamine 2000 for 24 h, followed by PEDV infection. At 72 h post-infection, RT-qPCR, WB, and plaque assay showed that when TRIF, MAVS, and STING in IPEC cells were silenced by corresponding siRNA, N gene transcription, N protein expression, and virus replication levels of PEDV were slightly increased compared with the control group (Figure 7D–F and Figure S3). These results further indicated that a variety of porcine innate immune signaling pathways exert the anti-PEDV function.

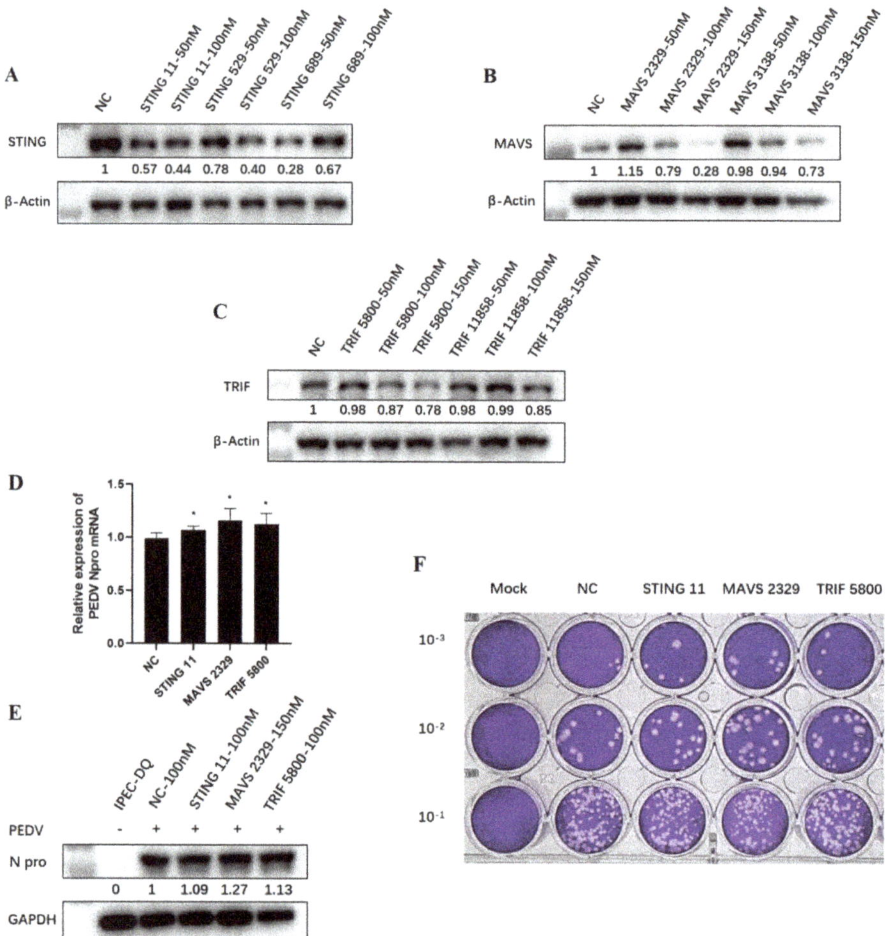

Figure 7. Effects of siRNA knockdown of endogenous TRIF, MAVS, and STING on PEDV replication in IPEC-DQ cells. (**A–C**) WB detection of TRIF, MAVS, and STING siRNA knockdown efficiency by different siRNAs. The gray values of each adaptor protein bands after normalization by β-Actin are shown below the corresponding protein bands. (**D**) The effect of signaling adaptor knockdown on PEDV replication was detected via RT-qPCR. * $p < 0.05$. (**E**) The impact of signaling adaptor knockdown on PEDV replication was detected via WB. The gray values of each N protein bands after normalization by GAPDH are shown below the corresponding protein bands. (**F**) The effect of adaptor knockdown on the level of supernatant PEDV at 72 h post-infection was detected via plaque assay.

4. Discussion

Porcine epidemic diarrhea (PED) is a swine disease with worldwide distribution which has important economic significance, posing a huge threat to the swine industry. At present, PEDV variants have emerged and evolved persistently [20]. Although there are several PED vaccines available on the market, protective immunity from vaccination depends on both effective innate immunity and adaptive immunity [21]. Innate immune response is the first line of host defense during infection and plays a crucial role in early recognition of and immune protection against virus infection. However, PEDV does not elicit a robust antiviral IFN response; the interaction between PEDV and host immunity is complex, and its mechanism has not been fully understood until recently [22]. To understand and maximize the potential of PEDV innate immunity, we examined the anti-PEDV innate immune response globally and completely by investigating the signaling adaptors which represent and reflect all currently known innate immune signaling pathways. Our results and findings indicated that multiple innate immune signaling pathways are involved in the recognition of and defense against PEDV infection.

Among different PRR families, three types of PRRs have been identified in the recognition of viral nucleic acids, including RLR detection of viral RNA in the cytoplasm [23], TLR recognition of viral RNA or DNA in the endosome [24], and CDR detection of viral DNA in the cytoplasm [25]. Here, we found and confirmed the significant roles of the RLR adaptor MAVS, the TLR3 adaptor TRIF, and the CDR adaptor STING in anti-PEDV response. Additionally, other innate signaling adaptors MyD88, RIPK2, ASC, and CARD9-BCL10-MALT1 (CBM complex) also appeared to play a role in anti-PEDV response. MyD88 is the common signaling adaptor for all TLRs except TLR3 [26]. RIPK2 is the signaling adaptor for NLRs NOD1 and NOD2, whereas ASC is the adaptor for other NLRs [27]. The signaling adaptor complex of CBM is for all CLRs [28]. Among these various PRRs, the roles of some PRRs such as NLRs and CLRs have never been investigated and need to be validated and investigated during PEDV infection.

Pigs at all life stages can be infected with PEDV; however, both the morbidity and mortality of suckling piglets under 7 days of age are very high. The mortality of fattening pigs and sows after infection is very low, and most infected pigs show recessive or transient diarrhea. Adult pigs are raised with mucosal immunity and often immunized with vaccines compared with newborn piglets. It is very likely that there is a difference in the PRR-adaptor cell signaling pathways between the intestinal epithelial cells of young vs. older pigs. Additionally, other resistance mechanisms may also be at play, which differs between young vs. old pigs.

We previously showed that multiple innate immune signaling pathways are involved in the recognition of and defense against porcine reproductive and respiratory syndrome virus (PRRSV), a member of *Arteriviridae* family [15]. Both PRRSV and PEDV harbor positive-sense single-stranded genomic RNA, which may confer similarities in the induced innate immune responses. Furthermore, the phenotype may be general to other viruses, that is, many virus infections are recognized and counteracted by several innate immune PRR-mediated signaling pathways. Subsequently, several intriguing questions arise: (1) What is the relative contribution of each PRR signaling pathway in the anti-PEDV innate immune response? (2) What is the relationship between different PRR signaling pathways in PEDV infection? (3) How does PEDV evade each of these PRR signaling pathways? Answering these questions will reveal an intricate and delicate interaction network between PEDV and host, which deserves further researches in the future.

In summary, our results suggest that multiple porcine PRR-mediated signaling pathways are involved in PEDV recognition and defense. These results deepen our understanding of PEDV innate immunity and help maximize the potential of innate immunity to control PEDV infection.

Supplementary Materials: The following supporting information can be downloaded at: https://www.mdpi.com/article/10.3390/v15081629/s1, Figure S1: The specific recognition of PEDV in Vero cells by PEDV N mAb; Figure S2: The genome sequencing results and characterization of STING, MAVS and TRIF KO cell clones; Figure S3: The quantitative column diagrams of the plaque assays in this study.

Author Contributions: Y.Z., Y.X., S.J., S.S., J.Z. (Jiajia Zhang), J.L. and Q.C. performed the experiments; W.Z., F.M. and N.C. contributed to data analysis; Y.Z. and J.Z. (Jianzhong Zhu) conceived the idea and wrote the paper. All authors have read and agreed to the published version of the manuscript.

Funding: This research was partly funded by the National Natural Science Foundation of China grant number [32172867; 32202818], the 111 Project grant number [D18007] and A Project Funded by the Priority Academic Program Development of Jiangsu Higher Education Institutions (PAPD).

Institutional Review Board Statement: Not applicable.

Informed Consent Statement: Not applicable.

Data Availability Statement: More data are available upon request.

Conflicts of Interest: The authors have no financial conflict of interest.

References

1. Jung, K.; Saif, L.J. Porcine epidemic diarrhea virus infection: Etiology, epidemiology, pathogenesis and immunoprophylaxis. *Vet. J.* **2015**, *204*, 134–143. [CrossRef]
2. Li, W.; Li, H.; Liu, Y.; Pan, Y.; Deng, F.; Song, Y.; Tang, X.; He, Q. New variants of porcine epidemic diarrhea virus, China, 2011. *Emerg. Infect. Dis.* **2012**, *18*, 1350–1353. [CrossRef] [PubMed]
3. Sun, M.; Ma, J.; Wang, Y.; Wang, M.; Song, W.; Zhang, W.; Lu, C.; Yao, H. Genomic and epidemiological characteristics provide new insights into the phylogeographical and spatiotemporal spread of porcine epidemic diarrhea virus in Asia. *J. Clin. Microbiol.* **2015**, *53*, 1484–1492. [CrossRef] [PubMed]
4. Jung, K.; Saif, L.J.; Wang, Q. Porcine epidemic diarrhea virus (PEDV): An update on etiology, transmission, pathogenesis, and prevention and control. *Virus. Res.* **2020**, *286*, 198045. [CrossRef] [PubMed]
5. Lin, F.; Zhang, H.; Li, L.; Yang, Y.; Zou, X.; Chen, J.; Tang, X. PEDV: Insights and Advances into Types, Function, Structure, and Receptor Recognition. *Viruses* **2022**, *14*, 1744. [CrossRef] [PubMed]
6. Li, S.; Yang, J.; Zhu, Z.; Zheng, H. Porcine Epidemic Diarrhea Virus and the Host Innate Immune Response. *Pathogens* **2020**, *9*, 367. [CrossRef]
7. Du, J.; Luo, J.; Yu, J.; Mao, X.; Luo, Y.; Zheng, P.; He, J.; Yu, B.; Chen, D. Manipulation of Intestinal Antiviral Innate Immunity and Immune Evasion Strategies of Porcine Epidemic Diarrhea Virus. *Biomed Res. Int.* **2019**, *2019*, 1862531. [CrossRef]
8. Beutler, B. Innate immunity: An overview. *Mol. Immunol.* **2004**, *40*, 845–859. [CrossRef]
9. Li, D.; Wu, M. Pattern recognition receptors in health and diseases. *Signal Transduct. Target. Ther.* **2021**, *6*, 291. [CrossRef]
10. Thompson, M.R.; Kaminski, J.J.; Kurt-Jones, E.A.; Fitzgerald, K.A. Pattern recognition receptors and the innate immune response to viral infection. *Viruses* **2011**, *3*, 920–940. [CrossRef]
11. Sun, M.; Yu, Z.; Luo, M.; Li, B.; Pan, Z.; Ma, J.; Yao, H. Screening Host Antiviral Proteins under the Enhanced Immune Responses Induced by a Variant Strain of Porcine Epidemic Diarrhea Virus. *Microbiol. Spectr.* **2022**, *10*, e0066122. [CrossRef] [PubMed]
12. Wong, L.Y.; Lui, P.Y.; Jin, D.Y. A molecular arms race between host innate antiviral response and emerging human coronaviruses. *Virol. Sin.* **2016**, *31*, 12–23. [CrossRef] [PubMed]
13. Zhang, Q.; Ke, H.; Blikslager, A.; Fujita, T.; Yoo, D. Type III Interferon Restriction by Porcine Epidemic Diarrhea Virus and the Role of Viral Protein nsp1 in IRF1 Signaling. *J. Virol.* **2018**, *92*, e01677-17. [CrossRef] [PubMed]
14. Zhang, K.; Lin, S.; Li, J.; Deng, S.; Zhang, J.; Wang, S. Modulation of Innate Antiviral Immune Response by Porcine Enteric Coronavirus. *Front. Microbiol.* **2022**, *13*, 845137. [CrossRef]
15. Xu, Y.; Ye, M.; Zhang, Y.; Sun, S.; Luo, J.; Jiang, S.; Zhang, J.; Liu, X.; Shao, Q.; Cao, Q.; et al. Screening of Porcine Innate Immune Adaptor Signaling Revealed Several Anti-PRRSV Signaling Pathways. *Vaccines* **2021**, *9*, 1176. [CrossRef]
16. Desmyter, J.; Melnick, J.L.; Rawls, W.E. Defectiveness of interferon production and of rubella virus interference in a line of African green monkey kidney cells (Vero). *J. Virol.* **1968**, *2*, 955–961. [CrossRef]
17. Kono, S.; Koase, M.; Sakata, H.; Shimizu, Y. Variable interferon productivity of Vero cells. Brief report. *Arch Gesamte Virusforsch* **1972**, *37*, 141–143. [CrossRef] [PubMed]
18. Jiang, S.; Ao, D.; Ni, J.; Chen, N.; Meurens, F.; Zhu, J. The signaling relations between three adaptors of porcine C-type lectin receptor pathway. *Dev. Comp. Immunol.* **2020**, *104*, 103555. [CrossRef]
19. Ao, D.; Li, S.; Jiang, S.; Luo, J.; Chen, N.; Meurens, F.; Zhu, J. Inter-relation analysis of signaling adaptors of porcine innate immune pathways. *Mol. Immunol.* **2020**, *121*, 20–27. [CrossRef]
20. Li, M.; Pan, Y.; Xi, Y.; Wang, M.; Zeng, Q. Insights and progress on epidemic characteristics, genotyping, and preventive measures of PEDV in China: A review. *Microb. Pathog.* **2023**, *181*, 106185. [CrossRef]
21. Dong, S.; Xie, C.; Si, F.; Chen, B.; Yu, R.; Li, Z. Immunization against porcine epidemic diarrhea virus and vaccine development. *Sheng Wu Gong Cheng Xue Bao* **2021**, *37*, 2603–2613. [PubMed]
22. Zhang, Y.; Chen, Y.; Zhou, J.; Wang, X.; Ma, L.; Li, J.; Yang, L.; Yuan, H.; Pang, D.; Ouyang, H. Porcine Epidemic Diarrhea Virus: An Updated Overview of Virus Epidemiology, Virulence Variation Patterns and Virus-Host Interactions. *Viruses* **2022**, *14*, 2434. [CrossRef]
23. Yoneyama, M.; Onomoto, K.; Jogi, M.; Akaboshi, T.; Fujita, T. Viral RNA detection by RIG-I-like receptors. *Curr. Opin. Immunol.* **2015**, *32*, 48–53. [CrossRef]
24. Kawai, T.; Akira, S. Toll-like receptor and RIG-I-like receptor signaling. *Ann. N. Y. Acad. Sci.* **2008**, *1143*, 1–20. [CrossRef] [PubMed]
25. Sparrer, K.M.; Gack, M.U. Intracellular detection of viral nucleic acids. *Curr. Opin. Microbiol.* **2015**, *26*, 1–9. [CrossRef]
26. Kawai, T.; Akira, S. TLR signaling. *Semin. Immunol.* **2007**, *19*, 24–32. [CrossRef] [PubMed]

27.   Platnich, J.M.; Muruve, D.A. NOD-like receptors and inflammasomes: A review of their canonical and non-canonical signaling pathways. *Arch. Biochem. Biophys.* **2019**, *670*, 4–14. [CrossRef]
28.   Geijtenbeek, T.B.; Gringhuis, S.I. Signalling through C-type lectin receptors: Shaping immune responses. *Nat. Rev. Immunol.* **2009**, *9*, 465–479. [CrossRef]

Article

Alternative Splicing of RIOK3 Engages the Noncanonical NFκB Pathway during Rift Valley Fever Virus Infection

Thomas Charles Bisom [1], Hope Smelser [1], Jean-Marc Lanchy [2] and J. Stephen Lodmell [2,3,*]

[1] Department of Chemistry and Biochemistry, University of Montana, Missoula, MT 59801, USA; thomas.bisom@umconnect.umt.edu (T.C.B.); hope.smelser@umconnect.umt.edu (H.S.)
[2] Division of Biological Sciences, University of Montana, Missoula, MT 59801, USA; jean-marc.lanchy@umontana.edu
[3] Center for Biomolecular Structure and Dynamics, University of Montana, Missoula, MT 59801, USA
* Correspondence: stephen.lodmell@umontana.edu

Abstract: Although the noncanonical NFκB pathway was originally identified as a cellular pathway contributing to lymphoid organogenesis, in the past 20 years, its involvement in innate immunity has become more appreciated. In particular, the noncanonical NFκB pathway has been found to be activated and even exploited by some RNA viruses during infection. Intriguingly, activation of this pathway has been shown to have a role in disrupting transcription of type 1 interferon (IFN), suggesting a rationale for why this response could be co-opted by some viruses. Rift Valley fever virus (RVFV) is a trisegmented ambisense RNA virus that poses a considerable threat to domestic livestock and human health. Previously, we showed the atypical kinase RIOK3 is important for mounting an IFN response to RVFV infection of human epithelial cells, and shortly following infection with RVFV (MP12 strain), RIOK3 mRNA is alternatively spliced to its X2 isoform that encodes a truncated RIOK3 protein. Alternative splicing of RIOK3 mRNA has an inhibitory effect on the IFN response but also stimulates an NFκB-mediated inflammatory response. Here, we demonstrate alternative splicing of RIOK3 mRNA is associated with activation of the noncanonical NFκB pathway and suggest this pathway is co-opted by RVFV (MP12) to enhance viral success during infection.

Keywords: noncanonical NFκB pathway; Rift Valley fever virus; RIOK3; innate immunity; alternative splicing

Citation: Bisom, T.C.; Smelser, H.; Lanchy, J.-M.; Lodmell, J.S. Alternative Splicing of RIOK3 Engages the Noncanonical NFκB Pathway during Rift Valley Fever Virus Infection. *Viruses* **2023**, *15*, 1566. https://doi.org/10.3390/v15071566

Academic Editors: Caijun Sun and Feng Ma

Received: 1 July 2023
Accepted: 14 July 2023
Published: 18 July 2023

1. Introduction

Careful co-ordination of interferon (IFN) and inflammatory responses to viral infection is essential for robust cellular immunity as well as avoidance of collateral tissue damage caused by overblown immune responses. Viruses have evolved many strategies to dysregulate discrete pathways to evade cellular responses to infection. Such dysregulation can be harmful to the host. For example, with the recent COVID-19 pandemic, exacerbated inflammatory responses coupled with a muted IFN response were frequently observed in the most severe clinical cases [1,2]. Therefore, understanding how viruses exploit innate immune pathways within the cell could be valuable for developing countermeasures to severe viral infections.

Rift Valley fever virus (RVFV; order *Bunyavirales*, family *Phenuiviridae*, genus *Phlebovirus* [3]) is an arthropod-borne RNA virus that poses a considerable threat to human health and agriculture. Transmission typically occurs from mosquitoes of the *Aedes* and *Culex* genera to humans and livestock, such as sheep, goats, and cattle [4–6]. Human infections can also occur by handling meat, blood, or milk or inhaling aerosolized droplets from fluids of infected animals [7]. Symptoms of human infection range from flu-like symptoms to hemorrhagic fever [8,9]. Hepatic complications, renal failure, and miscarriage have also been reported in humans [10,11], and these pathologies have likewise been observed in animal models of Rift Valley fever and in livestock [12–16]. Due to its virulence and risk of being transmitted through aerosolization, RVFV is also considered a threat to national security in some countries, including the United States, where it is considered a USDA/HHS Overlap Select Agent [17,18]. Understanding the host innate immunologic responses to RVFV infection is, therefore, valuable to global human health and security.

RVFV has an ambisense tripartite RNA genome and, following infection of a cell, genomic viral RNA is detected predominantly by the retinoic acid inducible gene I (RIG-I) intracellular receptor [19,20]. After detecting viral RNA, RIG-I associates with mitochondrial antiviral signaling protein (MAVS) and, following recruitment of TANK binding kinase 1 (TBK1) to MAVS and subsequent phosphorylation of interferon regulatory factor 3 (IRF3) by TBK1, a type 1 IFN response is initiated [21–23]. This results in transcription and translation of type 1 IFNs, such as IFNβ, which, following translation, activate the JAK/STAT pathway through autocrine and paracrine signaling [24,25]. This pathway then results in transcription of a subset of genes, the IFN stimulated genes, that induce an antiviral state in the cell [26–28].

Although the noncanonical NFκB pathway (also known as the alternative NFκB pathway) was originally identified as a cellular pathway involved in adaptive immunity and secondary lymphoid organogenesis [29–32], it has more recently become appreciated in cellular innate immunity for its role in transcription of inflammatory cytokines and chemokines during viral infection [33–37]. Intriguingly, it has also been shown to disrupt transcription of type 1 IFN [38]. During activation of the noncanonical NFκB pathway, an activating signal frees NFκB-inducing kinase (NIK; also known as MAP3K14) from ubiquitination by an E3 ubiquitin ligase complex comprised of TRAF3 (linking NIK to the complex), TRAF2, and cIAP1/2 and thereby spares NIK from proteasomal degradation [39–43]. NIK then phosphorylates IKKα [44]. Phosphorylated IKKα then phosphorylates p100 (also known as NFκB2) dimerized with RelB in the cytosol, leading to proteolytic processing of p100 to p52 [33,45–47]. The noncanonical NFκB dimer of p52 complexed with RelB then enters the nucleus for transcription of genes specific to the noncanonical NFκB pathway, such as CXCL13 (also knowns as BLC) and CCL19 (also known as ELC) [48]. Furthermore, while in the nucleus, the noncanonical NFκB dimer is thought to inhibit the IFN response in part by competing with the canonical (also known as classical) NFκB dimer for the κB site at the IFNβ promoter (Figure 1) [38].

Canonical (p50:RelA, p50:c-Rel, and p50:p50) and noncanonical (p52:RelB) NFκB dimers belong to the same family of NFκB transcription factors and share a Rel homology domain [49–52], but there are key differences in their transcriptional activities and in how and when these transcription factors become active [53–55]. For example, in order for canonical NFκB dimers to become transcriptionally active, they must be freed from members of inhibitor of κB (IκB) family of inhibitory proteins in the cytosol, such as IκBα [56–58]. Upon phosphorylation, IκB proteins are ubiquitinated and then degraded by the proteosome, allowing canonical NFκB dimers to enter the nucleus for transcription of target genes [59–61].

There are several other characteristic differences between the canonical and noncanonical NFκB pathways. Typically, activation of the canonical pathway occurs almost immediately upon immune stimulation, but the noncanonical pathway has a much slower onset and depends on de novo protein synthesis [32,33,62,63]. Distinct receptors activate the noncanonical and canonical pathways, and activation of the noncanonical pathway is more selective and exclusive to a subset of receptors [32,33]. The noncanonical pathway can be activated through binding of the LTβR receptor with ligands, such as LIGHT [64,65]. Activation of the canonical pathway is typified by binding of TNFα with TNFR1 receptor [66]. However, both canonical and noncanonical pathways are known to be interwoven, and activation of one pathway can sometimes result in or even rely on activation of the other. For example, when the noncanonical NFκB pathway was first described, expression (but not processing) of p100 was shown to depend on the canonical NFκB dimer, p50:RelA [65].

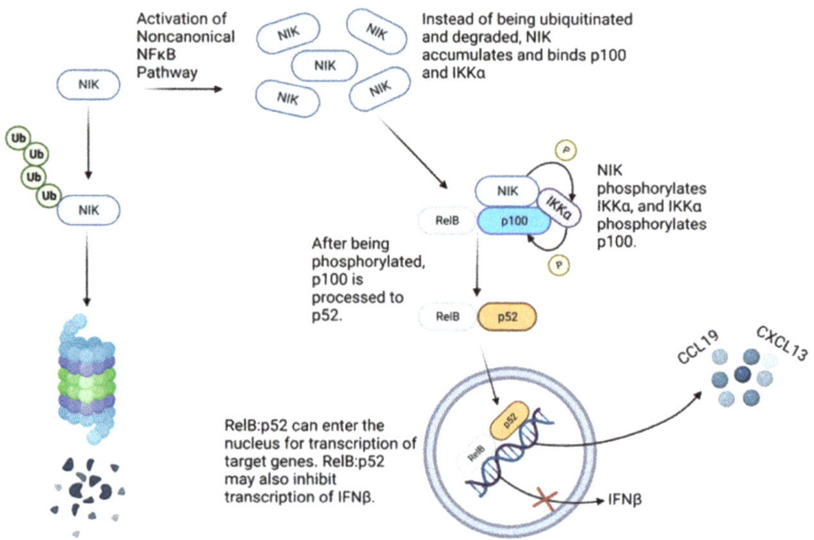

Figure 1. Noncanonical NFκB pathway overview. See text for details. Typically, the onset of the noncanonical NFκB pathway is delayed compared to that of the canonical NFκB pathway. In this study, however, we show engagement of the noncanonical pathway prior to or concomitant with the canonical pathway during RVFV MP12 infection.

In addition to a type 1 IFN response that is initiated by RIG-I detection of viral RNA, both canonical and noncanonical NFκB pathways can also be activated when RIG-I associates with MAVS [36,67]. We and others have shown the atypical kinase RIOK3 is important for regulating a RIG-I-induced IFN response [68–72]. However, shortly following RVFV infection, a shift toward alternative splicing of RIOK3 mRNA to its X2 isoform occurs, culminating in a diminished IFN response but an elevated NFκB-mediated inflammatory response in human epithelial cells [70,73]. Therefore, promoting this alternative splicing of RIOK3 mRNA could serve as a means for RVFV to evade a type 1 IFN response and potentially lead to a dysregulated innate immune response. Here, we show the noncanonical NFκB pathway is activated early during RVFV (MP12 strain) infection and that alternative splicing of RIOK3 occurs specifically during activation of the noncanonical NFκB pathway. Furthermore, the alternatively spliced isoform of RIOK3, called RIOK3 X2, aids in expression of p100, which diminishes the type 1 IFN response. Collectively, these observations suggest that RVFV infection and alternative splicing of RIOK3 mRNA accompany induction of the noncanonical NFκB pathway, which decreases IFN production and enhances viral multiplication.

2. Materials and Methods

2.1. Cell Culture, Viruses, and Infections

Vero 76, human embryonic kidney 293 (HEK 293), and hepatocarcinoma (HepG2) cells were obtained from ATCC (Manassas, VA, USA). Vero 76 and HEK 293 cells were cultured in Dulbecco's Modified Eagle Medium (DMEM) containing 10% fetal bovine serum (FBS) and penicillin + streptomycin (pen/strep) (Thermo Fisher Scientific, Waltham, MA, USA), and HepG2 cells were grown in 1:1 DMEM:F12 containing 10% FBS and pen/strep (Thermo Fisher Scientific). Since renal failure and hepatic complications have frequently been observed in severe cases of RVFV infection, HEK 293 and HepG2 cells were selected for use in this study. Experiments were initiated when cells were near confluency (70–90%). Furthermore, the attenuated biosafety level 2 (BSL2) laboratory RVFV strain MP12 (kindly provided by Brian Gowen Utah State University, Logan, UT, USA) was used for all viral infections. MP12 is an attenuated RVFV strain derived from the wt ZH548 isolate, and this strain was used because it can be handled in BSL2 containment. RVFV MP12 harbors mutations in all three of its genomic segments to mitigate its pathogenicity [74], but many of the virological attributes of MP12, such as its cytopathogenicity in cell culture, are still shared with RVFV ZH548, making MP12 a good surrogate for initial studies of RVFV. MP12 viral stocks were propagated in Vero 76 cells, and the second parental generation was used for infections in experiments. During experimental infections, cells were incubated with virus in DMEM or DMEM:F12 containing no FBS and no pen/strep for 1 h. Media were then exchanged with DMEM or DMEM:F12 containing 2% FBS and pen/strep, and the cells were incubated at 37 °C, 5% CO_2; see text for specific times when cells were harvested post-infection. Quantification of infectious dose was performed in TCID50 assays described by Smith et al. [75] with Vero 76 cells, and TCID50/mL values were obtained using the Reed and Muench Method. Manipulations of the viruses used in this study are compliant with both the Institutional Biosafety Committee at the University of Montana, Missoula, and NIH requirements in regard to their handling under BSL2 containment conditions.

2.2. Reagents

TNFα and LIGHT were purchased from BioLegend (San Diego, CA, USA). Prior to treating cells with TNFα or LIGHT, growth media were removed; cells were washed with PBS, and PBS was exchanged with fresh DMEM media containing 10 or 2% FBS and pen/strep. Then, when treating with TNFα, TNFα was added such that its final concentration in each well was 20 ng/mL, and when treating with LIGHT, LIGHT was added to each well to reach a final concentration of 50 ng/mL. After treating with either reagent, cells were then incubated at 37 °C, 5% CO_2; see text for incubation times used in each assay.

2.3. Plasmids, Oligonucleotides, and Transfections

Plasmid transfections were accomplished with Lipofectamine 2000 (Thermo Fisher Scientific). In these transfections, growth media were removed, cells were washed with PBS, and PBS in each well was exchanged with a volume of OptiMEM transfection media equal to half the volume of growth media. The cells were then incubated at 37 °C, 5% CO_2. While the cells were incubating, transfection master mixes were prepared. When preparing Lipofectamine master mixes, Lipofectamine was added to OptiMEM media and incubated at room temperature for 5–20 min. Lipofectamine was added to each master mix such that the ratio of μL Lipofectamine 2000 to μg of construct transfected would be ~2.5 per well. After incubation of Lipofectamine master mixes, these mixtures were added to DNA construct master mixes comprising construct DNA and OptiMEM, and the resulting transfection mixture was left to incubate at room temperature for 20 min to allow assembly of DNA: liposome complexes. Transfection mixtures were then dispensed into wells with cells incubated in OptiMEM media, and the mixtures were added to each well at a volume equal to half the volume of growth media. Before any downstream treatments were initiated, cells were incubated with the transfection mixtures at 37 °C, 5% CO_2 for at least 24 h to allow sufficient expression of the constructs. The RIOK3 X2 construct, which we previously showed expressed the predicted protein/peptide, was expressed in a phRL-CMV backbone, and the empty vector (EV) control plasmid described in the text was phRL-CMV backbone lacking any insert [70].

2.4. RNA Extractions and Reverse Transcription

When extracting RNA, cells were lysed with TRIzol (Thermo Fisher Scientific), and RNA was extracted from the reagent according to the manufacturer's protocol. To separate RNA into an aqueous layer, lysates were first treated with 0.2 mL chloroform per 1 mL TRIzol. Following vortex and incubation at room temperature for 10 min, the lysates were spun at 12,000× g and 4 °C for 15 min. Aqueous layers containing RNA were then transferred to new tubes, and to precipitate the RNA, 2.5 μL of 4 mg/mL glycogen and

0.5 mL isopropanol per 1 mL of TRIzol used were added. After vortex and incubation at room temperature for 10 min, RNA samples were then spun at 12,000× g and 4 °C for 10 min. Supernatant was then removed, and the RNA pellets were washed with 1 mL of 75% ethanol per 1 mL of TRIzol used. Washes were accomplished by brief vortex of the pellets in 75% ethanol and centrifugation at 7600× g and 4 °C for 5 min. The 75% ethanol solution was then removed, and the pellets dried at room temperature for 5–10 min. RNA was then resuspended in nanopure H_2O, incubated at 55–60 °C for 10–15 min, and stored at −80 °C or used immediately. Extracted RNA was then reverse transcribed to cDNA with Maxima H Minus Reverse Transcriptase (Thermo Fisher Scientific). Prior to reverse transcription, RNA in each sample was diluted to ~100 ng/μL. Diluted RNA was incubated with random hexamer primers and dNTPs at 65 °C for 5 min, chilled, and reverse transcription reagents (reverse transcription buffer and Maxima H Minus Reverse transcriptase) were then added. After addition of reverse transcription reagents, samples were incubated at 25 °C for 10 min., 50 °C for 30 min, and 85 °C for 5 min. Resulting cDNA was then either immediately used or stored at −20 °C.

2.5. RT-qPCR

RT-qPCR measurements were obtained using SYBR Green Master Mix (Thermo Fisher Scientific) and the CFX Connect Real-Time PCR Detection System (Bio-Rad, Hercules, CA, USA). First, cDNA was diluted 1:5 in nanopure H_2O. Measurements were obtained using a 384-well plate, and each well with sample contained 5 μL SYBR Green Master Mix, 1 μL forward + reverse 10 μM primer mix, and 4 μL diluted cDNA. cDNA was then amplified and measured under the following conditions: (1) 50 °C for 2 min, (2) 95 °C for 2 min, (3) 95 °C for 15 s, (4) 60 °C for 1 min, (5) plate read, (6) go to (3) 39×, (7) 65 °C for 31 s, (8) 65 °C for 5 s + 0.5 °C/cycle Ramp 0.5 °C/s, (9) plate read, (10) go to 8) 60×. Relative normalized expression was quantified in CFX Maestro (Bio-Rad) using the ΔΔCt method, and RNA was normalized to GAPDH (see Table S1 for qPCR primer sequences).

2.6. Gel Electrophoresis

Prior to electrophoresis, cDNA was amplified by PCR using InPhusion Flash Hi-Fidelity PCR Master Mix (Thermo Fisher Scientific). Each reaction mixture consisted of 10 μL Phusion Flash PCR Master Mix, 1 μL of 10μM forward primer, 1 μL of 10 μM reverse primer, 6 μL nanopure H_2O, and 2 μL cDNA. cDNA was then amplified under the following conditions: (1) 98 °C for 10 s, (2) 98 °C for 1 s, (3) 65 °C for 5 s, (4) 72 °C for 15 s, (5) go to (2) 29×, (6) 72 °C for 1 min, (7) 4 °C forever. Amplified cDNAs were then resolved on a 1% agarose gel at 80 V. After rocking gels for 10 min in a 1× TAE bath containing ethidium bromide, images were obtained using a Gel Doc XR + Molecular Imager (Bio-Rad). Figures show representative samples of biological duplicates or triplicates.

2.7. Western Blotting

Cells harvested for Western blotting were grown in 6-well plates. Prior to harvest, media was removed, and the cells were washed with PBS (Thermo Fisher Scientific). To remove cells from the plate, cell scraping or trypsin (Thermo Fisher Scientific) was used. When scraping cells, after PBS wash, 5 mL PBS were added per well while the 6-well plate was on ice. Cells were scraped, and suspended cells were transferred to conical centrifuge tubes kept on ice. A total of 5 mL of PBS were again added per well, and scraping and transfer of suspensions to centrifuge tubes was repeated. The cells were then centrifuged at ~1000 rpm, 4 °C for 5–10 min. PBS was then removed, and the cells were either lysed immediately or stored at −80 °C until lysis. When removing cells from 6-well plates with trypsin, 1 mL trypsin was added per well after washing with PBS. Cells were incubated at 37 °C, 5% CO_2 for 2–5 min until they no longer adhered to the plate(s), and the suspensions were then transferred to conical centrifuge tubes and were centrifuged at ~1000 rpm, 4 °C for ~5 min. Trypsin was removed, and cell pellets were resuspended in chilled PBS (10 mL/cell pellet). Resuspensions were centrifuged at ~1000 rpm, 4 °C for

5–10 min, and PBS was removed. Cells were either lysed immediately or stored at $-80\,^\circ$C until lysis. Cells were lysed with radioimmunoprecipitation buffer (RIPA; 10 mM Tris HCl (pH 8.0), 1 mM EDTA, 0.5 mM EGTA, 1% Triton X-100, 0.1% sodium deoxycholate, 0.1% SDS, 140 mM NaCl) containing Halt Protease Inhibitor Cocktail (Thermo Fisher Scientific).

Lysates were either used immediately or stored at $-80\,^\circ$C. Total protein concentrations in lysates were then measured with a BCA assay (Thermo Fisher Scientific) and diluted accordingly with RIPA to reach as equal concentrations as possible. Proteins were then resolved by denaturing polyacrylamide gel electrophoresis using Tris-Glycine SDS Running Buffer (25 mM Tris Base, 192 mM Glycine, 0.1% SDS) and 8% acrylamide Tris-Glycine pre-cast gels (NuSep, Germantown, MD, USA). Proteins were transferred from the gel to PVDF membrane (MilliporeSigma, Burlington, MA, USA) at 30 V for 960 min in transfer buffer (25 mM Tris base, 0.19 M glycine, 20% methanol). Membranes were blocked with non-fat milk (5% *w/v* non-fat dry milk in TBST) at room temperature for 1 h and subsequently washed with Tris-buffered saline Tween-20 (TBST; 20 mM Tris-HCl pH 7.5, 0.15 M NaCl, 0.1% Tween-20) 3 times, 5 min each time. Incubation in 1° antibodies was then performed at room temperature for 2 h or 4 $^\circ$C overnight. TBST washes were again performed 3 times, 5 min each time, and the membrane was incubated with horseradish peroxidase (HRP) 2° antibodies at room temperature for 1 h. After washing with TBST 3 times, 10 min each time, membranes were incubated in visualization solution (Solution A: 100 mM Tris-HCl pH 8.5, 2.5 mM luminol, 0.396 mM coumaric acid + Solution B: 100 mM Tris-HCl pH 8.5, 0.0192% H_2O_2) and visualized on an Image Reader LAS-3000 (FujiFilm, Greenwood, SC, USA). When visualizing proteins specific to more than one 1° antibody, the membrane was severed at a molecular weight distinguishing the two proteins and incubated separately with distinct 1° antibodies. The 1° antibodies used were rat anti-p100/p52, mouse anti-IκBα, mouse anti-β-tubulin, and mouse anti-GAPDH and were all purchased from BioLegend. The 2° antibodies used were HRP goat anti-rat IgG (BioLegend) and HRP goat anti-mouse IgG (Sigma-Aldrich, St. Louis, MO, USA). Where applicable, band intensities were quantified by ImageJ software [76]. See Table S2 for ratios of antibody/blocking solution used. Figures show representative samples of biological duplicates.

2.8. Statistical Analysis

All experiments were replicated in biological duplicate or triplicate. Western blots shown are representative examples of duplicated or triplicated experiments. Statistically significant differences in the quantitative RT-PCR data were determined using a two-tailed unpaired Student's *t*-test in GraphPad Prism 8. Error bars are reported as standard error of the mean

3. Results

3.1. The Noncanonical NFκB Pathway Is Activated Early in RVFV (MP12) Infection

As described above, alternative splicing of RIOK3 mRNA is observed upon RVFV infection. Moreover, the suppression of IFN expression and enhanced expression of a subset of inflammatory cytokines is observed when alternatively spliced RIOK3 mRNA is expressed [70,73]. Suppression of IFN and enhanced expression of certain inflammatory cytokines are also characteristics of activation of the NFκB noncanonical pathway [33]. Therefore, we sought to determine whether the noncanonical NFκB pathway is activated during RVFV infection and whether alternative splicing of RIOK3 mRNA has a role in this pathway. First, to specifically determine whether infection with RVFV (strain MP12) activated the noncanonical NFκB pathway, we measured expression of genes that are indicative of this pathway.

Since CXCL13 and CCL19 have been shown to be transcribed specifically during activation of the noncanonical NFκB pathway [48], mRNAs of these chemokines were measured at multiple timepoints following RVFV MP12 infection. HEK 293 cells were infected with MP12 at an MOI of 2, and cells were harvested at 6, 12, 18, 24, and 30 h post-infection (h.p.i.). CXCL13 and CCL19 mRNAs at these times were then measured by RT-qPCR, and throughout RVFV MP12 infection, relative expression (normalized to GAPDH) of these chemokines continued to increase, suggesting progressive activation of the noncanonical NFκB pathway (Figure 2a,b). Intriguingly, relative expression of NIK and p100 also increased throughout infection, and importantly, there was a statistically significant increase in p100 expression as early as 6 h.p.i. (Figure 2c,d). The same increase in relative expression of CXCL13, CCL19, NIK, and p100 was not observed in mock infected cells harvested at the same time points, and relative expression of CXCL13, CCL19, NIK, and p100 did not change significantly from 0 to 30 h in these cells (Figure S1). Furthermore, in HepG2 cells, relative normalized expression of CXCL13, NIK, and p100 also increased significantly in RVFV-MP12-infected cells, and although the difference in relative expression of CCL19 between infected and mock infected cells did not rise to statistical significance, CCL19 expression was reproducibly elevated in RVFV-MP12-infected cells (Figure S2).

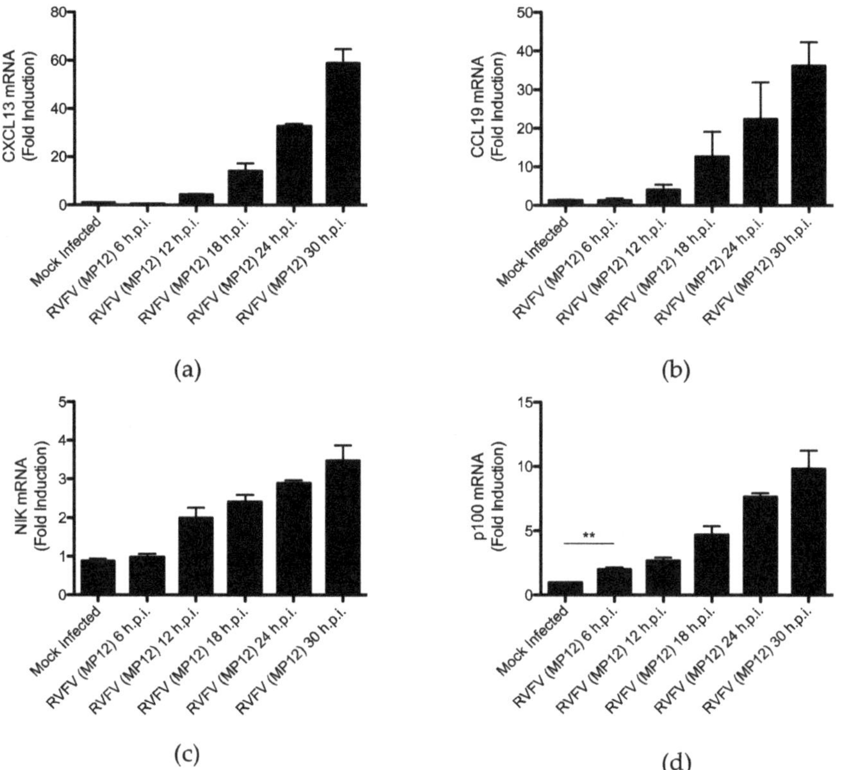

Figure 2. The noncanonical NFκB pathway is activated during RVFV MP12 infection. RVFV-infected cells were lysed at the times post-infection indicated, and mRNAs of the noncanonical NFκB pathway genes (**a**) CXCL13, (**b**) CCL19, (**c**) NIK, and (**d**) p100 were quantified by RT-qPCR, normalized to GAPDH mRNA levels. Plots are representative of the data as the mean value of 3 biological replicates +/− SEM. Student's *t*-test: ** $p < 0.01$.

Processing of p100 to p52 is a key step in activation of the noncanonical NFκB pathway. p100 inhibits RelB from translocating to the nucleus and thereby inhibits transcription of noncanonical NFκB genes, but when p100 is processed to p52 the RelB/p52 dimer is favored to enter the nucleus for transcription of noncanonical NFκB genes [32,77]. Notably, constitutive processing of p100 in most cell types is lacking, thereby limiting and strictly regulating activation of the noncanonical NFκB pathway. For example, in most cells, p100 predominates largely over p52, and when p100 is overexpressed it is scarcely converted to p52. Therefore, any noticeable increase in p52 protein levels is strongly indicative of activation of the noncanonical NFκB pathway [47,78]. To assess p100 processing, protein levels of p52 were measured during RVFV MP12 infection by Western blotting. HEK 293 cells were infected with MP12 for 6 and 24 h, and p52 levels were then visualized using an antibody that recognizes both p100 and its proteolytic product, p52. Levels of p52 appear to increase as early 6 h.p.i. and remain increased at 24 h.p.i., suggesting processing of p100 to p52 indeed occurs early following RVFV MP12 infection (Figure 3a). Quantification of band intensity further supports these results (Figure S3a).

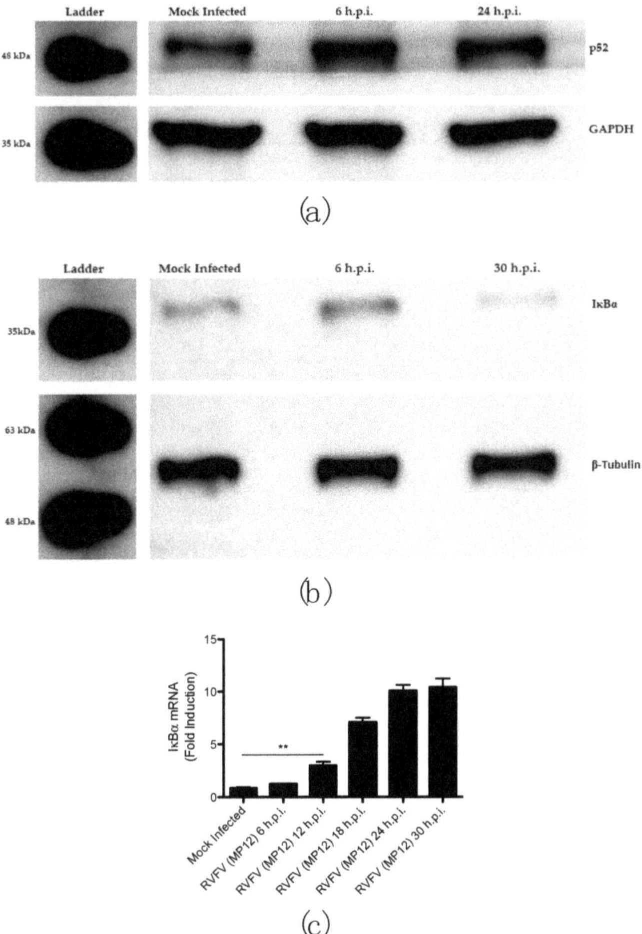

Figure 3. Activation of the noncanonical NFκB precedes activation of the canonical NFκB pathway during RVFV MP12 infection. Activation of the noncanonical pathway was assessed by (**a**) measuring p100 processing, and activation of the canonical pathway was assessed by (**b**) measuring degradation of IκBα via Western blotting. Activation of the canonical pathway was also measured, via RT-qPCR, by (**c**) assessing relative mRNA expression of IκBα normalized to GAPDH. Images in (**a,b**) are representative samples of biological duplicates. Furthermore, p52 levels were also assessed by Western blotting with the same samples shown in (**b**), and quantification of band intensities is shown in Figure S3b. In these samples, p52 is increased at 6 h.p.i. and is further increased at 30 h.p.i. The plot shown in (**c**) presents the data as the mean value of 3 biological replicates +/− SEM. Student's *t*-test: ** $p < 0.01$.

Activation of the canonical NFκB pathway during RVFV MP12 infection was also assessed by measuring protein levels of IκBα during infection. Since IκBα protein inhibits transcriptional activity of canonical NFκB dimers, IκBα is expected to decrease with activation of the canonical pathway. Therefore, similar to the appearance of p52 being a hallmark of activation of the noncanonical pathway, decreased IκBα is indicative of activation of the canonical pathway [59–61]. IκBα protein level was probed via Western blotting at 6 and 30 h.p.i., and intriguingly, in MP12-infected HEK 293 cells, a decrease in IκBα protein is not observed at 6 h.p.i. but is at 30 h.p.i., suggesting the canonical NFκB pathway is activated after the noncanonical pathway (Figure 3b). Normally, induction of the noncanonical pathway is delayed and activated after the canonical pathway [32,33]. Early activation of the noncanonical pathway and delayed onset of the canonical pathway was also observed in RVFV-MP12-infected HepG2 cells, in which degradation of IκBα was not observed at 6 h.p.i., but accumulation of p52 was observed (Figure S4).

Although IκBα protein level decreases with activation of the canonical NFκB pathway, IκBα mRNA has been shown to increase with activation of the canonical pathway, thereby contributing to the oscillatory nature of this pathway [79]. Therefore, IκBα mRNA was also measured by RT-qPCR throughout infection of HEK 293 cells with MP12, and it did not increase considerably above baseline until 12 h.p.i. (Figure 3c). Since p100 mRNA expression is significantly elevated by 6 h.p.i. and p52 was also shown to appear by this time, yet IκBα mRNA was not significantly elevated until 12 h.p.i., these data show that p100 processing (and thereby activation of the noncanonical pathway) precedes activation of the canonical pathway during RVFV MP12 infection. Collectively, these results suggest the noncanonical NFκB pathway is activated early in RVFV MP12 infection and even before the canonical NFκB pathway, implying the noncanonical pathway, and specifically p100 processing, may be co-opted by RVFV as a means to subvert the antiviral IFN response.

3.2. Alternative Splicing of RIOK3 Correlates with Activation of the Noncanonical NFκB Pathway

Our laboratory has shown alternative splicing of the atypical kinase RIOK3 occurs throughout RVFV infection [70,80]. Furthermore, alternative splicing of RIOK3 per se results in a diminished IFN response and elevated NFκB-mediated inflammatory response in epithelial cells [73], which are also characteristics of activation of the noncanonical NFκB pathway. It is notable that expression of the RVFV virulence factor NSs inhibits expression of IFN [81,82]. However, the isoform of RIOK3 expressed from the alternatively spliced mRNA can itself inhibit expression of IFN in the absence of NSs [73]. Furthermore, as shown above, activation of the noncanonical pathway occurs early in RVFV MP12 infection. Therefore, we sought to determine whether alternative splicing of RIOK3 is a programmed event during general activation of the noncanonical NFκB pathway.

To assess alternative splicing of RIOK3 as a function of activation of the noncanonical NFκB pathway, cDNA of RIOK3 mRNA was amplified by RT-PCR between exons 5 and 10, allowing up to three bands to be visualized in an agarose gel (Figure 4). The slowest migrating band corresponds to full-length canonically spliced RIOK3 mRNA; the middle band corresponds to mRNA of RIOK3's principal alternatively spliced isoform, RIOK3 X2, which lacks part of exon 8, and the fastest migrating band corresponds to mRNA of a RIOK3 X1/X2 hybrid that lacks all of exon 7 and part of exon 8. Notably, the alternatively spliced mRNAs contain a premature stop codon that would yield a truncated protein after translation.

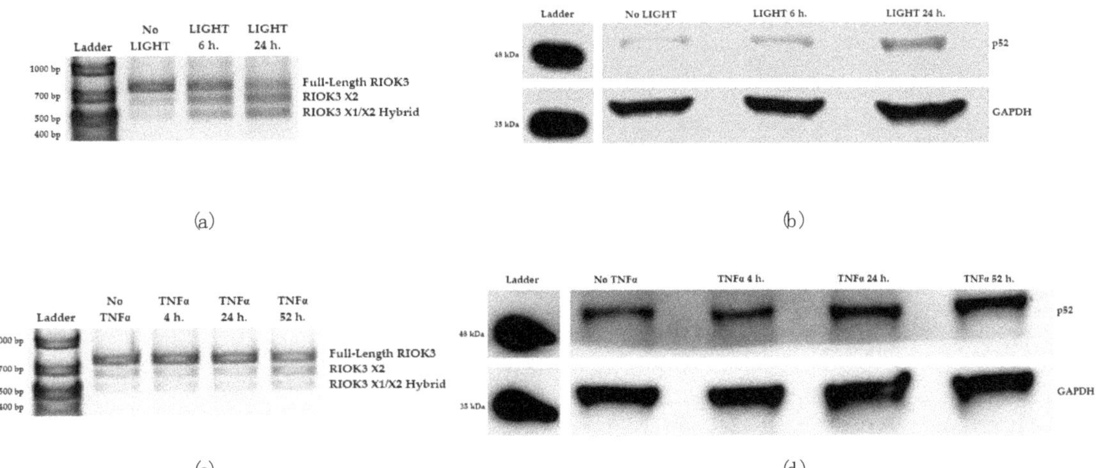

Figure 4. Alternative splicing of RIOK3 occurs specifically during activation of the noncanonical NFκB pathway. At 0, 6, and 24 h post-LIGHT treatment, (**a**) alternative splicing of RIOK3 and (**b**) activation of the noncanonical pathway were assessed. RIOK3 alternative splicing was assessed by RT-PCR and gel electrophoresis, and activation of the noncanonical pathway was shown through probing for p52 via Western blotting. At 0, 4, 24, and 52 h post-TNFα treatment, (**c**) alternative splicing of RIOK3 and (**b**) activation of the noncanonical pathway were also assessed in the same manner as in (**a,b**). Images in (**a–d**) show representative samples of biological duplicates or triplicates.

To examine the kinetics of RIOK3 mRNA splicing upon NFκB pathway activation, HEK 293 cells were treated with the noncanonical NFκB pathway agonist LIGHT for 6 and 24 h, and alternative splicing of RIOK3 was assessed as mentioned above. With LIGHT treatment, alternative splicing of RIOK3 was observed by 6 h and appeared to increase up to 24 h after LIGHT treatment (Figure 4a). This correlated with p100 processing, as indicated by an increase in p52 levels (Figure 4b). However, when treating HEK 293 cells with TNFα, which immediately activates the canonical NFκB pathway, alternative splicing of RIOK3 was not observed between 4 and 24 h post-TNFα treatment, suggesting alternative splicing is not programmed into activation of the canonical NFκB pathway (Figure S5).

Since the noncanonical NFκB pathway is known to also become activated following TNFα treatment but normally has delayed kinetics compared to the canonical pathway [32,33], we also assessed alternative splicing of RIOK3 at later times post-TNFα treatment. By 52 h post-TNFα treatment, alternative splicing of RIOK3 was observed (Figure 4c). The kinetics of appearance of RIOK3 alternative splicing corresponded with increased p52 levels, which appear to begin to increase at 24 h and become abundant by 52 h post-TNFα treatment (Figure 4d). These results suggest alternative splicing of RIOK3 might occur upon activation of the noncanonical NFκB pathway but not during activation of the canonical NFκB pathway.

Our laboratory recently showed that the splicing factor TRA2β plays an important role in RIOK3 splicing patterns. In fact, RIOK3 expression levels are governed by alternative splicing of TRA2β mRNA. TRA2β was shown to be important for canonical splicing of RIOK3 mRNA to full-length RIOK3 [80]. However, when TRA2β is alternatively spliced to incorporate a poison exon, nonfunctional mRNA isoforms of TRA2β with higher molecular weights are produced and rapidly degraded by nonsense-mediated decay [83,84]. Reduction in TRA2β was shown to result in alternative splicing of RIOK3 to RIOK3 X2 [80]. Therefore, TRA2β mRNA processing was also probed during LIGHT or TNFα treatment. During LIGHT treatment, TRA2β nonfunctional isoforms were observed 6 and 24 h post-treatment. During TNFα treatment, TRA2β nonfunctional isoforms are also observed by

52 h post-treatment (Figure 5a,b). These results corroborate our finding that alternative splicing of RIOK3 may in fact be part of the noncanonical NFκB response.

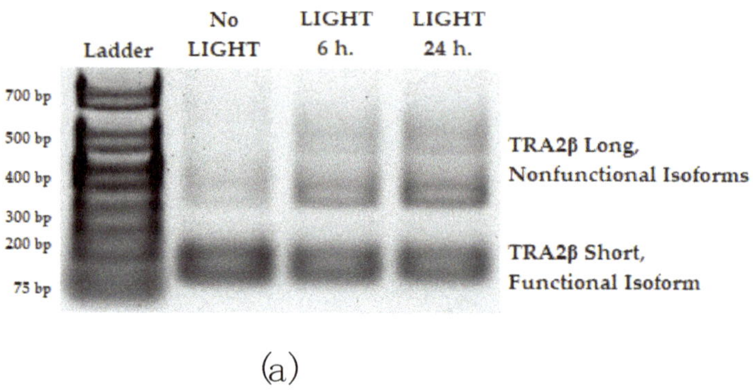

(a)

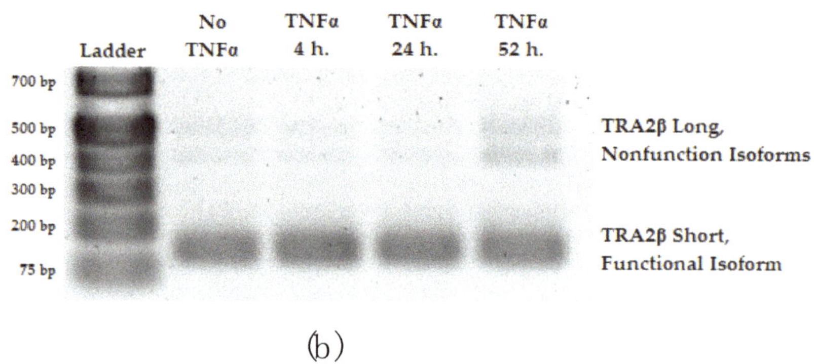

(b)

Figure 5. Alternative splicing of TRA2β to include a poison exon and produce nonfunctional isoforms also occurs specifically during activation of the noncanonical NFκB pathway. Alternative splicing of TRA2β was assessed during (**a**) LIGHT and (**b**) TNFα treatment using RT-PCR and gel electrophoresis. Images in (**a**,**b**) show representative samples of biological duplicates or triplicate.

3.3. Expression of RIOK3 X2 during RVFV (MP12) Infection Increases p100 Expression and Decreases IFNβ Expression

Since alternative splicing of RIOK3 occurs during RVFV infection and the splicing shift appears concurrently with activation of the noncanonical NFκB pathway, we then investigated whether RIOK3's alternatively spliced isoform, RIOK3 X2, has a direct effect on innate immune responses during RVFV MP12 infection.

To test this, RIOK3 X2 was overexpressed via transfection of an expression plasmid harboring the open reading frame for RIOK3 X2 in HEK 293 cells, and the cells were subsequently infected with RVFV MP12 at an MOI ~2. RT-qPCR was then used to quantify relative normalized expression of noncanonical NFκB genes CXCL13, CCL19, NIK, and p100. No differences in CXCL13 and CCL19 relative expression were observed, and although expression of NIK was elevated in infected cells transfected with RIOK3 X2 compared to infected cells transfected with an empty vector (EV) control plasmid, the increase was not statistically significant (Figure S6). However, there was a statistically significant elevation in relative expression of p100 in infected cells overexpressing RIOK3 X2, suggesting RIOK3 X2 expression enhances p100 expression and thus likely fuels the noncanonical NFκB pathway (Figure 6a).

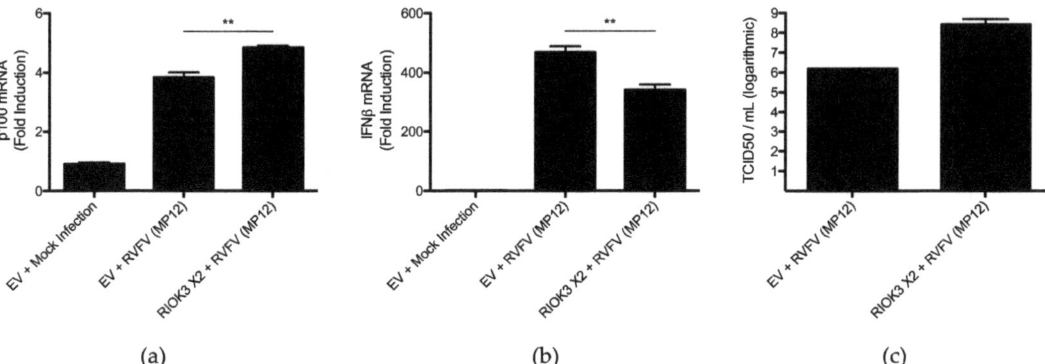

Figure 6. Overexpression of RIOK3 X2 in RVFV-MP12-infected cells results in increased p100 expression, decreased IFNβ expression, and increased RVFV MP12 TCID50/mL. Relative expression of (**a**) p100 and (**b**) IFNβ were measured by RT-qPCR, normalizing to GAPDH mRNA. TCID50/mL values were obtained from TCID50 assays using Vero 76 cells. In (**a–c**), HEK 293 cells were transfected with EV control plasmid or RIOK3 X2 for 24 h and subsequently infected with RVFV MP12 for 30 h. Plots present the data as the mean value of 3 biological replicates +/− SEM. Student's *t*-test: ** $p < 0.01$.

Furthermore, to assess whether RIOK3 X2's engagement of the noncanonical NFκB pathway during RVFV MP12 infection affects the IFN response, relative expression of IFNβ was also measured in MP12-infected HEK 293 cells overexpressing RIOK3 X2. Compared to infected cells transfected with control plasmid, there was a statistically significant decrease in IFNβ expression in infected cells overexpressing RIOK3 X2 (Figure 6b). Again, although it is known that RVFV NSs protein can suppress IFN expression, this experiment examined the effect of exogenous expression of RIOK3 X2 among parallel sets of RVFV-MP12-infected cells. Therefore, we conclude that overexpression of RIOK3 X2 during RVFV MP12 infection elevates p100 expression with a concomitant decrease in IFNβ expression.

Lastly, to directly determine whether the effects of RIOK3 X2 on innate immune responses could be beneficial for RVFV propagation in cell culture, virus titers in supernatants of MP12-infected HEK 293 cells transfected with RIOK3 X2 or MP12-infected HEK 293 cells transfected with control plasmids were quantified by TCID50 assays. TCID50/mL was ~2 logs higher in supernatant from infected cells overexpressing RIOK3 X2 compared to that of controls (Figure 6c). Therefore, expression of RIOK3 X2 during RVFV MP12 infection could be beneficial for the virus, and this might be due, at least in part, to RIOK3 X2 engaging the noncanonical NFκB pathway through supplementing the p100 pool in the infected cell and thereby diminishing the cell's type 1 IFN response.

4. Discussion

A hallmark of the noncanonical NFκB pathway is the appearance of p52, the proteolytically processed subunit of p100. Cellular expression of p52 is exquisitely under tight regulation by the cell, and even small changes in its steady-state levels are associated with activation or suppression of the noncanonical pathway [47,78]. When complexed with RelB, p52 plays a large role in disrupting the type 1 IFN response by competing with the canonical NFκB dimer for the κB binding site of the IFNβ promoter [38]. Here, we show alternative splicing of RIOK3 coincides with enhanced expression of p52's precursor, p100. Notably, alternative splicing of RIOK3 occurs concurrently with activation of the noncanonical NFκB pathway, and although it is possible that alternative splicing of RIOK3 may be initiated by activation or inhibition of a pathway associated with both TNFα and LIGHT treatments, results here demonstrate alternative splicing of RIOK3 is likely associated with the noncanonical NFκB pathway. Moreover, since overexpression of RIOK3 X2 results in decreased IFNβ expression and increased p100 expression, these results are also consistent with previous results [38] showing that the noncanonical NFκB pathway has an inhibitory effect on the IFN pathway. The link between the NFκB noncanonical pathway and IFN restriction is further emphasized by our finding that increased p100 protein during RVFV MP12 infection is processed to p52.

Since the noncanonical NFκB pathway has an inhibitory effect on the antiviral IFN response, exploiting and co-opting this pathway was proposed to be advantageous for some viruses [34,36,37]. Our results indicate that activation of the noncanonical NFκB pathway occurs early and *before* activation of the canonical NFκB pathway during RFVF MP12 infection, which would not be expected if the activation of the noncanonical pathway we observe was simply part of a coordinated innate immune response to the virus. That is, in a typical cellular innate immune response to viral infection, activation of noncanonical NFκB pathway would be expected to occur significantly later than activation of the canonical pathway. Furthermore, elevated infectious particles observed in RIOK3 X2 overexpressing cells that were shown to have elevated p100 expression further suggest the supportive role this pathway could have in RVFV replication and spread.

Here, we add RVFV to the short list of viruses that use this unusual strategy to enhance their success. Furthermore, we demonstrate that alternative splicing of key genes is instrumental in co-opting the noncanonical NFκB pathway during RVFV infection. Exactly how alternative splicing of RIOK3 is involved in the cell's activation of the noncanonical NFκB pathway and how RIOK3 X2 expression results in increased expression of p100 will need to be addressed in future studies. However, one potential mechanism for how RIOK3 X2 might be able to increase p100 expression could be through RIOK3 X2 mediating crosstalk between the canonical and noncanonical pathways. Full-length RIOK3 has been shown to inhibit the canonical NFκB pathway [85,86], while expression of RIOK3 X2 correlates with stimulation of this pathway [73]. Therefore, the increased expression of p100 observed here with overexpression of RIOK3 X2 could be a result of activating the canonical NFκB p50:RelA dimer to enter the nucleus to stimulate transcription of p100, which is needed for continued processing of p100 into p52. This would be supported by the finding that p100 expression depends on p50:RelA [65].

It is still unclear whether alternative splicing of RIOK3 alone can activate the noncanonical NFκB pathway or whether RIOK3 alternative splicing occurs downstream of the initial activation of this pathway. Here, we observed alternative splicing of RIOK3 after p52 begins to appear at 24 h post-TNFα treatment, so it is possible that alternative splicing of RIOK3 occurs after the initial activation of the noncanonical pathway. Moreover, alternative splicing in general has been shown to be important for regulating both canonical and noncanonical NFκB pathways [87,88], and intriguingly, it has also been found to be linked to dysregulation of NFκB-mediated immune responses during viral infection [89]. Therefore, further understanding of the onset of alternative splicing of RIOK3 during activation of the noncanonical NFκB pathway could be important for both understanding how innate immune responses are regulated and how they can be dysregulated during viral infection and other disease states.

The present findings help illuminate how cellular innate immune pathways can be targeted and co-opted for viral success (Figure 7). We show the noncanonical NFκB pathway is one innate immune pathway in particular that likely can be exploited by viruses such as RVFV to diminish the potent antiviral IFN response. Notably, alternative splicing of RIOK3, and likely other mRNAs, is an important event that facilitates the switch toward the noncanonical NFκB pathway to mitigate the IFN response. A deeper understanding of viral strategies to undermine cellular defenses could lead to new antiviral therapeutic strategies to combat infections in people and livestock.

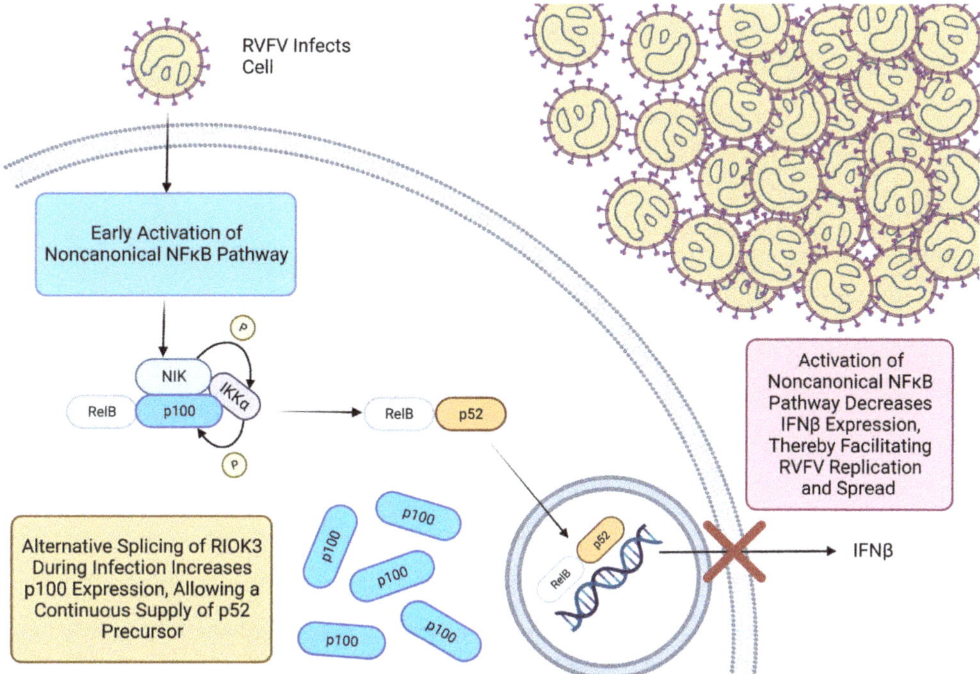

Figure 7. Alternative splicing of RIOK3 replenishes the p100 pool in the cell during RVFV MP12 infection, allowing continuous processing of p100 to p52 during infection.

Supplementary Materials: The following supporting information can be downloaded at: https://www.mdpi.com/article/10.3390/v15071566/s1, Figure S1: Relative normalized expression of (a) CXCL13, (b) CCL19, (c) NIK, and (d) p100 in HEK 293 cells mock infected for 0 to 30 h; Figure S2: Relative normalized expression of (a) CXCL13, (b) CCL19, (c) NIK, and (d) p100 in mock- and RVFV MP12-infected HepG2 cells; Figure S3: Relative p52 protein levels in (a) HEK 293 cells infected with RVFV MP12 for 6 and 24 h, (b) HEK 293 cells infected with RVFV MP12 for 6 and 30 h, (c) HepG2 cells infected with RVFV MP12 for 6 and 24 h, and (d) HepG2 cells infected with RVFV MP12 for 6 and 30 h; Figure S4: Activation of the (a) noncanonical NFκB pathway and (b) canonical NFκB pathway in HepG2 cells throughout RVFV MP12 infection; Figure S5: Alternative splicing of RIOK3 0-24 h TNFα treatment; Figure S6: Relative Normalized Expression of (a) CXCL13, (b) CCL19, and (c) NIK in mock- or RVFV MP12-infected HEK 293 cells transfected with EV or RIOK3 X2; Table S1:Primer sequences; Table S2: Ratios of antibodies per blocking solution used.

Author Contributions: T.C.B.: Conceptualization, Investigation, Visualization, Data Curation, Formal Analysis, Writing—Original Draft Preparation; H.S.: Conceptualization, Investigation, Visualization; J.-M.L.: Conceptualization, Investigation, Visualization; J.S.L.: Conceptualization, Supervision, Writing—Review and Editing, Funding Acquisition. All authors have read and agreed to the published version of the manuscript.

Funding: We gratefully acknowledge funding for this research from the following sources: NIH grants R03TR002937 to J.S.L., P30GM140963 (B. Bowler, P.I.), the UM Center for Translational Medicine, and NIH R41TR003929 (F. Astruc-Diaz, P.I.).

Institutional Review Board Statement: Not applicable.

Informed Consent Statement: Not applicable.

Data Availability Statement: Not applicable.

Acknowledgments: Plots were generated using GraphPad Prism 8, and illustrations were prepared using BioRender. Furthermore, we wish to thank the Center for Biomolecular Structure and Dynamics at the University of Montana for core support for this work. We also thank Philippe Diaz, Scott Samuels, and members of their laboratories for fruitful discussions and use of their equipment. This manuscript was written in loving memory of Shea Elizabeth Downey who touched so many people with her beautiful life.

Conflicts of Interest: The authors declare no conflict of interest.

References

1. De la Rica, R.; Borges, M.; Gonzalez-Freire, M. COVID-19: In the Eye of the Cytokine Storm. *Front. Immunol.* **2020**, *11*, 558898. [CrossRef] [PubMed]
2. Blanco-Melo, D.; Nilsson-Payant, B.E.; Liu, W.C.; Uhl, S.; Hoagland, D.; Moller, R.; Jordan, T.X.; Oishi, K.; Panis, M.; Sachs, D.; et al. Imbalanced Host Response to SARS-CoV-2 Drives Development of COVID-19. *Cell* **2020**, *181*, 1036–1045.e9. [CrossRef]
3. Adams, M.J.; Lefkowitz, E.J.; King, A.M.Q.; Harrach, B.; Harrison, R.L.; Knowles, N.J.; Kropinski, A.M.; Krupovic, M.; Kuhn, J.H.; Mushegian, A.R.; et al. Changes to taxonomy and the International Code of Virus Classification and Nomenclature ratified by the International Committee on Taxonomy of Viruses (2017). *Arch. Virol.* **2017**, *162*, 2505–2538. [CrossRef] [PubMed]
4. Linthicum, K.J.; Davies, F.G.; Kairo, A.; Bailey, C.L. Rift Valley fever virus (family Bunyaviridae, genus Phlebovirus). Isolations from Diptera collected during an inter-epizootic period in Kenya. *J. Hyg.* **1985**, *95*, 197–209. [CrossRef]
5. Linthicum, K.J.; Britch, S.C.; Anyamba, A. Rift Valley Fever: An Emerging Mosquito-Borne Disease. *Annu. Rev. Entomol.* **2016**, *61*, 395–415. [CrossRef] [PubMed]
6. Kwasnik, M.; Rozek, W.; Rola, J. Rift Valley Fever—A Growing Threat To Humans and Animals. *J. Vet. Res.* **2021**, *65*, 7–14. [CrossRef]
7. Nicholas, D.E.; Jacobsen, K.H.; Waters, N.M. Risk factors associated with human Rift Valley fever infection: Systematic review and meta-analysis. *Trop. Med. Int. Health* **2014**, *19*, 1420–1429. [CrossRef] [PubMed]
8. Ikegami, T.; Makino, S. The Pathogenesis of Rift Valley Fever. *Viruses* **2011**, *3*, 493–519. [CrossRef] [PubMed]
9. Javelle, E.; Lesueur, A.; Pommier de Santi, V.; de Laval, F.; Lefebvre, T.; Holweck, G.; Durand, G.A.; Leparc-Goffart, I.; Texier, G.; Simon, F. The challenging management of Rift Valley Fever in humans: Literature review of the clinical disease and algorithm proposal. *Ann. Clin. Microbiol. Antimicrob.* **2020**, *19*, 4. [CrossRef]
10. Al-Hazmi, M.; Ayoola, E.A.; Abdurahman, M.; Banzal, S.; Ashraf, J.; El-Bushra, A.; Hazmi, A.; Abdullah, M.; Abbo, H.; Elamin, A.; et al. Epidemic Rift Valley fever in Saudi Arabia: A clinical study of severe illness in humans. *Clin. Infect. Dis.* **2003**, *36*, 245–252. [CrossRef]
11. Baudin, M.; Jumaa, A.M.; Jomma, H.J.E.; Karsany, M.S.; Bucht, G.; Naslund, J.; Ahlm, C.; Evander, M.; Mohamed, N. Association of Rift Valley fever virus infection with miscarriage in Sudanese women: A cross-sectional study. *Lancet Glob. Health* **2016**, *4*, e864–e871. [CrossRef] [PubMed]
12. Cartwright, H.N.; Barbeau, D.J.; McElroy, A.K. Rift Valley Fever Virus Is Lethal in Different Inbred Mouse Strains Independent of Sex. *Front. Microbiol.* **2020**, *11*, 1962. [CrossRef] [PubMed]
13. Coetzer, J.A.; Ishak, K.G.; Calvert, R.C. Sequential development of the liver lesions in new-born lambs infected with Rift Valley fever virus. II. Ultrastructural findings. *Onderstepoort J. Vet. Res.* **1982**, *49*, 109–122. [PubMed]
14. Rippy, M.K.; Topper, M.J.; Mebus, C.A.; Morrill, J.C. Rift Valley fever virus-induced encephalomyelitis and hepatitis in calves. *Vet. Pathol.* **1992**, *29*, 495–502. [CrossRef]
15. McMillen, C.M.; Hartman, A.L. Rift Valley fever in animals and humans: Current perspectives. *Antiviral Res.* **2018**, *156*, 29–37. [CrossRef]
16. McMillen, C.M.; Boyles, D.A.; Kostadinov, S.G.; Hoehl, R.M.; Schwarz, M.M.; Albe, J.R.; Demers, M.J.; Hartman, A.L. Congenital Rift Valley fever in Sprague Dawley rats is associated with diffuse infection and pathology of the placenta. *PLoS Negl. Trop. Dis.* **2022**, *16*, e0010898. [CrossRef]
17. NIAID. NIAID Emerging Infectious Diseases/Pathogens. Available online: https://www.niaid.nih.gov/research/emerging-infectious-diseases-pathogens (accessed on 1 April 2023).
18. Grossi-Soyster, E.N.; LaBeaud, A.D. Rift Valley Fever: Important Considerations for Risk Mitigation and Future Outbreaks. *Trop. Med. Infect. Dis.* **2020**, *5*, 89. [CrossRef]
19. Habjan, M.; Andersson, I.; Klingstrom, J.; Schumann, M.; Martin, A.; Zimmermann, P.; Wagner, V.; Pichlmair, A.; Schneider, U.; Muhlberger, E.; et al. Processing of genome 5′ termini as a strategy of negative-strand RNA viruses to avoid RIG-I-dependent interferon induction. *PLoS ONE* **2008**, *3*, e2032. [CrossRef]
20. Elliott, R.M.; Weber, F. Bunyaviruses and the type I interferon system. *Viruses* **2009**, *1*, 1003–1021. [CrossRef]
21. Fitzgerald, K.A.; McWhirter, S.M.; Faia, K.L.; Rowe, D.C.; Latz, E.; Golenbock, D.T.; Coyle, A.J.; Liao, S.M.; Maniatis, T. IKKepsilon and TBK1 are essential components of the IRF3 signaling pathway. *Nat. Immunol.* **2003**, *4*, 491–496. [CrossRef]
22. Sharma, S.; tenOever, B.R.; Grandvaux, N.; Zhou, G.P.; Lin, R.; Hiscott, J. Triggering the interferon antiviral response through an IKK-related pathway. *Science* **2003**, *300*, 1148–1151. [CrossRef]

23. Brisse, M.; Ly, H. Comparative Structure and Function Analysis of the RIG-I-Like Receptors: RIG-I and MDA5. *Front. Immunol.* **2019**, *10*, 1586. [CrossRef] [PubMed]
24. Sato, M.; Tanaka, N.; Hata, N.; Oda, E.; Taniguchi, T. Involvement of the IRF family transcription factor IRF-3 in virus-induced activation of the IFN-beta gene. *FEBS Lett.* **1998**, *425*, 112–116. [CrossRef] [PubMed]
25. Velazquez, L.; Fellous, M.; Stark, G.R.; Pellegrini, S. A protein tyrosine kinase in the interferon alpha/beta signaling pathway. *Cell* **1992**, *70*, 313–322. [CrossRef]
26. Der, S.D.; Zhou, A.; Williams, B.R.; Silverman, R.H. Identification of genes differentially regulated by interferon alpha, beta, or gamma using oligonucleotide arrays. *Proc. Natl. Acad. Sci. USA* **1998**, *95*, 15623–15628. [CrossRef]
27. Muller, M.; Laxton, C.; Briscoe, J.; Schindler, C.; Improta, T.; Darnell, J.E., Jr.; Stark, G.R.; Kerr, I.M. Complementation of a mutant cell line: Central role of the 91 kDa polypeptide of ISGF3 in the interferon-alpha and -gamma signal transduction pathways. *EMBO J.* **1993**, *12*, 4221–4228. [CrossRef]
28. Hu, X.; Li, J.; Fu, M.; Zhao, X.; Wang, W. The JAK/STAT signaling pathway: From bench to clinic. *Signal Transduct. Target. Ther.* **2021**, *6*, 402. [CrossRef]
29. De Togni, P.; Goellner, J.; Ruddle, N.H.; Streeter, P.R.; Fick, A.; Mariathasan, S.; Smith, S.C.; Carlson, R.; Shornick, L.P.; Strauss-Schoenberger, J.; et al. Abnormal development of peripheral lymphoid organs in mice deficient in lymphotoxin. *Science* **1994**, *264*, 703–707. [CrossRef]
30. Alimzhanov, M.B.; Kuprash, D.V.; Kosco-Vilbois, M.H.; Luz, A.; Turetskaya, R.L.; Tarakhovsky, A.; Rajewsky, K.; Nedospasov, S.A.; Pfeffer, K. Abnormal development of secondary lymphoid tissues in lymphotoxin beta-deficient mice. *Proc. Natl. Acad. Sci. USA* **1997**, *94*, 9302–9307. [CrossRef] [PubMed]
31. Weih, F.; Caamano, J. Regulation of secondary lymphoid organ development by the nuclear factor-kappaB signal transduction pathway. *Immunol. Rev.* **2003**, *195*, 91–105. [CrossRef]
32. Sun, S.C. The noncanonical NF-kappaB pathway. *Immunol. Rev.* **2012**, *246*, 125–140. [CrossRef] [PubMed]
33. Sun, S.C. The non-canonical NF-kappaB pathway in immunity and inflammation. *Nat. Rev. Immunol.* **2017**, *17*, 545–558. [CrossRef] [PubMed]
34. Struzik, J.; Szulc-Dabrowska, L. Manipulation of Non-canonical NF-kappaB Signaling by Non-oncogenic Viruses. *Arch. Immunol. Ther. Exp.* **2019**, *67*, 41–48. [CrossRef] [PubMed]
35. Ruckle, A.; Haasbach, E.; Julkunen, I.; Planz, O.; Ehrhardt, C.; Ludwig, S. The NS1 protein of influenza A virus blocks RIG-I-mediated activation of the noncanonical NF-kappaB pathway and p52/RelB-dependent gene expression in lung epithelial cells. *J. Virol.* **2012**, *86*, 10211–10217. [CrossRef]
36. Liu, P.; Li, K.; Garofalo, R.P.; Brasier, A.R. Respiratory syncytial virus induces RelA release from cytoplasmic 100-kDa NF-kappa B2 complexes via a novel retinoic acid-inducible gene-Imiddle dotNF- kappa B-inducing kinase signaling pathway. *J. Biol. Chem.* **2008**, *283*, 23169–23178. [CrossRef]
37. Choudhary, S.; Boldogh, S.; Garofalo, R.; Jamaluddin, M.; Brasier, A.R. Respiratory syncytial virus influences NF-kappaB-dependent gene expression through a novel pathway involving MAP3K14/NIK expression and nuclear complex formation with NF-kappaB2. *J. Virol.* **2005**, *79*, 8948–8959. [CrossRef]
38. Jin, J.; Hu, H.; Li, H.S.; Yu, J.; Xiao, Y.; Brittain, G.C.; Zou, Q.; Cheng, X.; Mallette, F.A.; Watowich, S.S.; et al. Noncanonical NF-kappaB pathway controls the production of type I interferons in antiviral innate immunity. *Immunity* **2014**, *40*, 342–354. [CrossRef]
39. Liao, G.; Zhang, M.; Harhaj, E.W.; Sun, S.C. Regulation of the NF-kappaB-inducing kinase by tumor necrosis factor receptor-associated factor 3-induced degradation. *J. Biol. Chem.* **2004**, *279*, 26243–26250. [CrossRef]
40. Varfolomeev, E.; Blankenship, J.W.; Wayson, S.M.; Fedorova, A.V.; Kayagaki, N.; Garg, P.; Zobel, K.; Dynek, J.N.; Elliott, L.O.; Wallweber, H.J.; et al. IAP antagonists induce autoubiquitination of c-IAPs, NF-kappaB activation, and TNFalpha-dependent apoptosis. *Cell* **2007**, *131*, 669–681. [CrossRef]
41. Vince, J.E.; Wong, W.W.; Khan, N.; Feltham, R.; Chau, D.; Ahmed, A.U.; Benetatos, C.A.; Chunduru, S.K.; Condon, S.M.; McKinlay, M.; et al. IAP antagonists target cIAP1 to induce TNFalpha-dependent apoptosis. *Cell* **2007**, *131*, 682–693. [CrossRef]
42. Vallabhapurapu, S.; Matsuzawa, A.; Zhang, W.; Tseng, P.H.; Keats, J.J.; Wang, H.; Vignali, D.A.; Bergsagel, P.L.; Karin, M. Nonredundant and complementary functions of TRAF2 and TRAF3 in a ubiquitination cascade that activates NIK-dependent alternative NF-kappaB signaling. *Nat. Immunol.* **2008**, *9*, 1364–1370. [CrossRef]
43. Zarnegar, B.J.; Wang, Y.; Mahoney, D.J.; Dempsey, P.W.; Cheung, H.H.; He, J.; Shiba, T.; Yang, X.; Yeh, W.C.; Mak, T.W.; et al. Noncanonical NF-kappaB activation requires coordinated assembly of a regulatory complex of the adaptors cIAP1, cIAP2, TRAF2 and TRAF3 and the kinase NIK. *Nat. Immunol.* **2008**, *9*, 1371–1378. [CrossRef] [PubMed]
44. Ling, L.; Cao, Z.; Goeddel, D.V. NF-kappaB-inducing kinase activates IKK-alpha by phosphorylation of Ser-176. *Proc. Natl. Acad. Sci. USA* **1998**, *95*, 3792–3797. [CrossRef]
45. Senftleben, U.; Cao, Y.; Xiao, G.; Greten, F.R.; Krahn, G.; Bonizzi, G.; Chen, Y.; Hu, Y.; Fong, A.; Sun, S.C.; et al. Activation by IKKalpha of a second, evolutionary conserved, NF-kappa B signaling pathway. *Science* **2001**, *293*, 1495–1499. [CrossRef] [PubMed]
46. Xiao, G.; Fong, A.; Sun, S.C. Induction of p100 processing by NF-kappaB-inducing kinase involves docking IkappaB kinase alpha (IKKalpha) to p100 and IKKalpha-mediated phosphorylation. *J. Biol. Chem.* **2004**, *279*, 30099–30105. [CrossRef] [PubMed]
47. Xiao, G.; Harhaj, E.W.; Sun, S.C. NF-kappaB-inducing kinase regulates the processing of NF-kappaB2 p100. *Mol. Cell* **2001**, *7*, 401–409. [CrossRef]

48. Bonizzi, G.; Bebien, M.; Otero, D.C.; Johnson-Vroom, K.E.; Cao, Y.; Vu, D.; Jegga, A.G.; Aronow, B.J.; Ghosh, G.; Rickert, R.C.; et al. Activation of IKKalpha target genes depends on recognition of specific kappaB binding sites by RelB:p52 dimers. *EMBO J.* **2004**, *23*, 4202–4210. [CrossRef]

49. Sen, R.; Baltimore, D. Multiple nuclear factors interact with the immunoglobulin enhancer sequences. *Cell* **1986**, *46*, 705–716. [CrossRef]

50. Gilmore, T.D. NF-kappa B, KBF1, dorsal, and related matters. *Cell* **1990**, *62*, 841–843. [CrossRef]

51. Bours, V.; Azarenko, V.; Dejardin, E.; Siebenlist, U. Human RelB (I-Rel) functions as a kappa B site-dependent transactivating member of the family of Rel-related proteins. *Oncogene* **1994**, *9*, 1699–1702.

52. Ryseck, R.P.; Weih, F.; Carrasco, D.; Bravo, R. RelB, a member of the Rel/NF-kappa B family of transcription factors. *Braz. J. Med. Biol. Res.* **1996**, *29*, 895–903.

53. Bonizzi, G.; Karin, M. The two NF-kappaB activation pathways and their role in innate and adaptive immunity. *Trends Immunol.* **2004**, *25*, 280–288. [CrossRef] [PubMed]

54. Liu, T.; Zhang, L.; Joo, D.; Sun, S.C. NF-kappaB signaling in inflammation. *Signal Transduct. Target. Ther.* **2017**, *2*, 17023. [CrossRef] [PubMed]

55. Yu, H.; Lin, L.; Zhang, Z.; Zhang, H.; Hu, H. Targeting NF-kappaB pathway for the therapy of diseases: Mechanism and clinical study. *Signal Transduct. Target. Ther.* **2020**, *5*, 209. [CrossRef]

56. Baeuerle, P.A.; Baltimore, D. Activation of DNA-binding activity in an apparently cytoplasmic precursor of the NF-kappa B transcription factor. *Cell* **1988**, *53*, 211–217. [CrossRef]

57. Baeuerle, P.A.; Baltimore, D. I kappa B: A specific inhibitor of the NF-kappa B transcription factor. *Science* **1988**, *242*, 540–546. [CrossRef]

58. Baeuerle, P.A.; Baltimore, D. A 65-kappaD subunit of active NF-kappaB is required for inhibition of NF-kappaB by I kappaB. *Genes Dev.* **1989**, *3*, 1689–1698. [CrossRef]

59. Palombella, V.J.; Rando, O.J.; Goldberg, A.L.; Maniatis, T. The ubiquitin-proteasome pathway is required for processing the NF-kappa B1 precursor protein and the activation of NF-kappa B. *Cell* **1994**, *78*, 773–785. [CrossRef]

60. Traenckner, E.B.; Wilk, S.; Baeuerle, P.A. A proteasome inhibitor prevents activation of NF-kappa B and stabilizes a newly phosphorylated form of I kappa B-alpha that is still bound to NF-kappa B. *EMBO J.* **1994**, *13*, 5433–5441. [CrossRef]

61. Brown, K.; Gerstberger, S.; Carlson, L.; Franzoso, G.; Siebenlist, U. Control of I kappa B-alpha proteolysis by site-specific, signal-induced phosphorylation. *Science* **1995**, *267*, 1485–1488. [CrossRef] [PubMed]

62. Claudio, E.; Brown, K.; Park, S.; Wang, H.; Siebenlist, U. BAFF-induced NEMO-independent processing of NF-kappa B2 in maturing B cells. *Nat. Immunol.* **2002**, *3*, 958–965. [CrossRef] [PubMed]

63. Liang, C.; Zhang, M.; Sun, S.C. beta-TrCP binding and processing of NF-kappaB2/p100 involve its phosphorylation at serines 866 and 870. *Cell. Signal.* **2006**, *18*, 1309–1317. [CrossRef] [PubMed]

64. Mauri, D.N.; Ebner, R.; Montgomery, R.I.; Kochel, K.D.; Cheung, T.C.; Yu, G.L.; Ruben, S.; Murphy, M.; Eisenberg, R.J.; Cohen, G.H.; et al. LIGHT, a new member of the TNF superfamily, and lymphotoxin alpha are ligands for herpesvirus entry mediator. *Immunity* **1998**, *8*, 21–30. [CrossRef]

65. Dejardin, E.; Droin, N.M.; Delhase, M.; Haas, E.; Cao, Y.; Makris, C.; Li, Z.W.; Karin, M.; Ware, C.F.; Green, D.R. The lymphotoxin-beta receptor induces different patterns of gene expression via two NF-kappaB pathways. *Immunity* **2002**, *17*, 525–535. [CrossRef]

66. Hayden, M.S.; Ghosh, S. Regulation of NF-kappaB by TNF family cytokines. *Semin. Immunol.* **2014**, *26*, 253–266. [CrossRef] [PubMed]

67. Seth, R.B.; Sun, L.; Ea, C.K.; Chen, Z.J. Identification and characterization of MAVS, a mitochondrial antiviral signaling protein that activates NF-kappaB and IRF 3. *Cell* **2005**, *122*, 669–682. [CrossRef]

68. Feng, J.; De Jesus, P.D.; Su, V.; Han, S.; Gong, D.; Wu, N.C.; Tian, Y.; Li, X.; Wu, T.T.; Chanda, S.K.; et al. RIOK3 is an adaptor protein required for IRF3-mediated antiviral type I interferon production. *J. Virol.* **2014**, *88*, 7987–7997. [CrossRef]

69. Willemsen, J.; Wicht, O.; Wolanski, J.C.; Baur, N.; Bastian, S.; Haas, D.A.; Matula, P.; Knapp, B.; Meyniel-Schicklin, L.; Wang, C.; et al. Phosphorylation-Dependent Feedback Inhibition of RIG-I by DAPK1 Identified by Kinome-wide siRNA Screening. *Mol. Cell* **2017**, *65*, 403–415.e8. [CrossRef]

70. Havranek, K.E.; White, L.A.; Bisom, T.C.; Lanchy, J.M.; Lodmell, J.S. The Atypical Kinase RIOK3 Limits RVFV Propagation and Is Regulated by Alternative Splicing. *Viruses* **2021**, *13*, 367. [CrossRef]

71. Shen, Y.; Tang, K.; Chen, D.; Hong, M.; Sun, F.; Wang, S.; Ke, Y.; Wu, T.; Sun, R.; Qian, J.; et al. Riok3 inhibits the antiviral immune response by facilitating TRIM40-mediated RIG-I and MDA5 degradation. *Cell Rep.* **2021**, *35*, 109272. [CrossRef]

72. Takashima, K.; Oshiumi, H.; Takaki, H.; Matsumoto, M.; Seya, T. RIOK3-mediated phosphorylation of MDA5 interferes with its assembly and attenuates the innate immune response. *Cell Rep.* **2015**, *11*, 192–200. [CrossRef]

73. Bisom, T.C.; White, L.A.; Lanchy, J.M.; Lodmell, J.S. RIOK3 and Its Alternatively Spliced Isoform Have Disparate Roles in the Innate Immune Response to Rift Valley Fever Virus (MP12) Infection. *Viruses* **2022**, *14*, 2064. [CrossRef] [PubMed]

74. Vialat, P.; Muller, R.; Vu, T.H.; Prehaud, C.; Bouloy, M. Mapping of the mutations present in the genome of the Rift Valley fever virus attenuated MP12 strain and their putative role in attenuation. *Virus Res.* **1997**, *52*, 43–50. [CrossRef] [PubMed]

75. Smith, M.R.; Schirtzinger, E.E.; Wilson, W.C.; Davis, A.S. Rift Valley Fever Virus: Propagation, Quantification, and Storage. *Curr. Protoc. Microbiol.* **2019**, *55*, e92. [CrossRef] [PubMed]

76. Schneider, C.A.; Rasband, W.S.; Eliceiri, K.W. NIH Image to ImageJ: 25 years of image analysis. *Nat. Methods* **2012**, *9*, 671–675. [CrossRef]
77. Solan, N.J.; Miyoshi, H.; Carmona, E.M.; Bren, G.D.; Paya, C.V. RelB cellular regulation and transcriptional activity are regulated by p100. *J. Biol. Chem.* **2002**, *277*, 1405–1418. [CrossRef]
78. Betts, J.C.; Nabel, G.J. Differential regulation of NF-kappaB2(p100) processing and control by amino-terminal sequences. *Mol. Cell. Biol.* **1996**, *16*, 6363–6371. [CrossRef]
79. Ruland, J. Return to homeostasis: Downregulation of NF-kappaB responses. *Nat. Immunol.* **2011**, *12*, 709–714. [CrossRef]
80. White, L.A.; Bisom, T.C.; Grimes, H.L.; Hayashi, M.; Lanchy, J.M.; Lodmell, J.S. Tra2beta-Dependent Regulation of RIO Kinase 3 Splicing During Rift Valley Fever Virus Infection Underscores the Links Between Alternative Splicing and Innate Antiviral Immunity. *Front. Cell. Infect. Microbiol.* **2021**, *11*, 799024. [CrossRef]
81. Billecocq, A.; Spiegel, M.; Vialat, P.; Kohl, A.; Weber, F.; Bouloy, M.; Haller, O. NSs protein of Rift Valley fever virus blocks interferon production by inhibiting host gene transcription. *J. Virol.* **2004**, *78*, 9798–9806. [CrossRef]
82. Wuerth, J.D.; Weber, F. Phleboviruses and the Type I Interferon Response. *Viruses* **2016**, *8*, 174. [CrossRef]
83. Stoilov, P.; Daoud, R.; Nayler, O.; Stamm, S. Human tra2-beta1 autoregulates its protein concentration by influencing alternative splicing of its pre-mRNA. *Hum. Mol. Genet.* **2004**, *13*, 509–524. [CrossRef]
84. Leclair, N.K.; Brugiolo, M.; Urbanski, L.; Lawson, S.C.; Thakar, K.; Yurieva, M.; George, J.; Hinson, J.T.; Cheng, A.; Graveley, B.R.; et al. Poison Exon Splicing Regulates a Coordinated Network of SR Protein Expression during Differentiation and Tumorigenesis. *Mol. Cell* **2020**, *80*, 648–665.e9. [CrossRef] [PubMed]
85. Shan, J.; Wang, P.; Zhou, J.; Wu, D.; Shi, H.; Huo, K. RIOK3 interacts with caspase-10 and negatively regulates the NF-kappaB signaling pathway. *Mol. Cell. Biochem.* **2009**, *332*, 113–120. [CrossRef] [PubMed]
86. Fenner, B.J.; Scannell, M.; Prehn, J.H. Identification of polyubiquitin binding proteins involved in NF-kappaB signaling using protein arrays. *Biochim. Biophys. Acta* **2009**, *1794*, 1010–1016. [CrossRef]
87. Leeman, J.R.; Gilmore, T.D. Alternative splicing in the NF-kappaB signaling pathway. *Gene* **2008**, *423*, 97–107. [CrossRef]
88. Michel, M.; Wilhelmi, I.; Schultz, A.S.; Preussner, M.; Heyd, F. Activation-induced tumor necrosis factor receptor-associated factor 3 (Traf3) alternative splicing controls the noncanonical nuclear factor kappaB pathway and chemokine expression in human T cells. *J. Biol. Chem.* **2014**, *289*, 13651–13660. [CrossRef] [PubMed]
89. Ameur, L.B.; Marie, P.; Thenoz, M.; Giraud, G.; Combe, E.; Claude, J.B.; Lemaire, S.; Fontrodona, N.; Polveche, H.; Bastien, M.; et al. Intragenic recruitment of NF-kappaB drives splicing modifications upon activation by the oncogene Tax of HTLV-1. *Nat. Commun.* **2020**, *11*, 3045. [CrossRef]

Article

Synergistic Effect of Treatment with Highly Pathogenic Porcine Reproductive and Respiratory Syndrome Virus and Lipopolysaccharide on the Inflammatory Response of Porcine Pulmonary Microvascular Endothelial Cells

Xinyue Yao [1], Wanwan Dai [2], Siyu Yang [1], Zhaoli Wang [1], Qian Zhang [1], Qinghui Meng [3,*] and Tao Zhang [1,*]

[1] Beijing Key Laboratory of Traditional Chinese Veterinary Medicine, Animal Science and Technology College, Beijing University of Agriculture, No. 7 Beinong Road, Beijing 102206, China; zhangqianbua@163.com (Q.Z.)
[2] College of Veterinary Medicine, Shanxi Agriculture University, Taigu 030801, China
[3] Beijing Milu Ecological Research Center, Beijing Research Institute of Science and Technology, Beijing 100076, China
*   Correspondence: mengqinghui2006@163.com (Q.M.); zhangtao@bua.edu.cn (T.Z.)

Abstract: The highly pathogenic porcine reproductive and respiratory syndrome virus (HP-PRRSV) often causes secondary bacterial infection in piglets, resulting in inflammatory lung injury and leading to high mortality rates and significant economic losses in the pig industry. Microvascular endothelial cells (MVECs) play a crucial role in the inflammatory response. Previous studies have shown that HP-PRRSV can infect porcine pulmonary MVECs and damage the endothelial glycocalyx. To further understand the role of pulmonary MVECs in the pathogenesis of HP-PRRSV and its secondary bacterial infection, in this study, cultured porcine pulmonary MVECs were stimulated with a HP-PRRSV HN strain and lipopolysaccharide (LPS). The changes in gene expression profiles were analyzed through transcriptome sequencing, and the differentially expressed genes were verified using qRT-PCR, Western blot, and ELISA. Furthermore, the effects on endothelial barrier function and regulation of neutrophil trans-endothelial migration were detected using the Transwell model. HP-PRRSV primarily induced differential expression of numerous genes associated with immune response, including IFIT2, IFIT3, VCAM1, ITGB4, and CCL5, whereas LPS triggered an inflammatory response involving IL6, IL16, CXCL8, CXCL14, and ITGA7. Compared to the individual effect of LPS, when given after HN-induced stimulation, it caused a greater number of changes in inflammatory molecules, such as VCAM1, IL1A, IL6, IL16, IL17D, CCL5, ITGAV, IGTB8, and TNFAIP3A, a more significant reduction in transendothelial electrical resistance, and higher increase in neutrophil transendothelial migration. In summary, these results suggest a synergistic effect of HP-PRRSV and LPS on the inflammatory response of porcine pulmonary MVECs. This study provides insights into the mechanism of severe lung injury caused by secondary bacterial infection following HP-PRRSV infection from the perspective of MVECs, emphasizing the vital role of pulmonary MVECs in HP-PRRSV infection.

Keywords: porcine reproductive and respiratory syndrome virus; secondary infection; lipopolysaccharide; microvascular endothelial cells; inflammation; endothelial barrier; transendothelial cell migration

Citation: Yao, X.; Dai, W.; Yang, S.; Wang, Z.; Zhang, Q.; Meng, Q.; Zhang, T. Synergistic Effect of Treatment with Highly Pathogenic Porcine Reproductive and Respiratory Syndrome Virus and Lipopolysaccharide on the Inflammatory Response of Porcine Pulmonary Microvascular Endothelial Cells. *Viruses* **2023**, *15*, 1523. https://doi.org/10.3390/v15071523

Academic Editors: Caijun Sun and Feng Ma

Received: 31 May 2023
Revised: 3 July 2023
Accepted: 6 July 2023
Published: 8 July 2023

1. Introduction

Porcine reproductive and respiratory syndrome virus (PRRSV) is an enveloped, positive-sense RNA virus belonging to the family Arteriviridae [1]. The virus has undergone rapid genetic evolution, and in 2006, a highly pathogenic strain of PRRSV (HP-PRRSV) emerged in Jiangxi Province, China [2]. HP-PRRSV is characterized by more severe symptoms, higher morbidity, and increased mortality compared to PRRSV [3]. Notably, HP-PRRSV-infected pigs are often complicated by secondary bacterial infections, which are believed to contribute to the occurrence of severe clinical symptoms and pathological damage such as high fever and acute lung injury. Gram-negative bacteria, such as *Escherichia coli, Haemophilus parasuis, Vibrio cholerae, Bacillus bronchisepticum,* and *Pasteurella multocida,* frequently cause secondary infections of HP-PRRSV and can release high concentrations of lipopolysaccharide (LPS) in the piglet lungs [4,5], and even exist in dust within pig barns at various concentrations [6]. Qiao et al. investigated the secretion of pro-inflammatory cytokines in pulmonary alveolar macrophages (PAMs) stimulated with HP-PRRSV and LPS. Their results showed that HP-PRRSV induced the secretion of inflammatory cytokines, such as IL-1β and TNF-α, and the additional presence of LPS further increased their secretion. Moreover, the increase in cytokine secretion was more significant with a stronger virulent PRRSV strain [7]. Similarly, when pigs were inoculated with LPS and PRRSV in the trachea, they exhibited more severe respiratory symptoms compared to those inoculated with PRRSV alone [8].

Although PAMs are considered the main target cells for PRRSV and play a crucial role in maintaining lung homeostasis and defense, they alone cannot fully explain the pathogenesis of PRRSV. The lung, as a highly vascularized organ, contains massive microvascular endothelial cells (MVECs), which not only serve as the physical basis for gas–blood exchange but also act as a barrier against pathogens circulating through the bloodstream to reach the lungs or disseminate from the lung tissue to the rest of the body. Furthermore, MVECs can interact with lung resident cells, such as PAMs, and then regulate the recruitment and exudation of neutrophils and other immune cells. Consequently, MVECs are a key control point in various pathological changes, such as inflammation and fever [9], and their involvement in the pathogenesis of many viruses and bacterial toxins has been well documented [10,11].

Previous studies in our laboratory have shown that porcine pulmonary MVECs are susceptible to HP-PRRSV [12], and that HP-PRRSV infection damages the structure and function of the endothelial surface glycocalyx [13]. Additionally, MVECs exhibit a certain non-specific immune response to HP-PRRSV infection [14]. However, the functional changes in porcine pulmonary MVECs during secondary bacterial infection following HP-PRRSV infection remain unclear.

This experiment aimed to further understand the role of MVECs in severe lung injury caused by HP-PRRSV-mediated secondary bacterial infection. Porcine pulmonary MVECs were cultured in vitro and stimulated with HP-PRRSV HN strain and/or LPS. We analyzed the differential gene expression using transcriptome sequencing technology and validated the differential expression of partial inflammatory molecules at the mRNA and protein levels. We also examined changes in endothelial barrier function and the transendothelial migration (TEM) of neutrophils using the Transwell model. The results of this study explain the severe inflammatory damage in the lung caused by HP-PRRSV-mediated secondary bacterial infection and consolidate the key role of porcine pulmonary MVECs in the pathogenesis of HP-PRRSV.

2. Materials and Methods

2.1. Virus

The HP-PRRSV HN strain used in the experiment was generously donated by Dr. Zhanzhong Zhao from the Beijing Institute of Animal Husbandry and Veterinary Medicine, Chinese Academy of Agricultural Sciences. The virus was cultured and propagated with Marc-145 cells, and its titer was determined to be $10^{-6.339}$ TCID$_{50}$/0.1 mL. Normal Marc-145 cells were treated using the same method and the cell lysate was collected as a control. Both the virus lysate and the control lysate were stored at $-80\,^{\circ}$C.

2.2. Cells

Porcine pulmonary MVECs were isolated from approximately 15-day-old SPF British Large White pigs (purchased from Beijing SPF Pig Breeding Management Center). All procedures were approved by the Animal Protection Committee of Beijing Agricultural

University (No. BUA_ZT2022012) and performed according to the previous method with some modifications [15]. Briefly, the lung edge tissue was taken, minced after removal of the pleura, digested with 0.2% type II collagenase solution (Worthington, LS004176, Lakewood, NJ, USA), and dissociated into a single cell suspension, which was plated and incubated. The primary culture was preliminarily purified using the differential attachment method. Then, CD31-positive cells were separated using a high-throughput magnetic bead sorter (Thermo Fisher Scientific, Vantaa, Finland), and the purified MVECs were cultured in endothelial medium (M&C gene technology, L2305005, Beijing, China) containing 5% serum (Aoqing Biotech, AQmv09900, Beijing, China) and 0.25% ECGS (M&C gene technology, CC019, Beijing, China). Immunofluorescence staining for factor VIII was carried out for further identification, and cells from passage 3–5 were used for experiments.

Neutrophils were isolated from the peripheral blood of adult pigs obtained from the Beijing Fifth Meat Joint Processing Plant using the Neutrophil Isolation Kit (Solarbio, P4140, Beijing, China) following the provided instructions. The isolated neutrophils were adjusted to a density of 1×10^8 cells/mL, stored at $-80\,^{\circ}\mathrm{C}$, and used within one month. Neutrophils were recovered and resuspended in RPMI 1640 medium (Aoqing Biotech, AQ11875, Beijing, China) containing 5% serum at 1×10^8 cells/mL, and cell viability was determined by staining with trypan blue solution (Solarbio, T8070, Beijing, China) before being used for experiments.

2.3. Transcriptome Sequencing

Porcine pulmonary MVECs grown to confluence in 10 cm Petri dishes were divided into four groups: control group, PRRSV group, LPS group, and PRRSV-LPS group, and each group had three replicates. In the PRRSV group and PRRSV-LPS group, cells were exposed to 3 mL of HP-PRRSV HN lysate for 1 h, and the control group and the LPS group were exposed to an equal amount of the control lysate for 1 h. After washing with D-Hank's solution and incubated in a maintenance medium containing 2% serum for 36 h, the LPS group and the PRRSV-LPS group were treated with LPS at a final concentration of 5 μg/mL (Solarbio, L8880, Beijing, China), and the other two groups were treated with an equal volume of maintenance medium. Over another 24 h incubation, total cellular RNA was extracted from each group. Transcriptome sequencing analysis was carried out by Wekemo Tech Group Co., Ltd. (Shenzhen, China)and the procedure is briefly described as follows.

The RNA concentration and integrity were determined using Nanodrop and an Agilent 2100 bioanalyzer, and enriched using magnetic beads connected with oligo-thymine to capture the polyadenylated tail. The enriched mRNA was then fragmented randomly using divalent cations in NEB Fragmentation Buffer, followed by NEB general library construction. The library was preliminarily quantified using a Qubit 2.0 Fluorometer, and the fragment length was assessed using an Agilent 2100 bioanalyzer. Accurate quantification of the library's effective concentration was performed using qRT-PCR to ensure a concentration higher than 2 nM. High-throughput sequencing was carried out using the Illumina system. After filtering the raw data, the sequences were aligned to the pig genome "https://ftp.ncbi.nlm.nih.gov/genomes/all/GCF/000/003/025/GCF_000003025.6_Sscrofa11.1/GCF_000003025.6_Sscrofa11.1_genomic.fna.gz (accessed on 4 January 2022)" using HISAT2 software. The aligned sequences were then assembled and quantified using String Tie. To identify differentially expressed genes, the threshold criteria of Fold Change > 1.5 and adjusted p-value (Padj) < 0.05 were applied. Functional enrichment analysis of the differentially expressed gene sets was performed using GO functional enrichment analysis and KEGG pathway enrichment analysis with cluster Profiler 3.14.3 software.

2.4. Primer Design and qRT-PCR

To validate the results from RNA sequencing, qRT-PCR was performed to detect the expression of eight differentially expressed genes related to the inflammatory response. The primer sequences were obtained from the data on the "https://www.ncbi.nlm.nih.gov (accessed on 1 June 2022)"website (Table 1) and were synthesized by Sangon Biotech

(Shanghai) Co., Ltd. The total cellular RNA extracted in Section 2.3 was reverse-transcribed into cDNA with a cDNA synthesis kit (Tsingke, TSK301, Beijing, China). The total reaction system was 20 μL, including 2 μL cDNA, 2 μL 10 M primer F/R, 10 μL 2 × Mix, and 6 μL ddH$_2$O. According to the instructions of SYBR^{TM} Green Master Mix (Applied Biosystems^{TM}, 00791640, Carlsbad, CA, USA) and the Tm value of the primers, the reaction program was set as follows: pre-denaturation at 95 °C for 2 min, denaturation at 95 °C for 15 s, annealing at 60 °C for 15 s, extension at 72 °C for 1 min, and a total of 40 cycles. The relative quantification of gene expression was performed using the $2^{-\Delta\Delta Ct}$ method.

Table 1. Gene primer sequence for qRT-PCR detection.

Genes	Primer Sequences (5′-3′)
IL1A	F: AAC CTG GAT GAG GCA GTG AAA R: AGC ACT CAC AAA CAG TCG GG
CXCL8	F: AAT ACG CAT TCC ACA CCT TT R: TGT TGT TGT TGC TTC TCA GT
ITGB8	F: ACT GGG CCA AAG TGA AGA AAA C R: ATC CTC TTG AGC ACA CCA TCC
ITGAV	F: CAG CGC GTC TTC GAT GTT TC R: CCG GTG AGA AGA CCA GTC AC
VCAM1	F: ACG CTT GAC GTG AAA GGA AG R: CAC CCC GAT GGC AGG TAT TA
CCL5	F: CAT GGC AGC AGT CGT CTT TAT C R: AAG TTT GCA CGA GTT CAG GC
TNFAIP3	F: ATC CGA CCC CTA CCG TGA C R: GGT GCT CTA CAA GGC CTC TC
CXCL2	F: GAT GCT AAA CAA GAG CAG TGC C R: CCC AGG GGC TAT TTG CTT CTC

2.5. ELISA

Porcine pulmonary MVECs were divided into groups as described in Section 2.3 and treated with the virus lysate or the control lysate. After incubation in a maintenance medium for 36 h and 60 h, the LPS group and the PRRSV-LPS group were exposed to LPS at a final concentration of 5 μg/mL for 12 h. The culture supernatant was collected from each group and stored at −80 °C for ELISA detection. Total protein was extracted using RIPA lysate (Beyotime, P0013B, Shanghai, China) for Western blot detection.

The frozen cell supernatants were thawed and centrifuged at 12,000 rpm for 20 min. The supernatants were used to measure IL-1α and IL-8 according to the kit instructions (RayBiotech, ELP-IL1a-1, ELP-IL8-1, Atlanta, GA, USA).

2.6. Western Blot

The protein samples were separated using SDS-PAGE (Beyotime, P0012AC, Shanghai, China). The stacking gel was run at 70 V, and the separating gel at 120 V. The proteins were transferred to a PVDF membrane (Merk Milipore, IPVH00010, MA, USA) at a constant voltage of 110 V for 80 min. After blocking with 5% skimmed milk solution (Beyotime, P0216, Shanghai, China) for 1 h, the membrane was incubated with β-actin antibody (Proteintech, 66009, Rosemont, IL, USA) at 1:20000 dilution and VCAM-1 antibody (Santa Cruz, sc-18864, Dallas, TX, USA) at a 1:1000 dilution overnight at 4 °C. After washing, a secondary antibody solution at a 1:20,000 dilution was added and incubated at room temperature for 2 h. Finally, a chemiluminescence kit (Beyotime, P0018S, Shanghai, China) was used to visualize the protein bands using a gel imaging instrument (Tanon, Tanon-5200, Shanghai, China).

2.7. Transendothelial Electrical Resistance (TEER)

To assess the impact of PRRSV-LPS stimulation on the barrier function of MVECs, cells were seeded in the insert of the Transwell culture system (Biofil, TCS004024, Guangzhou,

China). TEER was measured using a Millicell-ERS endothelial resistance instrument (Milipore, MERS00002, MA, USA). After the resistance stabilized, MVECs were divided into groups and treated as described in Section 2.3. At 0, 36, 48, 60, 72, and 96 h post-infection with the HP-PRRSV HN strain, the resistance values were measured, respectively. TEER was calculated according to the following formula.

$$\text{Teer } (\Omega \cdot \text{cm}^2) = \text{Resistance value} \times \text{Transwell membrane area}$$

2.8. TEM Assay

MVECs were seeded in the insert of the Transwell plate, and divided into groups, and treated as described in Section 2.7. At 12 h after LPS stimulation, 2×10^5 neutrophils were added to the insert of Transwell plates. After 4 h incubation, the neutrophils in the lower chamber were collected, stained with trypan blue, and counted with a cytometer, and their mobility was calculated according to the following formula.

$$\text{Mobility} = (\text{neutrophil number in the lower chamber}/2 \times 10^5) \times 100\%$$

2.9. Data Analysis

For transcriptome sequencing data, p values were corrected by the Benjamini–Hochberg method [16]. The grayscale of protein bands from Western blotting was quantified using Image J software. All data are expressed as "mean ± standard deviation", and GraphPad Prism 8.0 software was used for data visualization and one-way analysis of variance. A p value less than 0.05 was considered statistically significant.

3. Results

3.1. Cell Characteristics

The CD31$^+$ MVECs purified from the primary cell culture of porcine lungs exhibited a polygonal appearance. They were routinely passaged and grew to confluence in about 5 days (Figure 1A). They could be stably propagated up to passage 7 before entering cellular senescence. Immunofluorescence staining of factor VIII was positive, with approximately 97% of cells exhibiting positive staining (Figure 1B,C). Additionally, cryopreserved porcine peripheral blood neutrophils were recovered and assessed using Trypan blue staining, with approximately 98% viability observed. Neutrophil purity was evaluated by Wright stain to be about 95%, and the cells displayed characteristic segmented nuclei, a prominent feature of neutrophils.

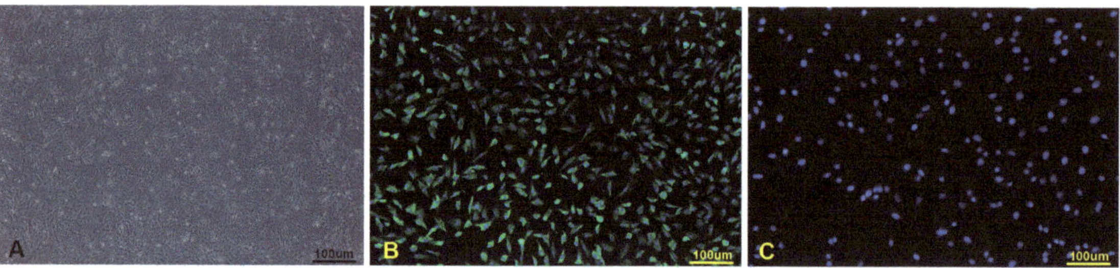

Figure 1. Microscopic morphology of porcine pulmonary MVECs and immunofluorescence staining for factor VIII. (**A**): Microscopic morphology of CD31$^+$ MVECs; (**B**): immunofluorescence staining for factor VIII; (**C**): negative control for immunofluorescence for factor VIII. (bar = 100 μm).

3.2. Transcriptome Changes of Porcine Pulmonary MVECs

To investigate the response of porcine pulmonary MVECs to HP-PRRSV and/or LPS stimulation, the transcriptome expression was analyzed by RNA-seq. The results showed that HP-PRRSV HN and LPS alone induced the differential expression of

101 (Figure 2A) and 175 (Figure 2B) genes, respectively, and their combined application resulted in the differential expression of 177 genes (Figure 2C). Venn analysis of the differentially expressed genes from individual treatments revealed only 20 genes in common (Figure 2D), indicating significant differences in the functional changes. Further analysis showed that compared to individual stimulation with HP-PRRSV HN or LPS, their combined treatment induced differential expression of 171 (Figure 3A) and 172 (Figure 3B) genes, respectively. Among these, only 21 genes were in common (Figure 3C). These findings suggest that their combined stimulation induces additional functional changes, and their respective roles are distinct.

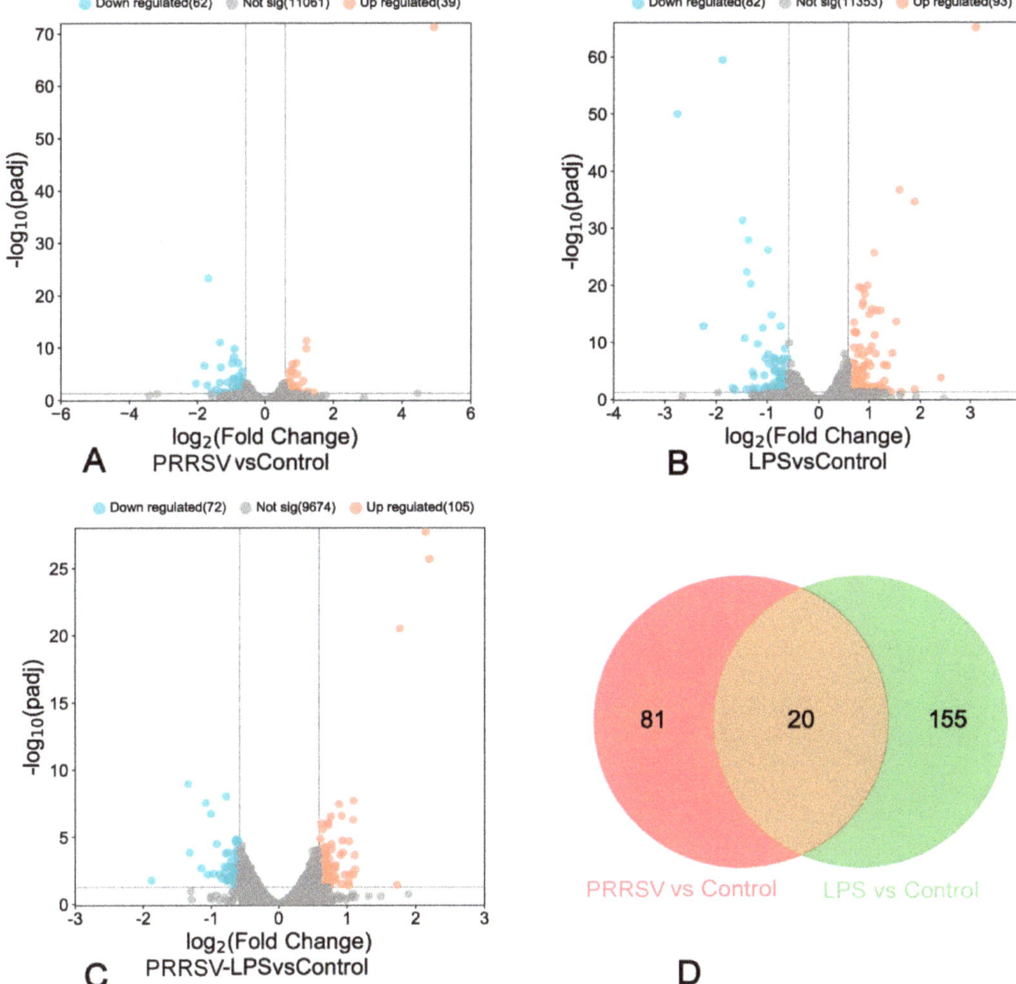

Figure 2. Volcano plot and Venn diagram illustrating differentially expressed genes in porcine pulmonary MVECs induced by HP-PRRSV and/or LPS. (**A**): Volcano plot of the PRRSV group versus control group; (**B**): volcano plot of the LPS group versus control group; (**C**): volcano plot of the PRRSV-LPS group versus control group; (**D**): Venn diagram of the HP-PRRSV group versus control group and the LPS group versus control group.

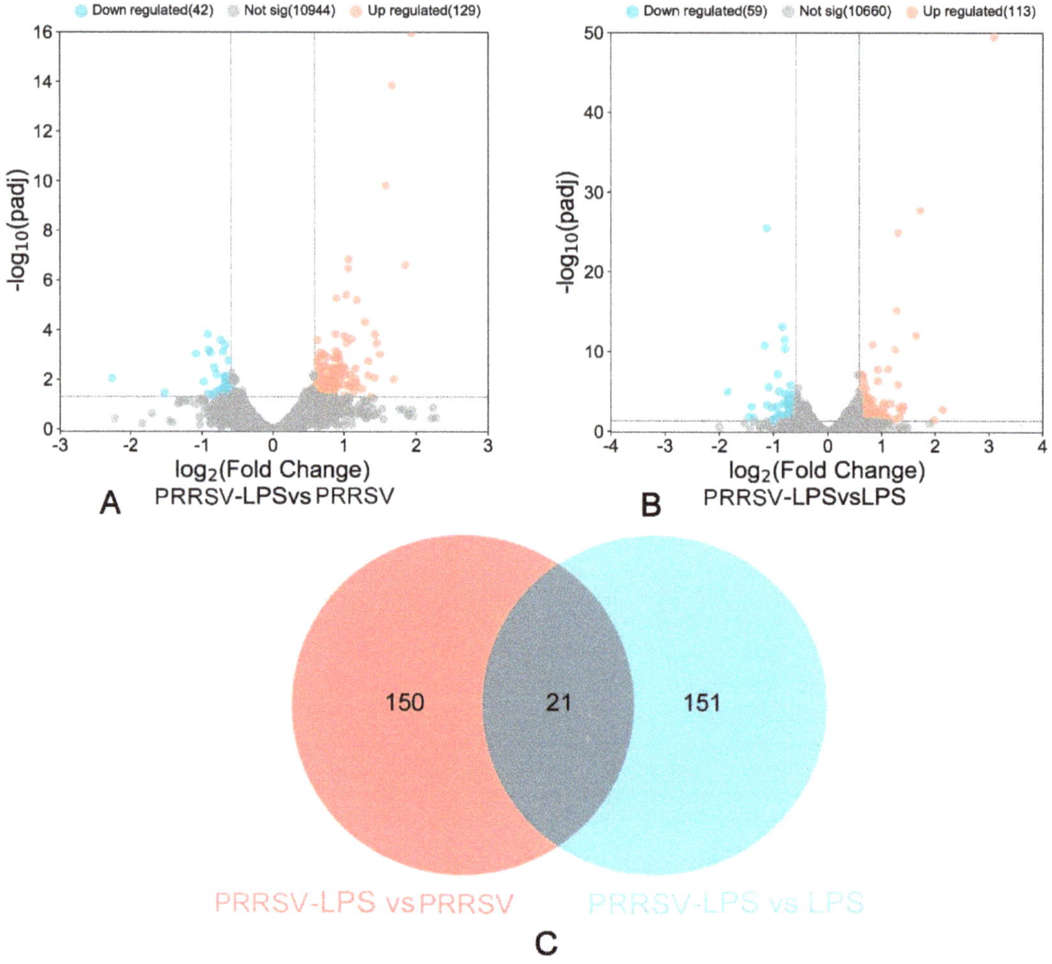

Figure 3. Volcano plot and Venn diagram demonstrating differentially expressed genes induced by combined stimulation compared to individual stimulation. (**A**): Volcano plot of the PRRSV-LPS group versus PRRSV group; (**B**): volcano plot of the PRRSV-LPS group versus LPS group; (**C**): Venn diagram of the PRRSV-LPS group versus PRRSV group and the PRRSV-LPS group versus LPS group.

3.3. GO and KEGG Functional Analysis of Differentially Expressed Genes

To understand the characteristics of the functional changes in porcine pulmonary MVECs induced by HP-PRRSV and/or LPS, gene ontology (GO) enrichment analysis was performed on the differentially expressed genes, and the top 20 significantly enriched GO items were selected to generate a histogram. The results revealed that in both individual and combined stimulation, the most significantly enriched genes were mainly associated with biological processes, although their specific items differed significantly. HP-PRRSV infection mainly induced responses to virus, cellular responses to type I interferon, type I interferon signaling pathway, negative regulation of viral gene replication, positive regulation of tyrosine phosphorylation of STAT protein, and positive regulation of JAK-STAT cascade (Figure 4A). The altered genes were involved not only in the viral immune response, such as CCL5, IFIT2, IFIT3, and SOCS3, but also in the inflammatory response, such as VCAM1, TNFSF15, ITGB4, and HES1 (Figure 4B). LPS mainly induced blood vessel development, positive regulation of smooth muscle cell proliferation, positive regulation of smooth muscle cell migration, cellular response to acid chemical, and cellular response to extracellular stimulus (Figure 4C). The altered genes primarily related to the inflammatory response, such as IL6, IL16, CXCL8, CXCL14, TNFSF15, and ITGA7 (Figure 4D). Furthermore, in comparison to HP-PRRSV alone, the biological processes caused by the combined stimulation mainly included connective tissue development, cardiovascular system development, blood vessel remodeling, regulation of signaling receptor activity, receptor ligand activity, and receptor regulator activity (Figure 4E). Their combined stimulation resulted in more changes in genes related to inflammatory responses, including IL-1α, CXCL2, CXCL8, ITGA1, and ITGB8 (Figure 4F), suggesting that LPS can induce a stronger inflammatory response in HP-PRRSV-pretreated MVECs. In comparison to LPS alone, its combined stimulation with HP-PRRSV mainly induced biological processes such as extracellular structure organization, cellular response to extracellular stimulus, extracellular structure organization, and extracellular matrix organization (Figure 4G). A large number of inflammatory genes, such as CCL 5, VCAM-1, ITGB 8, IGTAV, IL-6, and IL17D, were regulated (Figure 4H), indicating that the LPS-induced inflammatory response is more pronounced when MVECs are pretreated with HP-PRRSV.

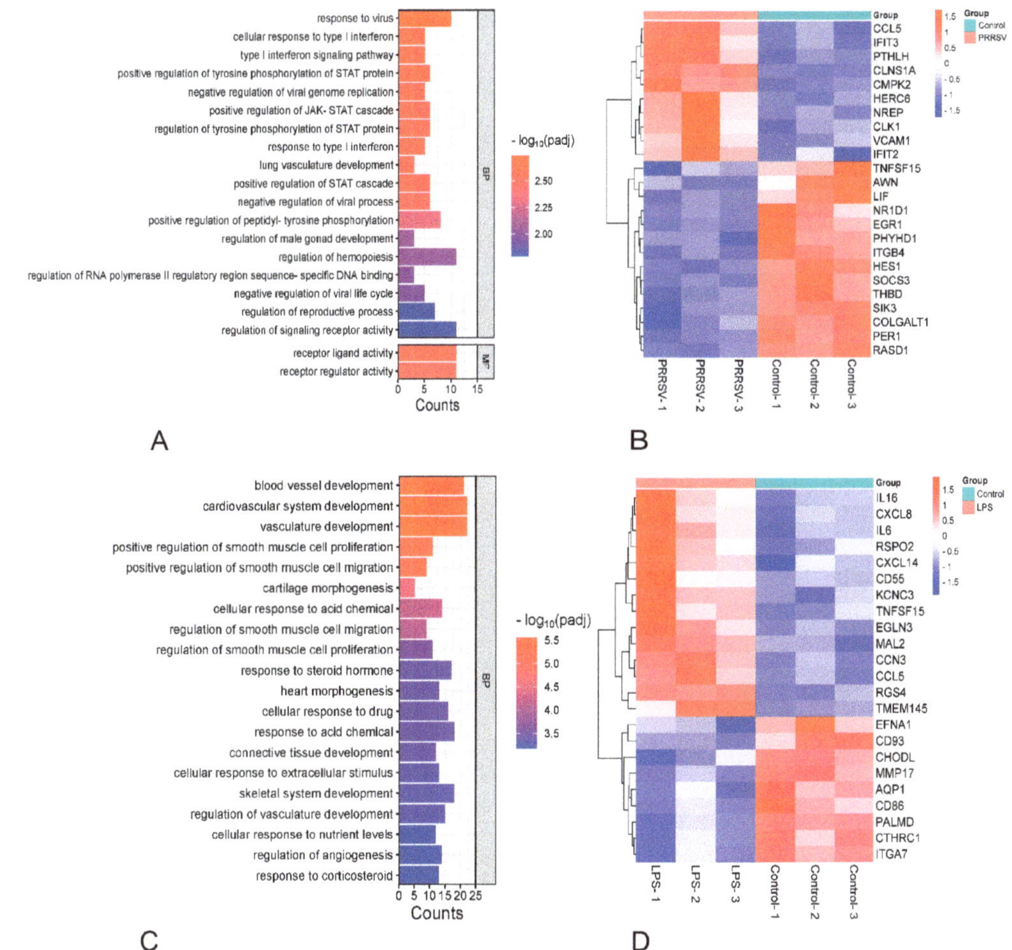

Figure 4. *Cont.*

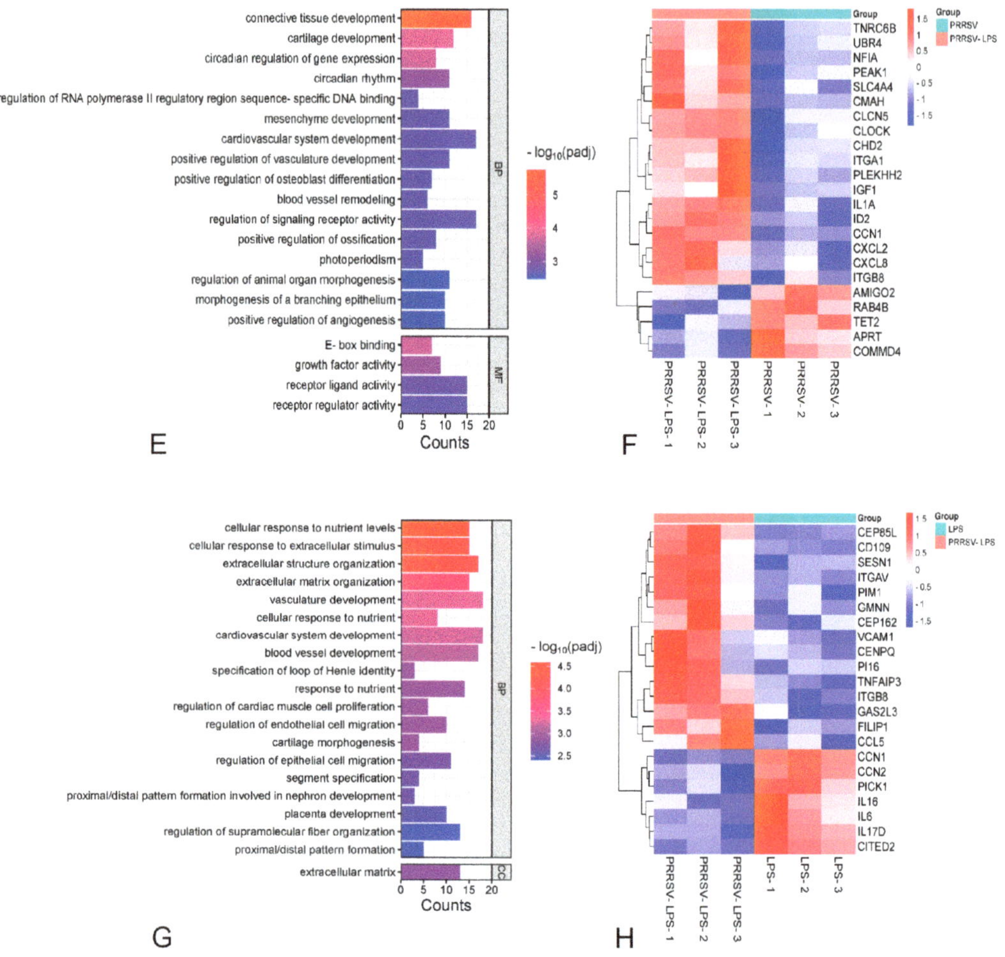

Figure 4. Histogram of GO enrichment analysis and heatmap of primarily differential genes. (**A**,**B**): PRRSV group versus control group, (**C**,**D**): LPS group versus control group, (**E**,**F**): PRRSV-LPS versus PRRSV group, (**G**,**H**): PRRSV-LPS group versus LPS group.

To clarify the signaling pathway associated with HP-PRRSV-mediated secondary bacterial infection, KEGG analysis was performed on the differentially expressed genes induced by combined stimulation. In comparison to LPS alone, the enriched pathways mainly included the MAPK signaling pathway, TNF signaling pathway, C-type lectin receptor signaling pathway, and IL-17 signaling pathway (Figure 5B). Similarly, compared to HP-PRRSV alone, the enriched pathways primarily comprised the PPAR signaling pathway, MAPK signaling pathway, cytokine–cytokine receptor interaction, and RIG-I-like receptor signaling pathway (Figure 5A). The results show that numerous inflammatory pathways are induced by their combined stimulation.

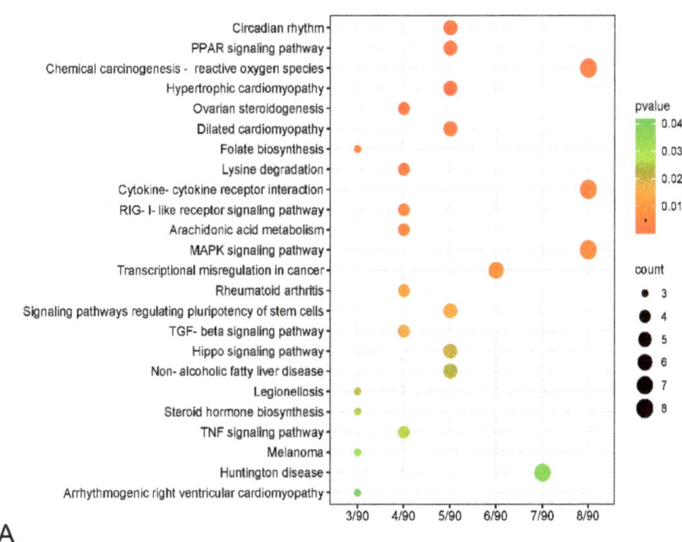

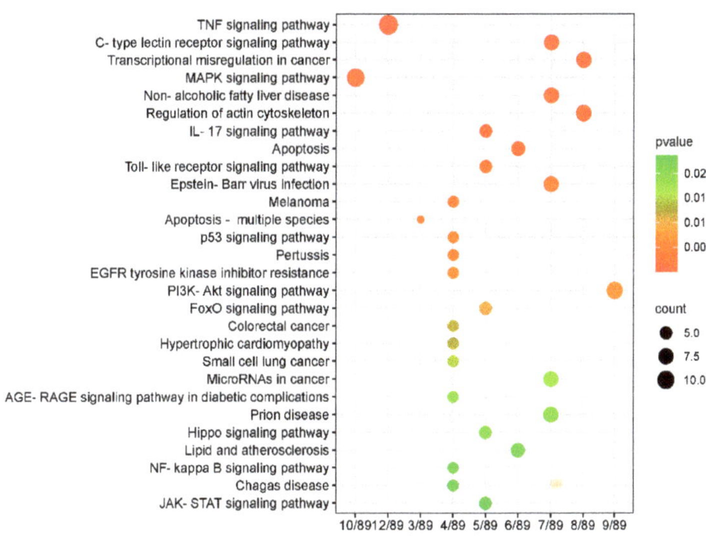

Figure 5. Histogram of KEGG enrichment analysis. (**A**): The PRRSV-LPS group versus the PRRSV group; (**B**): the PRRSV-LPS group versus the LPS group.

3.4. Validation of Differentially Expressed Genes by qRT-PCR

Given that many differentially expressed genes caused by HP-PRRSV and LPS stimulation, either alone or in combination, are related to immune and inflammatory responses, eight genes that have been well studied in immune and inflammatory responses with definitive roles were selected for qRT-PCR analysis to confirm the RNA-sequencing data. The results showed that the combined stimulation induced significantly upregulated mRNA expression of IL-1α, CXCL8, CXCL2, and ITGB8 compared to PRRSV stimulation alone, and that of ITGAV, CCL5, VCAM1, and TNFAIP3 compared to LPS stimulation alone (Figure 6).

These findings were consistent with the RNA-seq results, indicating the reliability of the transcriptome sequencing results.

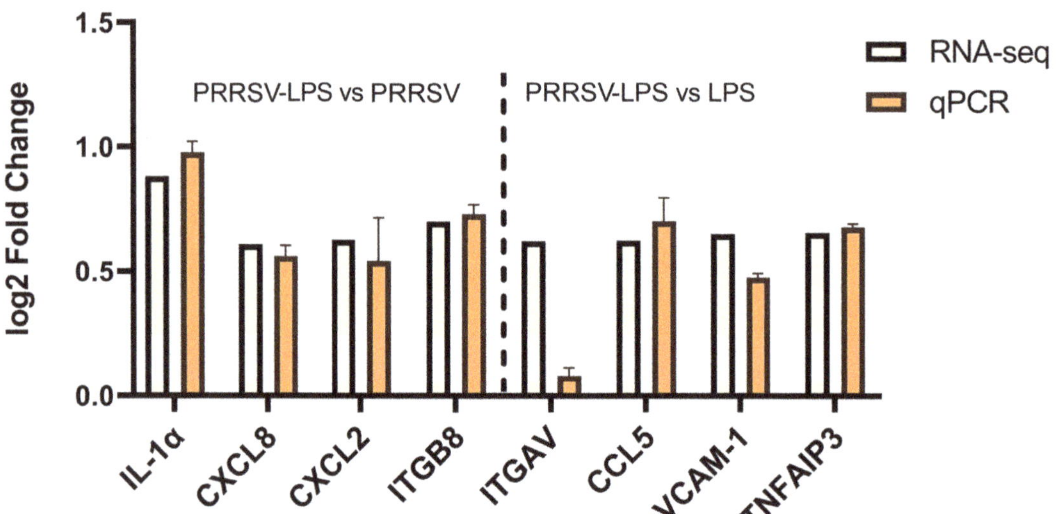

Figure 6. The mRNA expression ratios of representative inflammatory molecules by qRT-PCR and RNA sequencing.

3.5. Changes of Three Differentially Expressed Molecules at the Protein Level

To evaluate the effect of the combined stimulation on the protein level of key inflammatory genes, the production of IL-1α, CXCL8, and VCAM-1 was examined by ELISA and Western blot. There was no detectable IL-1α at 48 h, while its concentration in the PRRSV-LPS group was significantly higher than that in the LPS group at 72 h ($p < 0.01$) (Figure 7A). The concentrations of CXCL8 in the PRRSV-LPS group were significantly higher than those in the PRRSV and LPS groups at 48 h and 72 h ($p < 0.01$) (Figure 7B). The expression of VCAM-1 in the PRRSV-LPS group was significantly higher than that in the LPS group at 48 h and 72 h ($p < 0.01$), and its increase compared to the PRRSV group had no significant difference ($p > 0.05$) (Figure 7C,D).

3.6. Reduced TEER of Porcine Pulmonary MVECs by HP-PRRSV and LPS Stimulation

In addition to the expression of inflammatory molecules, disruption of the endothelial barrier is another characteristic of inflammatory response. In this study, the impact of HP-PRSSV and LPS stimulation on the TEER of porcine pulmonary MVECs was evaluated using a Transwell model. The results showed that the TEER reached a plateau stage at 48 h following the seeding of MVECs (Figure 8A), indicating the formation of a confluent monolayer. The stimulation of HP-PRRSV and LPS, either alone or in combination, significantly lowered the TEER compared to the control group (Figure 8B), and the TEER in the PRRSV + LPS group was significantly lower than that in the PRRSV or LPS group. The results suggest that HP-PRRSV HN and LPS exert a synergistic destruction of the endothelial barrier.

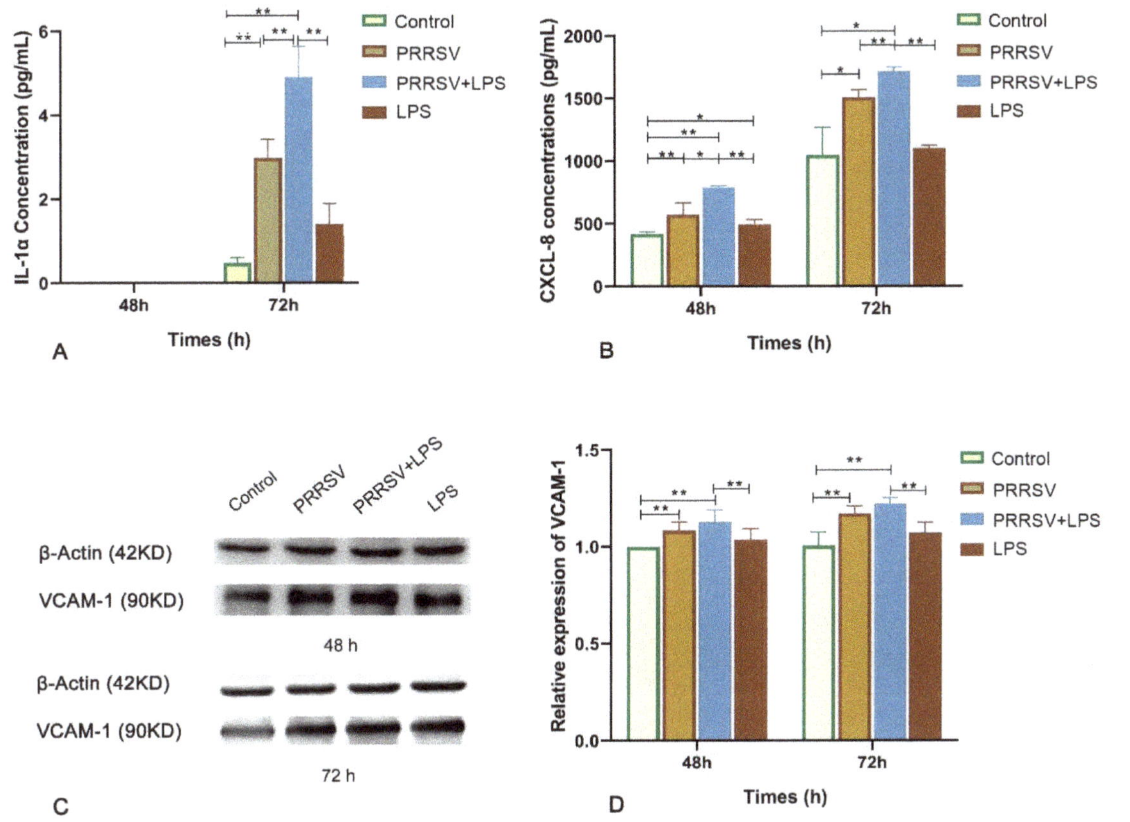

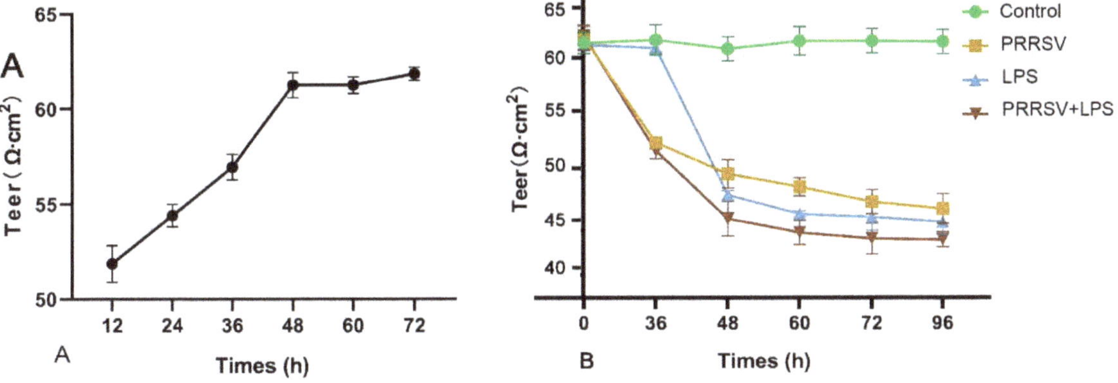

Figure 7. Effect of PRRSV and LPS stimulation on the production of, IL-1α (**A**), CXCL-8 (**B**), and VCAM-1 (**C,D**) in porcine pulmonary MVECs at the protein level. * indicates $p < 0.05$; ** indicates $p < 0.01$.

Figure 8. TEER changes of porcine pulmonary MVECs (**A**) induced by effects of HP-PRRSV and LPS stimulation (**B**).

3.7. Increased TEM of Neutrophils by HP-PRRSV and LPS Stimulation

Given the evidence that stimulation with HN-PRRSV and LPS induces differential expression of numerous inflammatory factors and disrupts the endothelial barrier, the Transwell model was further used to investigate their effect on TEM of neutrophils. The results showed a significant increase of neutrophil TEM in response to the stimulation of PRRSV and/or LPS ($p < 0.01$), and the PRRSV-LPS group exhibited a significantly higher migration rate compared to the PRRSV and LPS group ($p < 0.01$) (Figure 9), suggesting a remarkable synergistic effect of the combined stimulation.

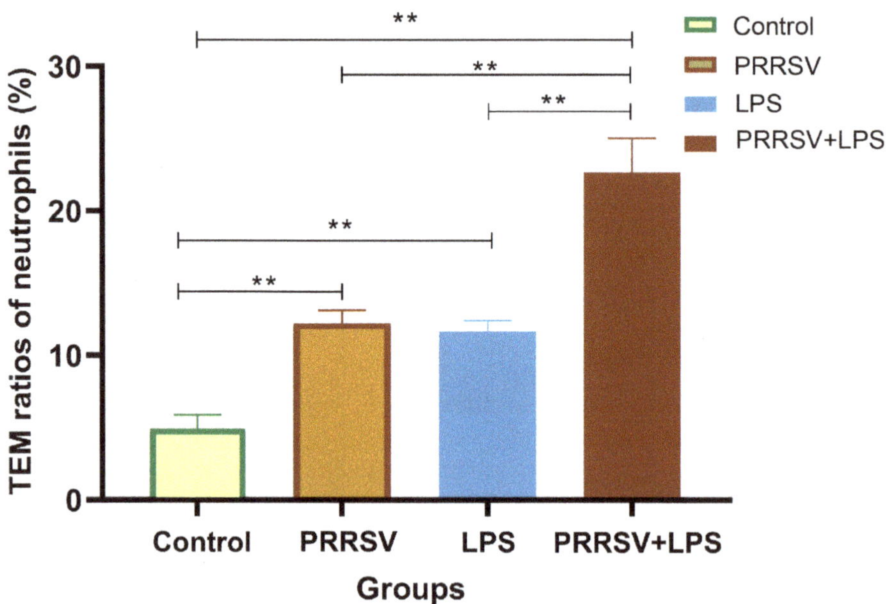

Figure 9. Effect of HP-PRRSV and LPS on neutrophil transendothelial migration. ** indicates $p < 0.01$.

4. Discussion

To explore the mechanism underlying the severe lung damage and high mortality rate in piglets caused by HP-PRRSV-mediated secondary bacterial infection, this study analyzed the transcriptome expression, TEER of the cell monolayer, and neutrophil TEM in porcine pulmonary MVECs treated with HP-PRRSV HN strain and LPS. The finding revealed distinct response patterns of porcine pulmonary MVECs to HP-PRRSV and LPS stimulation. HP-PRRSV infection resulted in differential expression of genes related to antiviral immunity and inflammatory response, while LPS stimulation mainly caused differential expression of genes associated with the inflammatory response. Interestingly, their combined stimulation increased the expression of inflammatory factors induced by LPS and exacerbated the functional damage caused by HP-PRRSV infection. Moreover, the combined stimulation significantly intensified the decrease in the TEER of MVECs and synergistically increased the neutrophil TEM. These results suggest that the combined stimulation of HP-PRRSV and LPS has the potential to synergistically enhance the production of inflammatory molecules, induce endothelial barrier dysfunction, and increase neutrophil TEM in porcine pulmonary MVECs. These more severe responses of MVECs may be one of the important mechanisms contributing to the development of severe lung lesions and increased mortality in piglets resulting from HP-PRRSV-mediated secondary bacterial infection.

MVECs have diverse biological functions and serve as crucial points in various physiological and pathological responses, exhibiting specific reactions to different stimuli [17,18]. This study also revealed distinct response patterns of porcine pulmonary MVECs to HP-PRRSV and LPS stimulation. The latter mainly caused differential expression of genes related to inflammatory response, while the former induced differential expression of many inflammatory genes as well as numerous non-specific immune genes, which suggests that porcine pulmonary MVECs not only are involved in the inflammatory lung injury caused by HP-PRRSV infection but also play an important role in the immune response to HP-PRRSV. Moreover, compared with HP-PRRSV or LPS stimulation alone, their combined stimulation resulted in the upregulation of more inflammatory molecules, including interleukins, chemokines, and adhesion molecules, suggesting that the inflammatory response of porcine pulmonary MVECs to HP-PRRSV-mediated secondary bacterial infection is more intense than that to either infection alone. Specifically, HP-PRRSV infection mainly promoted the effects of LPS on the MAPK signaling pathway, TNF signaling pathway, c-type lectin receptors, etc. However, LPS stimulation exacerbated the effects of HP-PRRSV infection on cytokine–cytokine receptor interaction, the RIG-I-like receptor signaling pathway, and so on.

The transcriptome sequencing results were confirmed at the mRNA and protein levels, while different treatment times were selected for the detection between the protein levels and the mRNA levels. In terms of mRNA detection, the treatment times of HP-PRRSV and LPS were 60 h and 24 h, respectively. For protein detection, the HP-PRRSV treatment times were 48 h and 72 h, and the LPS treatment time was 12 h. The results showed that the combined PRRSV-LPS stimulation had stronger effects on the increased production of three inflammatory molecules than either individual stimulation, although their production altered to some extent with the treatment time. Specifically, the secretions of IL-1α and CXCL8 were significantly higher in the PRRSV-LPS group compared to either the PRRSV group or the LPS group, while there was no significant difference between the PRRSV-LPS group and the LPS group at the mRNA level. This inconsistency may be attributed to the longer treatment time of LPS at the protein level, which brought stronger promoting factors for following LPS stimulation. IL-1α is a biologically active precursor released during tissue injury and necrotic cell death [19]. CXCL8, also known as interleukin 8 (IL-8), plays a central role in mediating inflammatory responses by attracting neutrophils to the site of inflammation, releasing vasoactive substances, and causing tissue immune damage [20]. VCAM-1 is involved in processes such as inflammatory response, immune response, and lymphocyte homing [21]. These three inflammatory molecules have been documented to be involved in PRRSV infection [22–24]. In this study, porcine pulmonary MVECs were demonstrated to be an important source tissue of inflammatory molecules.

Endothelial barrier dysfunction is an important marker of inflammatory injury in MVECs, and TEER measurement is a common method for evaluating endothelial barrier function [25]. Glycocalyx is the main component of the microvascular endothelial barrier [26]. Our previous study has demonstrated that HP-PRRSV infection leads to the destruction of the endothelial glycocalyx in porcine pulmonary MVECs [13], suggesting a potential impairment of their barrier function. This present study demonstrated that HP-PRRSV infection resulted in a decrease in the TEER of porcine pulmonary microvascular endothelial cell monolayer, although its effect was less pronounced compared to that of LPS stimulation alone. Furthermore, their combined stimulation led to a more substantial decline in TEER, indicating that the endothelial barrier dysfunction was exacerbated.

The neutrophil TEM experiment revealed that while HP-PRRSV HN and LPS stimulation individually increased neutrophil mobility to a similar extent, their combined stimulation significantly augmented it. Both the increased production of inflammatory factors and impaired endothelial barrier function may be the contribution of enhanced neutrophil TEM. During the process of neutrophils crossing MVECs, there is a complex interaction between them, and the function of neutrophils that migrate across the endothelium monolayer may be altered [27]. The combined stimulation of HP-PRRSV and LPS exacerbated the dysfunction of porcine pulmonary MVECs, likely impairing their regulation of neutrophil function. The functional status of neutrophils across impaired endothelial barrier deserves further study.

5. Conclusions

This study demonstrated the synergistic effect of the combined stimulation of HP-PRRSV and LPS on the inflammatory injury of porcine pulmonary MVECs. The findings of this research shed light on the underlying mechanism of severe lung injury caused by a secondary bacterial infection following HP-PRRSV infection, with a particular emphasis on the vital role of pulmonary MVECs in HP-PRRSV infection.

Author Contributions: T.Z. and Q.M. conceived the work. X.Y., W.D., S.Y. and Z.W. carried out the experiments. X.Y. analyzed the data and wrote the initial draft. W.D. and Q.Z. edited the manuscript. All authors commented on the manuscript. All authors have read and agreed to the published version of the manuscript.

Funding: This study was supported by the National Natural Science Foundation of China (No. 32273050, No. 31672597), Beijing Municipal Natural Science Foundation (No. 6202003, 5212003).

Institutional Review Board Statement: The protocol for the primary culture of porcine pulmonary MVECs was approved by the Animal Care and Protection Committee of Beijing University of Agriculture (No. BUA_ZT2022012).

Informed Consent Statement: Not applicable.

Data Availability Statement: The datasets generated for this study are available on request to the corresponding author.

Conflicts of Interest: The authors declare no conflict of interest.

References

1. Meulenberg, J.J. PRRSV, the virus. *Vet. Res.* **2000**, *31*, 11–21. [CrossRef] [PubMed]
2. Han, J.; Zhou, L.; Ge, X.; Guo, X.; Yang, H. Pathogenesis and control of the Chinese highly pathogenic porcine reproductive and respiratory syndrome virus. *Vet. Microbiol.* **2017**, *209*, 30–47. [CrossRef] [PubMed]
3. Li, Z.; He, Y.; Xu, X.; Leng, X.; Li, S.; Wen, Y.; Wang, F.; Xia, M.; Cheng, S.; Wu, H. Pathological and immunological characteristics of piglets infected experimentally with a HP-PRRSV TJ strain. *BMC Vet. Res.* **2016**, *12*, 230. [CrossRef]
4. Zhang, J.; Wang, J.; Zhang, X.; Zhao, C.; Zhou, S.; Du, C.; Tan, Y.; Zhang, Y.; Shi, K. Transcriptome profiling identifies immune response genes against porcine reproductive and respiratory syndrome virus and Haemophilus parasuis co-infection in the lungs of piglets. *J. Vet. Sci.* **2022**, *23*, e2. [CrossRef] [PubMed]
5. Xiangjin, Y.; Zeng, J.; Li, X.; Zhang, Z.; Din, A.U.; Zhao, K.; Zhou, Y. High incidence and characteristic of PRRSV and resistant bacterial Co-Infection in pig farms. *Microb. Pathog.* **2020**, *149*, 104536. [CrossRef] [PubMed]
6. Zejda, J.E.; Barber, E.; Dosman, J.A.; Olenchock, S.A.; McDuffie, H.H.; Rhodes, C.; Hurst, T. Respiratory health status in swine producers relates to endotoxin exposure in the presence of low dust levels. *J. Occup. Med.* **1994**, *36*, 49–56.
7. Qiao, S.; Feng, L.; Bao, D.; Guo, J.; Wan, B.; Xiao, Z.; Yang, S.; Zhang, G. Porcine reproductive and respiratory syndrome virus and bacterial endotoxin act in synergy to amplify the inflammatory response of infected macrophages. *Vet. Microbiol.* **2011**, *149*, 213–220. [CrossRef]
8. Labarque, G.; Van Reeth, K.; Van Gucht, S.; Nauwynck, H.; Pensaert, M. Porcine reproductive-respiratory syndrome virus infection predisposes pigs for respiratory signs upon exposure to bacterial lipopolysaccharide. *Vet. Microbiol.* **2002**, *88*, 1–12. [CrossRef]
9. Dalal, P.J.; Muller, W.A.; Sullivan, D.P. Endothelial Cell Calcium Signaling during Barrier Function and Inflammation. *Am. J. Pathol.* **2020**, *190*, 535–542. [CrossRef]
10. Varga, Z.; Flammer, A.J.; Steiger, P.; Haberecker, M.; Andermatt, R.; Zinkernagel, A.S.; Mehra, M.R.; Schuepbach, R.A.; Ruschitzka, F.; Moch, H. Endothelial cell infection and endotheliitis in COVID-19. *Lancet* **2020**, *395*, 1417–1418. [CrossRef]
11. Reyes, L.; Getachew, H.; Dunn, W.A.; Progulske-Fox, A. Porphyromonas gingivalis W83 traffics via ICAM1 in microvascular endothelial cells and alters capillary organization in vivo. *J. Oral Microbiol.* **2020**, *12*, 1742528. [CrossRef]
12. Li, P.; Yang, Z.; Ma, S.; Hu, G.; Dong, H.; Zhang, T. Susceptibility of porcine pulmonary microvascular endothelial cells to porcine reproductive and respiratory syndrome virus. *J. Vet. Med. Sci.* **2020**, *82*, 1404–1409. [CrossRef] [PubMed]
13. Song, X.; Wu, Y.; Wu, X.; Hu, G.; Zhang, T. Effects of highly pathogenic porcine reproductive and respiratory syndrome virus infection on the surface glycoprofiling of porcine pulmonary microvascular endothelial cells. *Viruses* **2022**, *14*, 2569. [CrossRef] [PubMed]

14. Wu, Y.; Song, X.; Cui, D.; Zhang, T. IFIT3 and IFIT5 play potential roles in innate immune response of porcine pulmonary microvascular endothelial cells to highly pathogenic porcine reproductive and respiratory syndrome virus. *Viruses* **2022**, *14*, 1919. [CrossRef] [PubMed]
15. Wu, Y.; Wang, Z.; Hu, G.; Zhang, T. Isolation and culture of rat intestinal mucosal microvascular endothelial cells using immunomagnetic beads. *J. Immunol. Methods* **2022**, *507*, 113296. [CrossRef]
16. Benjamini, Y.; Hochberg, Y. Controlling the False Discovery Rate: A Practical and Powerful Approach to Multiple Testing. *J. R. Stat. Soc. Ser. B Stat. Methodol.* **1995**, *57*, 289–300. [CrossRef]
17. Jin, Y.; Ji, W.; Yang, H.; Chen, S.; Zhang, W.; Duan, G. Endothelial activation and dysfunction in COVID-19: From basic mechanisms to potential therapeutic approaches. Signal Transduct. *Target. Ther.* **2020**, *5*, 293. [CrossRef]
18. Rommel, M.G.E.; Milde, C.; Eberle, R.; Schulze, H.; Modlich, U. Endothelial-platelet interactions in influenza-induced pneumonia: A potential therapeutic target. *Anat. Histol. Embryol.* **2020**, *49*, 606–619. [CrossRef]
19. Lopalco, G.; Maggio, M.C.; Sota, J.; Dinarello, C.A. Interleukin 1α: A comprehensive review on the role of IL-1α in the pathogenesis and treatment of autoimmune and inflammatory diseases. *Autoimmun. Rev.* **2021**, *20*, 102763.
20. Russo, R.C.; Garcia, C.C.; Teixeira, M.M.; Amaral, F.A. The CXCL8/IL-8 chemokine family and its receptors in inflammatory diseases. *Expert Rev. Clin. Immunol.* **2014**, *10*, 593–619. [CrossRef]
21. Troncoso, M.F.; Ortiz-Quintero, J.; Garrido-Moreno, V.; Sanhueza-Olivares, F.; Guerrero-Moncayo, A.; Chiong, M.; Castro, P.F.; García, L.; Gabrielli, L.; Corbalán, R.; et al. VCAM-1 as a predictor biomarker in cardiovascular disease. *Biochim. Biophys. Acta Mol. Basis Dis.* **2021**, *1867*, 166170. [CrossRef] [PubMed]
22. Wu, Y.; Song, X.; Li, P.; Wang, Z.; Zhao, Z.; Zhang, T. Highly pathogenic porcine reproductive and respiratory syndrome virus-induced inflammatory response in porcine pulmonary microvascular endothelial cells and effects of herbal ingredients on main inflammatory molecules. *Int. Immunopharmacol.* **2023**, *118*, 110012. [CrossRef] [PubMed]
23. De Lamache, D.D.; Moges, R.; Siddiq, A.; Allain, T.; Feener, T.D.; Muench, G.P.; McKenna, N.; Yates, R.M.; Buret, A.G. Immunomodulating properties of Tulathromycin in porcine monocyte-derived macrophages infected with porcine reproductive and respiratory syndrome virus. *PLoS ONE* **2019**, *14*, e0221560.
24. Amarilla, S.P.; Gómez-Laguna, J.; Carrasco, L. A comparative study of the local cytokine response in the lungs of pigs experimentally infected with different PRRSV-1 strains: Upregulation of IL-1α in highly pathogenic strain induced lesions. *Vet. Immunol. Immunopathol.* **2015**, *164*, 137–147. [CrossRef] [PubMed]
25. Srinivasan, B.; Kolli, A.R.; Esch, M.B.; Abaci, H.E.; Shuler, M.L.; Hickman, J.J. TEER measurement techniques for in vitro barrier model systems. *J. Lab. Autom.* **2015**, *20*, 107–126. [CrossRef]
26. Zhao, F.; Zhong, L.; Luo, Y. Endothelial glycocalyx as an important factor in composition of blood-brain barrier. *CNS Neurosci. Ther.* **2021**, *27*, 26–35. [CrossRef]
27. Filippi, M.D. Neutrophil transendothelial migration: Updates and new perspectives. *Blood J. Am. Soc. Hematol.* **2019**, *133*, 2149–2158. [CrossRef]

Review

The Host Cytoskeleton Functions as a Pleiotropic Scaffold: Orchestrating Regulation of the Viral Life Cycle and Mediating Host Antiviral Innate Immune Responses

Meilin Li, Dingkun Peng, Hongwei Cao, Xiaoke Yang, Su Li [ID], Hua-Ji Qiu *[ID] and Lian-Feng Li *[ID]

State Key Laboratory for Animal Disease Control and Prevention, Harbin Veterinary Research Institute, Chinese Academy of Agricultural Sciences, Harbin 150069, China
* Correspondence: qiuhuaji@caas.cn (H.-J.Q.); lilianfeng@caas.cn (L.-F.L.)

Abstract: Viruses are obligate intracellular parasites that critically depend on their hosts to initiate infection, complete replication cycles, and generate new progeny virions. To achieve these goals, viruses have evolved numerous elegant strategies to subvert and utilize different cellular machinery. The cytoskeleton is often one of the first components to be hijacked as it provides a convenient transport system for viruses to enter the cell and reach the site of replication. The cytoskeleton is an intricate network involved in controlling the cell shape, cargo transport, signal transduction, and cell division. The host cytoskeleton has complex interactions with viruses during the viral life cycle, as well as cell-to-cell transmission once the life cycle is completed. Additionally, the host also develops unique, cytoskeleton-mediated antiviral innate immune responses. These processes are also involved in pathological damages, although the comprehensive mechanisms remain elusive. In this review, we briefly summarize the functions of some prominent viruses in inducing or hijacking cytoskeletal structures and the related antiviral responses in order to provide new insights into the crosstalk between the cytoskeleton and viruses, which may contribute to the design of novel antivirals targeting the cytoskeleton.

Keywords: host cytoskeleton; viral replication cycle; host–virus interactions; orchestrated crosstalk

Citation: Li, M.; Peng, D.; Cao, H.; Yang, X.; Li, S.; Qiu, H.-J.; Li, L.-F. The Host Cytoskeleton Functions as a Pleiotropic Scaffold: Orchestrating Regulation of the Viral Life Cycle and Mediating Host Antiviral Innate Immune Responses. *Viruses* **2023**, *15*, 1354. https://doi.org/10.3390/v15061354

Academic Editors: Caijun Sun and Feng Ma

Received: 16 May 2023
Revised: 8 June 2023
Accepted: 9 June 2023
Published: 12 June 2023

1. Introduction

As an essential part of maintaining the normal function of cells, the cytoskeleton plays important roles in the activities of life, including endocytosis, cell division, intracellular transport, motility, force transmission, reactions to external forces, adhesion and preservation, and cell shape adaptation. The cytoskeleton is mainly composed of three types of cytoskeletal polymers, including actin filaments (AFs), microtubules (MTs), and intermediate filaments (IFs) [1]. AFs, MTs, and IFs constitute a complex network involved in the functions of eukaryotic cells, providing cells with the ability to perform multiple functions uniformly [2]. These proteins assemble into different structures to play broad roles [3], and form highly structured and dynamic networks. An intricate network of components is capable of swift adaptation in response to both external and internal stimuli, enabling precise regulation within minutes [1]. These three cytoskeletal proteins have different functions, but they are mutually regulated and work together to complete vital movement [4]. Information related to the cytoskeleton is summarized (Table 1 and Figure 1).

When the normal physiological activities of cells are disturbed, the cytoskeleton also undergoes remarkable changes accordingly, and abnormal conditions usually occur when the cells themselves propagate out of control or are disturbed by exogenous substances [5]. Cytoskeletal alterations contribute to the spread and migration of cancer cells [6]. Changes in cytoskeletal proteins in passively infected cells can also affect the infection process of microorganisms, such as viruses, bacteria, and parasites [5].

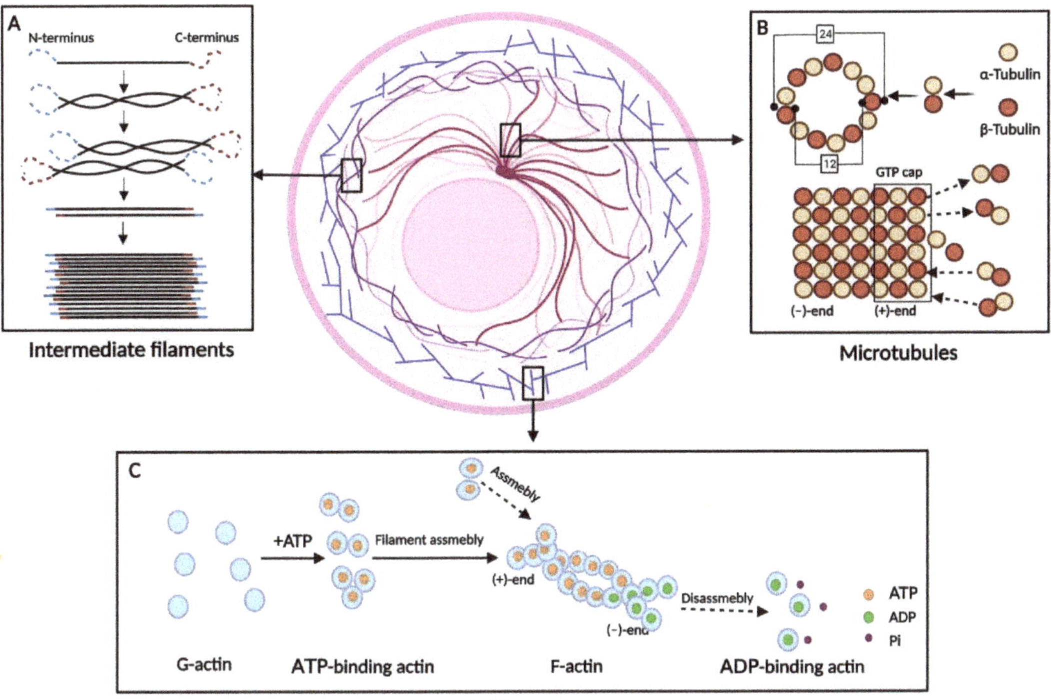

Figure 1. Schematic diagram of the distribution and structure of the cytoskeleton in the cell. (**A**) Intermediate filaments are formed by the spiral aggregation of monomers into dimers, followed by the aggregation of two dimers into a tetramer; finally, eight tetramers assemble to create a unit-length filament. (**B**) Microtubules (MTs) are composed of α- and β-heterodimers, which assemble into a hollow tube structure. The elongation of MTs occurs via the addition of heterodimers, resulting in the formation of a GTP cap at the positive end of the MTs. (**C**) Actin filaments are formed via a multistep process that begins with the binding of G-actin monomers to ATP. Following this step, G-actin monomers associate to form unstable dimers or trimers, which then elongate the filament. At the positive end, ATP-binding actin will be assembled and the ATP will gradually be hydrolyzed into ADP and Pi, once a G-actin is added to the filament. The minus end of an F-actin filament often contains the actin molecules in an ADP-binding form.

The cytoskeleton plays an active role in the viral life cycle. The process involves the virus invading the cell, traveling to the replication site, localizing the viral components to the proper assembly site after replication, and transporting them to the viral budding site. Since cortical actin affects the deformation of the cell membrane [7], virus entry into the cell must be regulated by actin [8]. Considering the capacity of MTs to facilitate intracellular transportation, their potential involvement in the viral life cycle is worth exploring. Many viral replicative mechanisms have been studied, and it has been found that microtubules can be used by viruses to transfer materials [9]. Upon viral infection, vimentin is crucial for stress response and signal transduction in cells [10]. This process assists the virus in propagating once it enters the cell [11]. In addition, viruses can use the cytoskeleton to spread from one cell to another and form a connecting channel between the two cells, which plays a significant role in their pathogenesis [12]. Recent research has confirmed that the cytoskeleton regulates the signaling pathway of IFN [13–15].

Current data show that the functions of the cytoskeleton are diverse. On one hand, the virus hijacks the cytoskeleton to complete its life cycle; on the other hand, the cytoskeleton assists cells to complete the process of innate immunity. This article reviews the functions of the cytoskeleton in viral colonization and propagation, including the process of virus invasion and host antiviral response.

Table 1. Structure and functions of the cytoskeleton.

Cytoskeleton Types	Main Members	Polymer Formation	Functions	References
Actin filaments (AFs)	β-Actin γ-Actin	G-actin forms an unstable dimer or trimer, and then the filaments are elongated by the addition of monomers.	Muscle contraction/ Maintenance of cell surface shape/ Deformable movement/ Cytokinesis	[1,16–18]
Microtubules (MTs)	α-Tubulin β-Tubulin	α- and β-Tubulin form a heterodimer, which is continuously extended. Thirteen extended tubulin protofilaments form a hollow tube.	Maintaining cell shape/ Transport of substances/ Assistant in mitosis	[1,19–24]
Intermediate filaments (IFs)	Acidic Keratins Basic Keratins Vimentin Lamins	IFs arise from the monomers spiraling around each other to form dimers. Two dimers aggregate to a tetramer and eight tetramers to a unit-length filament.	Maintaining cell morphology/ Signal transduction/ Involved in cellular stress	[1,25–29]

2. Physiological Functions of the Cytoskeleton on Normal Conditions

AFs are the major structural components of cells, and actin is the most abundant protein in many eukaryotic cells [30]. Actin is a 42-kDa protein with 375 amino acids and is highly conserved across a variety of species. It has six isoforms and more than 60 proteins [31], and only β-actin and γ-actin are expressed in most mammalian cell types [32]. The monomeric form of actin, known as G-actin, is the most fundamental structure for actin to perform its biological activities. Through a double or triple helix, G-actin creates the dimer or trimer F-actin, which is 7 nm in diameter. G-actin and F-actin serve different purposes, and the formation between them changes continually, preserving a relative balance in the absence of a stimulus. Actin has an ATP-binding region at its center, which binds to ATP in order to aggregate. The positive end of F-actin continuously binds to G-actin with ATP, extending into filament. ATP at the negative end will hydrolyze into ADP and Pi, resulting in the depolymerization of F-actin [1]. The equilibrium between the two formations is destroyed when the cellular activities change. For example, AFs prefer to polymerize when they are required to maintain cell morphological stability, whereas they typically depolymerize when they are required for cell deformation and movement [33]. Actin-binding proteins (ABPs) regulate F-actin [34]. For example, the actin-related protein (Arp2/3) is an ABP that drives G-actin polymerization to form F-actin, which can be activated by GTP depletion [35].

Actin, regarded as the most dynamic among the three major cytoskeletal proteins, is capable of rapid and significant structural alterations within minutes, which crucially contribute to determining cellular morphology [1]. Actin is involved in many physiological processes including cell motility, division, differentiation, senescence, death cell motility, division, differentiation, senescence, and death [36]. In all eukaryotic cells, actin regulates most cellular functions, including intercellular adhesion, cell motility, and cell division. The actin cytoskeleton is located in the cytoplasmic side of the plasma membrane and consists of a filamentous F-actin network that interfaces with the plasma membrane via surface receptors. Cortical actin is involved in all events related to the expression and presentation of membrane and cell surface molecules, the formation and movement of endocytic and phagocytic vesicles, viral entry, exocytosis, and viral export [16]. Actin regulates DNA repair, chromatin remodeling condensation, and gene transcription in the nucleus [17]. Moreover, nuclear actin has been identified as a crucial component of chromatin remodeling complexes that regulate gene expression. It interacts with all three RNA polymerases and plays a critical role in transcription initiation and elongation [37].

Long tubular organelles, known as MTs, are essential in eukaryotic cells and play a significant role in the cell cycle. These genes are highly conserved in many species. α-Tubulin is composed of 450 amino acid residues and β-tubulin is composed of 455 amino acid residues, which have a molecular weight of approximately 55 kDa. It has an average outer diameter of 24 nm and an inner diameter of 12 nm [1], and its structure is composed of α-tubulin and β-tubulin heterodimers assembled into a hollow polymer [19]. α-Tubulin and β-tubulin are sequentially arranged to form a single fiber, and 13 such fibers are arranged to form a hollow tubular structure [1]. The slower ends of polymerization and dissociation are the negative end of MTs, and this portion of α-tubulin is exposed. Faster polymerization and dissociation occur at the positive end of the microtubule exposed to β-tubulin [19]. Since a GTP cap exists at the microtubule positive ends, polymerization and depolymerization can be quickly completed in the cell, and are responsible for microtubule mass formation and dynamic interactions with different subcellular structures [20]. It also points to the positive end of the plasma membrane, which contributes to the intracellular trafficking of MT-bound vesicles. Owing to the growth and shortening of the positive ends, microtubule dynamics are generated [21]. MTs are altered by GTP hydrolysis for energy. GTP can be bound to α-tubulin, whereas GTPase is present on β-tubulin. Microtubule-binding proteins (MBPs) can directly or indirectly bind to MTs to regulate their dynamics, assembly, disassembly, and stability [22]. In contrast, the negative ends are hardly involved in depolymerization and polymerization, and determine the geometry of MTs network. Therefore, they are usually stably anchored at the microtubule nucleation sites [23]. The development of various forms of microtubule-organizing centers (MTOCs) results in a highly polymerized tubulin-containing structure, while γ-tubulin is highly polymerized [38]. In most cells, MTs are radially distributed around the cell from the center of MTOCs, with the positive pole pointing toward the cell membrane [1]. Microscopic motor proteins are divided into dynein and kinesin. Dynein transports intracellular material toward MTOCs, whereas kinesin transports intracellular materials toward the cell membrane. Intracellular vesicles can then be transported to different organelles through MTs, which is of great significance for efficient functioning [24]. The functional roles of MTs can be categorized into three distinct areas: cell motility, signal transduction regulation, and intracellular transportation [1].

IFs are fundamental building blocks of the cellular architecture and are generated from a vast array of proteins encoded by at least 70 genes [25]. The molecular weight of the protein is 52–58 kDa. Among these, vimentin is relatively conserved across various species, with a molecular weight ranging from 52 to 58 kDa. The structure of IFs consists of an N-terminus, a central α-helical rod domain, and a C-terminus of varying lengths [26]. During the interaction, the N- and C-termini of the monomer remain unwound, whereas the middle segment forms a parallel dimer, subsequently forming an antiparallel tetramer. Eight of these tetramers form a single filament with a compressed diameter of 10 nm [39]. IFs are classified into five categories based on their structural composition and sequence homology. Types I and II are acidic and basic keratins that form a heteropolymeric structure comprising 54 distinct subtypes of type III IFs, including vimentin [27], whereas type IV IFs are primarily expressed in the nervous system and contain three neurofilament heteropolymers (NF-L/M/H): synemin, internexin, and nestin. Type V IFs are nuclear filaments called lamins, and consist of A/C, B1, and B2 lamins [1]. Vimentin interacts with a range of proteins and performs crucial biological functions in IFs networks. Lysosomal and aggregate localization, cell migration, and various organelles and cellular components can be fixed within a specific range using the vimentin framework [28]. Vimentin, which is critical for multiple cellular functions, can sense and respond to cellular stress, including oxidative stress [29]. Vimentin can bind to the NF-κB sites, thereby changing the immune response [15], which is also critical for intracellular signal transduction.

3. Pathological Roles of the Cytoskeleton on Abnormal Conditions

3.1. Neoplasm and Cancer

The cytoskeleton is essential for cancer progression, and contributes to the metastasis and spread of tumor cells by maintaining the cell shape, movement, and other functions. Actin remodeling can promote tumor invasive growth and tumor cell proliferation in the skeletal muscle [40]. In addition, Rho small GTPases belong to the Ras superfamily of GTPases, which regulate a wide array of cellular processes related to their key roles controlling the cytoskeleton. Rho-GTPase is an important player in key signaling pathways that regulate cell migration, such as cytoskeletal dynamics, the assembly and disassembly of cell–cell connections, directional sensing, and integrin–matrix adhesion [41]. Rho-GTPases, their modulators, and effectors are involved in several aspects of cancer progression [42]. As an illustration, epithelial mesenchymal transition (EMT) is the process through which epithelial cells transform into mesenchymal cells. Once epithelial cells undergo EMT, they reorganize the cytoskeleton and change the signalling programs that define the cell shape and reprogram gene expression, and individual cells become more aggressive as a result [43]. EMTs are also influenced by changes in the cytoskeleton, such as altered intermediate filament composition caused by the inhibition of cytokeratin and the activation of vimentin [44]. This enables cell motility in response to changes in the structure of IFs, possibly as a result of the interaction of vimentin with motor proteins [45].

3.2. Passive Infection with Bacteria, Viruses, or Parasites

Viruses, bacteria, parasites, and other microorganisms infect cells, which then produce corresponding countermeasures. The most significant change that occurs as a result of this conversion is the recombination of actin [46]. Bacterial infections and inflammation can disrupt the epithelial barrier, and host cell cytoskeletal changes in the host cell can directly mediate bacterial invasion into the intracellular environment [47]. In the intracellular niche, some bacteria then utilize the host cytoskeletal network to spread from cell to cell [48]. The binding of the *Listeria monocytogenes* surface proteins (InlA) to cell receptors promotes two posttranslational modifications of E-cadherin, primarily comprising host kinase Src phosphorylation followed by ubiquitination by the E3 ubiquitin ligase Hakai. This actin undergoes polymerization, which is a key molecular event required for virus entry into the cell [49]. The dissemination process of the intracellular pathogen *Shigella* primarily relies on actin assembly at the bacterial pole, propelling the pathogen throughout the infected cells [50]. Vimentin plays a role in facilitating bacterial transport, leading to subsequent immune inflammatory responses [51]. Viral infection alters normal cytoskeletal functions to optimize viral replication and virions production. Rabies virus (RABV) causes dendrite damage and actin depolymerization due to a reduction in actin fragments in nerve cells [52]. It also regulates the gene expression of cytoskeleton-related proteins and disrupts biological pathways that require cytoskeletal proteins [53]. Plasmodium and other intracellular parasites can use host factors such as hemoglobin S and C to modify and reshape the actin cytoskeleton network, thus changing the cargo transport mode of the organism and protecting patients from infection (Table 2) [54].

Table 2. Pathological roles of the cytoskeleton.

Types of Pathogeneses	Changes in the Cytoskeleton	The Effects of the Changes	Pathological Roles	References
Cancers	Depolymerization and polymerization of actin	Contributing to cell migration	Devoting to cancer cells spread and replicate quickly	[41–45]
	Depolymerization, polymerization and modification of microtubules	Participating in cell movement through signal transduction and as a transport structure		
	Interaction of vimentin with actin and microtubules.	Contributing to cell–matrix adhesion and migration		
	Activation of vimentin expression, and interaction of vimentin with motor proteins	Aims to enhance cell motility, which is conducive to the process of epithelial–mesenchymal transition (EMT)		
Intracellular bacteria infected	Actin is recruited and interacts with actin regulatory factors Arp2/3	Leading to bacterial engulfment and internalization in a membrane-bound vacuole	Promoting the infection of intracellular bacteria	[46–51,55]
	Microtubule depolymerization and the activity of the Rho family of enzymes that control microtubules are affected and interfered with by bacterial production of *Clostridium difficile* toxin A (TcdA)	Participating in bacterial transportation and the consequential immune-inflammatory responses		
	Vimentin is expressed on the cell surface, secreted and located extracellularly	Contributing to stress reaction; vimentin can be both pro- and anti-bacterial, favoring bacterial invasion in some contexts, but also involved in bacterial-induced inflammation regulation		
Viruses infected	Actin depolymerizes and polymerizes, and kinetoproteins are recruited.	Contributing to entry and internalization	Assisting the virus to complete its life cycle	[15,52,53]
	Microtubule and motor proteins interact with viral proteins, microtubule depolymerization and polymerization, motor proteins are changed	Transporting viral components, formation of replicative organelles		
	The vimentin expression is changed	Contributing to viral replication and signaling		
Parasites infected	Plasmodium can promote actin polymerization in vitro	Inhibiting the movement of cargo vesicles to the erythrocyte plasma membrane	Promoting severe *Plasmodium falciparum* malaria infection	[54]

3.3. Pathological Process

The links between viral infection, cell morphology and changes in the actin cytoskeleton were determined using the description of the transformation [56]. The syncytium is formed by the fusion of cells of one or more species, which requires the rupture and reconnection of adjacent cell membranes. This entire process involves the support of the actin cytoskeleton beneath the cell membrane [57]. The pathogen responsible for COVID-19, severe acute respiratory syndrome coronavirus 2 (SARS-CoV-2), induces syncytia formation, which can increase the spread of the virus and facilitate the elimination of immune cells [58].

The disruption is also associated with neurodegenerative diseases. For example, infection with mouse hepatitis virus (MHV) induces tau phosphorylation through a mechanism dependent on glycogen synthase kinase-3β, which disrupts the stabilizing capacity of MTs, potentially leading to brain damage [59]. In Middle East respiratory syndrome coronavirus (MERS-CoV) and SARS-CoV-2 infections, troponin attaches to AFs and the level of troponin in the heart muscle of patients increases [60].

The influence of the cytoskeleton on viruses has been found to be significant. Here, what occurs when viruses hijack the cytoskeleton from various perspectives is discussed.

4. Multiple Engagements of the Cytoskeleton in Viral Life Cycle by Targeting Various Stages

AFs and MTs provide structural support for virus entry into the cell, the spatial configuration of endosomal membranes, intracellular transit, and recycling back to the cell surface, which are driven by different motor proteins. Vimentin regulates the transcription and translation of viral proteins [54]. In conclusion, the viral life cycle is greatly aided by alterations in the cytoskeleton and these effects vary depending on the virus species. The roles of cytoskeletal modifications are described in Figure 2 according to the stages of the viral life cycle.

4.1. Entry and Internalization

Viruses can enter cells via a variety of pathways, including membrane fusion and endocytosis. Viruses are usually captured by the pseudopodia of the cell, bind to their receptors, and enter the cell via membrane fusion [61]. In a recent report on SARS-CoV-2, cortical actin accumulation was observed in the plasma membrane of infected cells, suggesting the role of actin in virion entry, release, and transmission [62].

Actin and its regulators play an equally important role in endocytosis. When the human respiratory syncytial virus (RSV) and herpes simplex virus type 1 (HSV-1) infect, the viral capsid is surrounded by F-actin in synaptosomes, and actin is transiently depolymerized to form vesicles [63,64]. Many viruses enter cells via endocytosis with the help of clathrin, which requires actin for energy. The main mechanism by which RABV particles enter cells is clathrin-dependent [65], and viral particles enter cell inputs with elongated structures and an incomplete clathrin coating, which are dependent on actin for internalization [66]. The entry of virion-containing pits is hindered by actin disruption after pharmaceutical pretreatment with an actin-depolymerizing agent, such as latrunculin B or cytochalasin D, which does not prevent coated pit formation. The experimental phenomena of impeded infection demonstrate that the completion of the viral invasion process cannot be supported by clathrin on its own without actin to provide support [66]. Upon the arrival of the virus in the cell body, clathrin recruitment is initiated, and viruses undergo actin-mediated cell surfing to entry-specific sites. Notably, viral surfing continues during clathrin recruitment in pH-dependent viruses such as vesicular stomatitis virus (VSV) [67]. Surfing occurs along filopodia and AFs as they move toward endocytic hot spots. The movement of AFs in cell-surface protrusions, also known as F-actin reverse flow, involves myosin motors [68]. Myosin II is present in cellular processes that promote viral movement, which may affect the retrograde flow of F-actin from the filament group. Myosin II is the major ATPase involved in viral cell surfing [69].

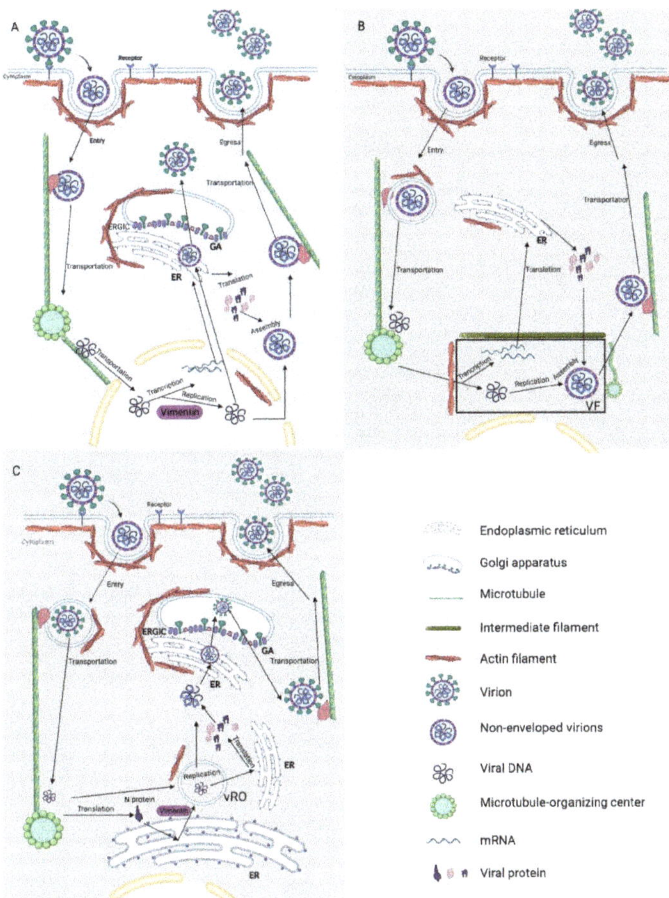

Figure 2. Schematic representation of different viruses using the cytoskeleton to complete their life cycle. (**A**) Similar to HSV, the process of cell entry requires actin rearrangement. After the viral genome enters the nucleus, replication and transcription begin to generate the nucleic acids and proteins required for viral assembly. This process is regulated by nuclear actin and vimentin. In the nucleus, viral proteins combine with DNA to form non-enveloped virions. Upon completion of assembly, virions either travel via ERGIC to produce mature virions or use microtubules for transport to the cell membrane and subsequent budding. (**B**) For poxviruses and African swine fever virus, the process of entry into cells requires actin rearrangement. The viral genome and related proteins are concentrated in the viral factory and begin to replicate and transcribe to generate the nucleic acids and proteins required for virus assembly. Actin/vimentin and microtubules are gathered around the viral factory. A large number of microtubule-organizing centers (MTOCs) are also present in the surrounding area. Subsequently, assembly is completed in the viral factory for microtubule-dependent transport out of the cell. (**C**) The cytoskeletal regulation of plus-strand RNA virus replication strategies is demonstrated. The process of entry into the cell requires actin rearrangement with subsequent transport to the ER region using microtubules. The viral genome is translated into the N protein, which is involved in the formation of vROs. Viral RNA replication is predominantly within vROs. Viral proteins assemble with RNA to generate non-enveloped virions, which undergo ERGIC to generate mature virions and are transported out of the cell in a microtubule-dependent manner. ER, Endoplasmic reticulum; GA, Golgi apparatus; ERGIC, ER/Golgi intermediate compartment structure; VF, virus factory; vROs, virus replication organelles.

In addition to the precise control of endocytosis by cortical actin, cell signal transduction is involved in the remodeling of the cytoskeleton after binding to the virus. During viral infection, Rho-GTPase signaling plays an important role in entry [70]; it is involved in regulating the actin structure, cytoskeleton assembly and remodeling, and mediating the phagocytosis of phagocytes with nucleating/elongation factors [71]. The glycoproteins of RSV and human parainfluenza virus type 3 (PIV-3) interact with RhoA to mediate viral entry [72]. Rac1 and Cdc42 are members of the RhoA-GTPase family involved in HSV-1 entry into neuronal and non-neuronal cells [73]. Another major signaling pathway is the phosphatidylinositol 3-kinase (PI3K)/protein kinase B (AKT), which plays a crucial role in cytoskeletal rearrangement. Dengue virus type 2 (DENV-2) infection induces AKT phosphorylation, leading to Rho activation and actin reorganization in Huh7 cells. The PI3K/AKT pathway is involved in DENV-2 infection in a Rho-GTPase and actin-dependent manner, and DENV-2 uses this signaling cascade to efficiently replicate in cells [74]. The third category is that most recently reported, showing that a disturbed actin cytoskeleton initiates the activation of pattern recognition receptors (PRRs). These sensor proteins are found in the cell membrane, nucleus, and cytoplasm. Many PRRs recognize certain viral or host-derived nucleic acids and, upon detection, cause the transcriptional activation of cytokines such as type I IFNs. Furthermore, PRRs are associated with cytoskeletal conversion. For instance, the retinoic acid-inducible gene I (RIG-I)-like receptor (RLR) pathway is activated following viral entry via actin rearrangement, a mechanism frequently linked to innate immunity [13].

No direct evidence has been reported regarding the association of IFs with viral invasion and transportation. Recent research has demonstrated that the establishment of human cytomegalovirus (CMV) infection is contingent upon the presence of an intact vimentin network, and that the cell tropism of CMV is contingent upon the integrity of the vimentin cytoskeleton [75].

4.2. Transport

MTs, AFs, and motor proteins are essential for generating the mechanical forces that drive the deformation and scission of cellular membranes. This mechanical activity facilitates the sorting of endosomal cargo and the generation of transport intermediates, enabling efficient intracellular transport processes. Virus transport by AFs and MTs often occurs shortly after entry into the cell membrane when it needs to traverse the inner layer of the cell membrane composed of microfilaments to inject its nucleic acid into the cell, or when the virus has completed replication and needs to be released from the cell membrane. Viral microfilament transport is a unidirectional movement that occurs after microfilament polymerization [76]. It has been demonstrated that the capsid protein and ICP0 of HSV-1 can interact with host MTs. Tubulin complex EB1 mediates the interaction between the viral capsid and the positive end of the microtubule, allowing the virus to undergo retrograde transport along the MTs upon entry [77]. ICP0 is a viral E3 ligase that destabilizes and unbundles MTs in Vero cells to aid in viral assembly and egress [78]. The destruction of AFs may hinder the assembly and egress of infectious bronchitis virus (IBV) [79]. Viruses can use MT trajectories as pathways for virions or as essential materials for virus assembly [80]. Viral proteins mediate the directional movement of virions along the MTs, which is important for viral transport out of the cell [81].

In fact, there are a few examples of viruses directly using MTs, more often using dynein and kinesin to transport viruses and cargo. Dynein interacts with the HSV-1 pUL37, and the key to its binding is the presence of a proline-rich domain in pUL37, and the interaction allows microtubules to smoothly transport the viral capsid [82]. The pUS9 of HSV-1 appears to use five arginine residues in its domain to bind to the host motor Kinesin-1 and contributes to anterograde axonal transport [83].

Interestingly, virions not only use motors for transport, but also dynein and kinesin on MTs to promote early infection [84]. The mature capsid (CA) core of human immunodeficiency virus 1 (HIV-1) encapsulates the viral genomic RNA, enzymes, and other viral proteins. The dynein motor complex is not only involved in the intracellular transport of the CA core, but may also be involved in the uncoating process of the CA core. After the downregulation of dynein expression, HIV-1 cDNA levels decrease, confirming that the HIV-1 reverse transcription is affected. However, the blockage of the CA take-off process by the virus caused by the downregulation of dynein expression appears to be transient, suggesting that uncoating is delayed rather than completely impaired [85]. Another study showed that multiple viral proteins directly interact with actin during HIV-1 infection, suggesting that HIV-1 may be anchored to cortical actin for reverse transcription and intracellular migration [86]. In conclusion, the retrograde transport of the HIV-1 CA core can effectively utilize dynein, while simultaneously using actin for transport.

4.3. Replication, Transcription, and Translation

During the replication of a positive-stranded RNA virus, viral replication organelles (vROs) composed of bilayer membrane-like structures are formed. The viral genomic RNA is wrapped in vROs to avoid the genome being degraded by other substances in the cytoplasm [87]. Live-cell imaging and sensors are used to monitor viral infections and replication, which shows that perinuclear inclusion in the SARS-CoV-2-infected cells is positive for dsRNA. Double-membrane vesicles (DMV), a form of vROs, are separated by reconstituted vimentin and encased in an IF cage. Drugs that inhibit vimentin also inhibit viral replication [88]. This cage is also surrounded by MTs that are excluded from the dsRNA-containing region, suggesting that IFs and MTs may serve to scaffold or confine the vROs compartment. Interestingly, bundles of cytoskeletal filaments have also been observed in the tomograms of infected cells in close proximity to vROs [62].

Notably, the formation of specialized replication organelles, or 'viral factories', has been observed in a range of viruses, including RNA viruses and large DNA viruses, such as poxvirus and African swine fever virus (ASFV). For these viruses, viral factories are typically organized around the periphery of the nucleus and serve as sites for efficient viral replication and assembly [83]. A study on vaccinia virus (VACV) has demonstrated that mRNA structures appear to be aligned on MTs, implying that MTs track connected mRNAs and cores. Accordingly, intact MTs are required for the typical punctate organization of viral mRNAs [89]. Early in infection, MTs retract toward the nucleus, rounding cell aggregates and bringing organelles close to the nucleus. Small early factories moved to the nuclear periphery in an MT-dependent manner to form larger factories [90].

Nuclear actin exists in the cell nucleus as a skeleton protein that participates in transcription, transcriptional regulation, and chromatin remodeling, and can control the nuclear expression of viral genes in the replication stages [91]. The movement of the HSV-1 capsid in the nucleus requires the participation of nuclear actin [92].

4.4. Assembly and Egress

VACV and ASFV assembly is consistent with viral replication and requires the involvement of a cytoskeleton-involved cytoskeleton-related perinuclear virus assembly factory [93]. Such cage structures, known as MTOCs, have been described above; their formation is important for the assembly of viral materials [94].

Newly synthesized viral proteins and nucleic acids in coronaviruses are transported to the actin-rich ER/Golgi intermediate compartment structure (ERGIC), which is adjacent to the endoplasmic reticulum and Golgi apparatus, facilitating the transfer of the virus from the endoplasmic reticulum to the Golgi apparatus, where the proteins undergo post-translational modification and complete the assembly stage [95]. Centrosomes are also important for the assembly of viruses; RNA viruses that replicate in the nucleus, such as retroviruses, can bud through centrosomes. For example, Foamy viruses must first accumulate in MTOCs and subsequently acquire an intact envelope via ERGIC to form intact virions [96].

Coronavirus-infected cells have been observed by electron microscopy, and actin parallel to the cell edge appears to be thickened [97]. The enhanced presence of actin can assist in providing a bending force to expel the progeny viral particles to the exterior [98]. Viral proteins are associated with the capsid proteins of viral particles, thus facilitating their ability to target sites of nuclear viral egress [99].

5. The Cytoskeleton Mediates Virus Transmission and Spread from Cell to Cell

Cell-to-cell transmission significantly boosts the effectiveness of viral transmission by concentrating the release of viral particles at the points of cell–cell contact [100]. It protects against antibodies that partially neutralize viruses [101], and under certain circumstances, overrides the inhibitory effects of specific antiviral restriction factors [102]. This viral transmission method may also affect the etiology and course of the infection [103]. It has been established that the cytoskeleton plays a major role in the transmission of viruses across cells (Figure 3).

5.1. Direct Transmission

The cells are linked by an open membrane channel called a tunnel nanotube (TNT) [104]. It can transmit a variety of items over great distances, including communication substances, genetic materials, and viruses. TNT can move not only small molecules, such as calcium ions, but also large molecules such as proteins, peptides, and organelles inside the cell [105]. With the help of this novel direct communication technique, the physiological and pathological aspects of various cell communication processes may be better understood, while also learning about novel long-distance communication mechanisms [105]. TNTs mostly consist of actin and MTs [106]. Despite ongoing research on TNT synthesis, the cytoskeleton plays a crucial role [104]. Moreover, Rho-GTPases play crucial signaling roles in this process [107]. It has been demonstrated that coronavirus, influenza virus, and HIV-1 cause TNT to develop and be transmitted between cells [107,108].

Virological synapse (VS) is a specialized site for the formation of virus-infected immune cells in contact with each other, and is a channel for the formation of contact between cells. The VS formation involves F-actin polymerization, depolymerization, and Rho-GTPase signaling [109]. Virions can be secreted from one cell to another via junctions [110]. After cell–cell contact, the cytoskeleton of infected cells rapidly polarizes to cell–cell junctions to form special sites at which different proteins are linked for virus transmission [111]. SARS-CoV-2 can spread between dendritic cells and target cells and invade nerve cells via connections similar to VS [112]. HIV-1 cell-to-cell transmission substantially increases the efficiency of viral transmission by concentrating the release of viral particles at the site of cell–cell interactions [113]. HIV-1 envelope proteins, such as Gag on the infected donor cells and CD4 on the uninfected target cells, interact to form VS, which requires actin support [114]. It has been suggested that during HIV infection viral particles are transported by MTs to cell-cell contacts, where they pass though the core region of the synapse and enter the target cell [110].

The structural continuity of tissues is maintained by three distinct types of cell–cell junction: desmosomes, tight and adherens junctions [115]. These junctions provide both extracellular and intracellular connections between neighboring cells, linking different elements of the cytoskeleton to form a cohesive structural network. In addition to their structural roles, these junctions are involved in regulating tissue integrity and controlling the diffusion of ions, solutes, and microorganisms through tissues [115]. It has been shown that hepatitis C virus and retroviruses may enter via tight junctions, human papilloma virus (HPV) may enter via adherens junctions, and HIV may modify gap junctions for entry [116].

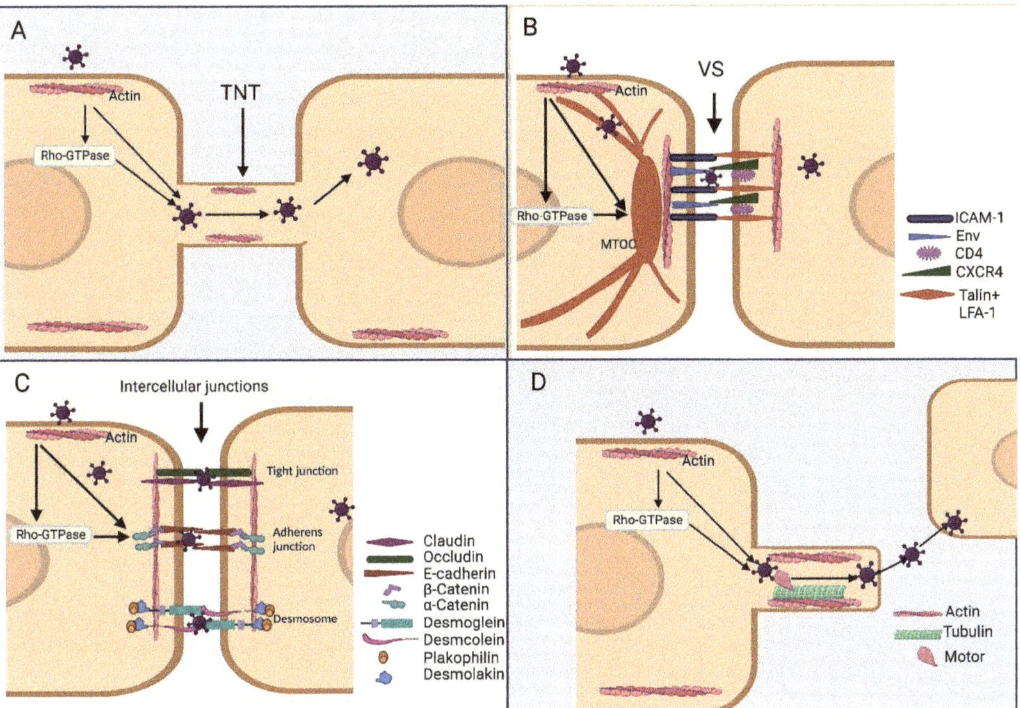

Figure 3. Schematic representation of the cell-to-cell spread of different viruses. (**A**) After viral infection, tunnel nanotubes are generated, and actin is required for this structure. The Rho receptor plays an important role during this period. TNT, tunnel nanotube. (**B**) During HIV infection, viral material is mobilized along microtubules to the site of cell–cell contact, where it traverses the central region of the synapse and enters the target cell. Upon arrival at the virological synapse (VS), LFA-1-talin complexes, and CD4 and CXCR4 are recruited in an actin-dependent manner. The VS is a specialized structure formed at the site of cell–cell contact, which facilitates the efficient transmission of HIV between infected and uninfected cells. VS, virological synapse. Env, HIV envelope protein; CXCR4, CXC chemokine receptor 4; ICAM-1, DC cognate ligands; LFA-1, lymphocyte function-associated antigen. (**C**) Intercellular junctions play a crucial role in maintaining the structural integrity and function of tissues. The three main types of intercellular junction are tight junctions, adherens junctions, and desmosomes. Tight junctions are formed by the transmembrane proteins claudin and occludin, which interact to form homotypic claudin–claudin and occludin–occludin complexes between adjacent cells. These transmembrane proteins bind directly to cytoplasmic adaptor proteins, which in turn associate with the actin cytoskeleton to maintain cell–cell adhesion and regulate the diffusion of ions and solutes through the tissue. Adherens junctions, on the other hand, are mediated by the transmembrane protein E-cadherin, which forms a ternary complex with the β-catenin and α-catenin. This complex binds to F-actin in a force-dependent manner, thereby regulating cell–cell adhesion and maintaining tissue integrity. Desmoglein and desmcolein bind to each other to form the core region. (**D**) Intracellular viruses can enter other cell-like comets due to slingshot-like structures formed by the intracellular microtubule and actin. Motor proteins are needed to empower this process.

5.2. Indirect Transmission

Many viruses use comets formed from actin to advance cytoplasmic viruses to the cell periphery or outside the infected cells [117]. Viral proteins may use actin-formed comet propulsion to pass through actin-enhanced cell junctions and be transported to neighboring cells [117]. The comet structure is essentially a slingshot structure formed by actin, which uses the elastic force of actin to push the virus out of the cell and facilitate its spread.

6. The Cytoskeleton Is Involved in the Immune Responses to Viral Infections

The cytoskeleton is also involved in innate immunity. Viral DNA is recognized by cyclic GMP-AMP synthetase (cGAS) [118], and viral RNA is recognized by retinoic acid-RLRs [119]. The recruitment of downstream molecules such as STING and MAVS results in the activation of downstream pathways. It can control gene expression, and its outcome is correlated with IFN production and expression [118,119].

The phosphorylation of RIG-I at Ser 8 and MDA5 at Ser 88 prevents RLR activation [120]. Once this site is dephosphorylated, RLRs are activated by the RNA viruses. The dephosphorylation of these sites by cellular protein phosphatase-1 PP1α or PP1γ is critical for RLR activation in response to viral infection [121], and virus-mediated perturbations of the actin cytoskeleton have been extensively documented to trigger RLR dephosphorylation via the PP1–R12C phosphatase complex [13]. Spire homolog 1 (Spir-1, also known as SPIRE1) has actin-binding domains, through which it nucleates actin filaments [122]. It has been demonstrated that Spir-1 stimulates innate immune signaling upon dsRNA sensing. Through a diphenylalanine motif, Spir-1 specifically contributes to the activation of interferon regulatory factor 3 (IRF3) and is also required for direct contact between Spir-1 and the VACV virulence factor K7. Spir-1 has been demonstrated to reduce VACV and ZIKV replication and/or dissemination, and is thus a virus restriction factor [123].

One example of a guanine nucleotide exchange factor (GEF) specific to RhoA, known as GEF-H1, is localized and confined to the MTs. This sequestration is associated with the precise temporal and spatial activation of Rho-GTPases [124]. Inactive GEF-H1 binds to the dynein motor complex on MTs, and GEF-H1 is activated and released from MTs upon cellular interactions, contributing to the recognition of intracellular pathogens. GEF-H1 can function in the RLR pathway in conjunction with MAVS and TANK-binding kinase 1 (TBK1); the inhibitor of nuclear factor-kappa B (IκB) kinase epsilon (IKKε) complexes to enhance the phosphorylation of IRF3 and the activation of the *ifnb1* promoter [125].

Recently, vimentin has been reported to play a role in many vital immune responses processes, and it has been described as a ligand for some PRRs. Vimentin expression may depend on IFNs [14], and viral infection may promote vimentin promoter activity. Vimentin overexpression is accompanied by enhanced viral replication, and the inhibition of IRF3 and TBK1 phosphorylation. Vimentin has been suggested to suppress the production of type I IFNs by targeting IRF3 or its associated binding partners, including TBK1 and inhibitors of IκB kinase epsilon (IKKε) [17]. During a viral infection, TBK1, IKK, and IRF3 form a complex. Once activated, TBK1 and IKK phosphorylate IRF3 to enhance its nuclear translocation. Both TBK1 and IKK possess an N-terminal kinase domain (KD) or interaction with IRF3 [126]. Vimentin and IRF3 bind to the KD domain of TBK1 or IKK, which may prevent the formation of the TBK1–IKK–IRF3 complex and the nuclear translocation of IRF3 [15,126]. The above three examples of RLR signaling pathways being affected are summarized in Figure 4.

Nucleotide oligomerization domain-containing protein 2 (NOD2) is an important receptor involved in cellular innate immunity, and vimentin is an NOD2-interacting protein in mammalian cells. A recent study has suggested that NOD2 interaction with vimentin is important for its ability to respond to the signals downstream of NF-κB [127].

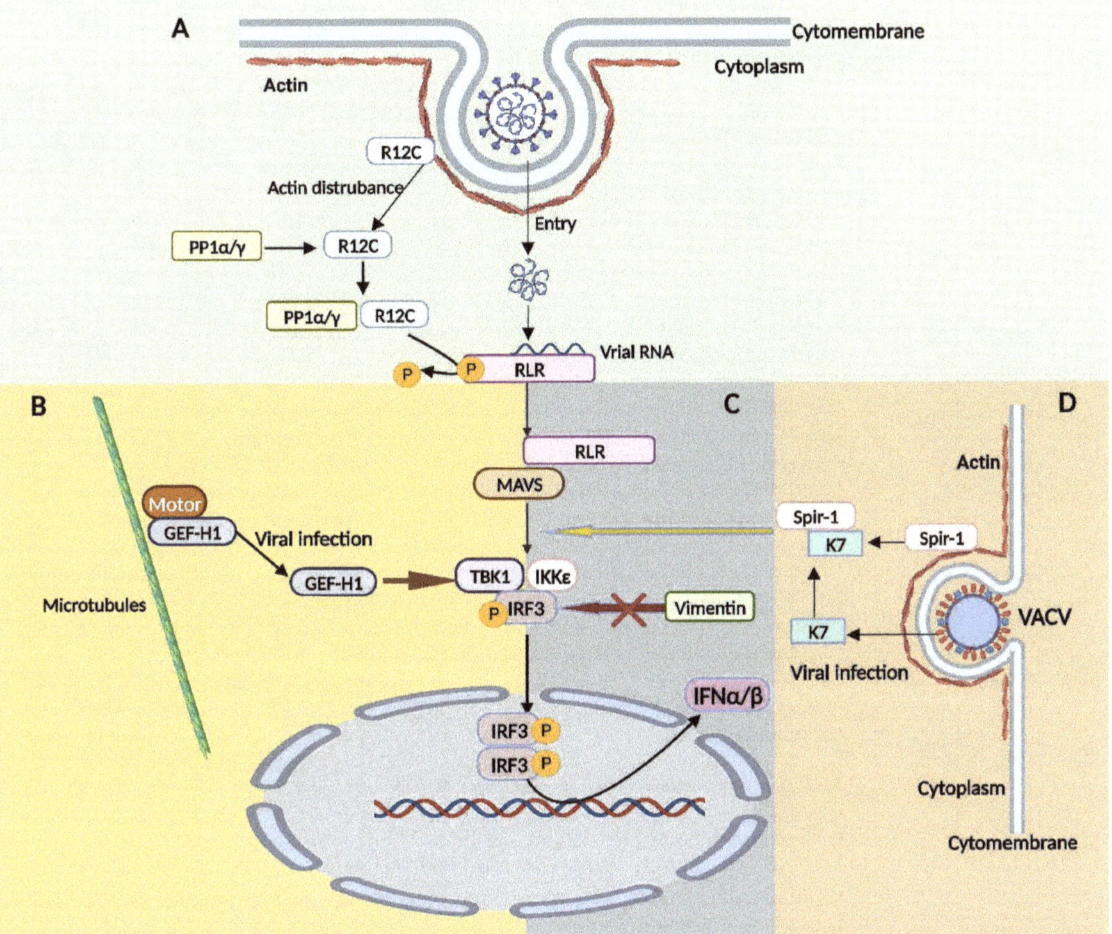

Figure 4. Involvement of the cytoskeleton in the immune response to viral infections. The inclusion of actin, the microtubule, and vimentin affects the regulation of the RLR signaling pathway in interferon production. (**A**) R12C can be released from the actin phosphatase complex during viral infection and participate in the binding of RLRs to PP1α/γ. (**B**) GEF-H1 can be released from the microtubule motor protein complex during viral infection and affect the phosphorylation of IRF3. (**C**) Vimentin can regulate TBK1 and IKKε, and affect the phosphorylation and nuclear import of IRF3. (**D**) The actin nucleating factor Spir-1 interacts with the K7 protein of vaccinia virus upon infection, promoting the IRF3 activation downstream of MAVS and upstream of IKK or TBK1. RLRs, retinoic acid-inducible gene I (RIG-I)-like receptor. R12C, a regulatory subunit of the protein phosphatase 1 (PP1). GEF-H1, guanine nucleotide exchange factor. IRF3, interferon regulatory factor. TBK1, TANK-binding kinase 1. IKKε, IκB kinase epsilon.

7. Conclusions and Prospects

In this review, we provide a comprehensive overview of the structure and biological functions of the cytoskeleton, focusing on changes in the cytoskeleton during viral infection and whether these changes affect the virus life cycle. Organisms are complex and the interactions between the cytoskeleton and viruses are ingenious. The cytoskeleton also resists the processes of invasion and replication through the cytoskeleton-mediated innate immune response upon viral infection.

For example, after infection with certain RNA viruses, actin is rearranged and R12C is released to regulate RLRs. This reveals that the cytoskeleton is like a warehouse containing a large number of signal regulatory substances. Once stimulated, a large number of enzymes and proteins are released during the process of depolymerization. Some of these substances can be used by cells to generate innate immunity to initiate signaling processes. Currently, there are no data to confirm whether other PRRs, such as cGAS-STING, have similar upstream signaling regulators. Obtaining these data is recommended for the study of broad-spectrum antiviral drugs.

In addition, we showed that the virus uses motor proteins to transport virions to the correct replication site during the invasion process, and the example of GEF-H1 demonstrates that the binding of motor proteins to the virus simultaneously engages GEF-H1 in the antiviral response. It is unclear whether the cytoskeleton regulates both the viral life cycle and the innate immune processes. Currently, no exact data are available, and this is worthy of further investigation.

Viral infection causes cytoskeleton alterations that can be either particular or general. During viral invasion, the membrane deforms and actin rearranges. R12C, which specifically binds to RLRs, can be released by actin disturbance via a non-specific pathway [13]. Vimentin, for instance, can participate in the invasion of viruses as a co-receptor (specific) [128] and impede the entry of HPV16-PsV, demonstrating that it can prevent virus–receptor contact through steric hindrance (non-specific) [129]. Additionally, upon viral infection, vimentin expression is increased, which may be influenced by IFN receptor 1 (IFNAR1), a nonspecific factor triggered by almost all viral infections, can inhibit TBK1 and IKKε during the process of enhanced vimentin expression [15]. Hence, the differences between non-specific and specific cytoskeleton-mediated activation need to be clarified.

In conclusion, the cytoskeleton is involved in almost all physiological processes in the cell, and therefore its roles in the process of virus invasion are inevitable. In the future, we will continue to study the cytoskeleton to provide insights into the design and development of antivirals.

Author Contributions: Conceptualization, H.-J.Q. and L.-F.L.; writing—original draft preparation, M.L.; writing—review and revision, M.L., D.P., H.C., X.Y., and S.L.; figure preparation, M.L.; manuscript revision and supervision, H.-J.Q. and L.-F.L.; and funding acquisition, L.-F.L. All authors have read and agreed to the published version of the manuscript.

Funding: The National Natural Science Foundation of China (no. 32072855) and the Natural Science Foundation of Heilongjiang Province of China (no. YQ2022C043).

Institutional Review Board Statement: Not applicable.

Informed Consent Statement: Not applicable.

Data Availability Statement: All data generated or analyzed during this study are included in this published article.

Conflicts of Interest: The authors declare no conflict of interest.

References

1. Hohmann, T.; Dehghani, F. The cytoskeleton—A complex interacting meshwork. *Cells* **2019**, *8*, 362. [CrossRef]
2. Fletcher, D.A.; Mullins, R.D. Cell mechanics and the cytoskeleton. *Nature* **2010**, *463*, 485–492. [CrossRef] [PubMed]
3. Abouelezz, A.; Almeida-Souza, L. The mammalian endocytic cytoskeleton. *Eur. J. Cell Biol.* **2022**, *101*, 151222. [CrossRef]
4. Seetharaman, S.; Etienne-Manneville, S. Cytoskeletal crosstalk in cell migration. *Trends Cell Biol.* **2020**, *30*, 720–735. [CrossRef] [PubMed]
5. Seo, D.; Gammon, D.B. Manipulation of the host cytoskeleton by viruses: Insights and mechanisms. *Viruses* **2022**, *14*, 1586. [CrossRef] [PubMed]
6. Li, X.; Wang, J. Mechanical tumor microenvironment and transduction: Cytoskeleton mediates cancer cell invasion and metastasis. *Int. J. Biol. Sci.* **2020**, *16*, 2014–2028. [CrossRef]
7. Molinie, N.; Rubtsova, S.N.; Fokin, A.; Visweshwaran, S.P.; Rocques, N.; Polesskaya, A.; Schnitzler, A.; Vacher, S.; Denisov, E.V.; Tashireva, L.A.; et al. Cortical branched actin determines cell cycle progression. *Cell Res.* **2019**, *29*, 432–445. [CrossRef]

8. Wang, I.H.; Burckhardt, C.J.; Yakimovich, A.; Greber, U.F. Imaging, tracking and computational analyses of virus entry and egress with the cytoskeleton. *Viruses* **2018**, *10*, 166. [CrossRef]

9. Buchwalter, R.A.; Ogden, S.C.; York, S.B.; Sun, L.; Zheng, C.; Hammack, C.; Cheng, Y.; Chen, J.V.; Cone, A.S.; Meckes, D.G., Jr.; et al. Coordination of Zika virus infection and viroplasm organization by microtubules and microtubule-organizing centers. *Cells* **2021**, *10*, 3335. [CrossRef]

10. Dutour-Provenzano, G.; Etienne-Manneville, S. Intermediate filaments. *Curr. Biol.* **2021**, *31*, R522–R529. [CrossRef] [PubMed]

11. Zhang, Y.; Zhao, S.; Li, Y.; Feng, F.; Li, M.; Xue, Y.; Cui, J.; Xu, T.; Jin, X.; Jiu, Y. Host cytoskeletal vimentin serves as a structural organizer and an RNA-binding protein regulator to facilitate Zika viral replication. *Proc. Natl. Acad. Sci. USA* **2022**, *119*, e2113909119. [CrossRef] [PubMed]

12. Ding, G.; Shao, Q.; Yu, H.; Liu, J.; Li, Y.; Wang, B.; Sang, H.; Li, D.; Bing, A.; Hou, Y.; et al. Tight junctions, the key factor in virus-related disease. *Pathogens* **2022**, *11*, 1200. [CrossRef] [PubMed]

13. Acharya, D.; Reis, R.; Volcic, M.; Liu, G.; Wang, M.K.; Chia, B.S.; Nchioua, R.; Gross, R.; Munch, J.; Kirchhoff, F.; et al. Actin cytoskeleton remodeling primes RIG-I-like receptor activation. *Cell* **2022**, *185*, 3588–3602. [CrossRef] [PubMed]

14. de Rivero Vaccari, J.P.; Minkiewicz, J.; Wang, X.; de Rivero Vaccari, J.C.; German, R.; Marcillo, A.E.; Dietrich, W.D.; Keane, R.W. Astrogliosis involves activation of retinoic acid-inducible gene-like signaling in the innate immune response after spinal cord injury. *Glia* **2012**, *60*, 14–21. [CrossRef]

15. Liu, H.; Ye, G.; Liu, X.; Xue, M.; Zhou, Q.; Zhang, L.; Zhang, K.; Huang, L.; Weng, C. Vimentin inhibits type I interferon production by disrupting the TBK1-IKKepsilon-IRF3 axis. *Cell Rep.* **2022**, *41*, 111469. [CrossRef] [PubMed]

16. Bovellan, M.; Romeo, Y.; Biro, M.; Boden, A.; Chugh, P.; Yonis, A.; Vaghela, M.; Fritzsche, M.; Moulding, D.; Thorogate, R.; et al. Cellular control of cortical actin nucleation. *Curr. Biol.* **2014**, *24*, 1628–1635. [CrossRef]

17. Kloc, M.; Chanana, P.; Vaughn, N.; Uosef, A.; Kubiak, J.Z.; Ghobrial, R.M. New insights into cellular functions of nuclear actin. *Biology* **2021**, *10*, 304. [CrossRef]

18. Pollard, T.D. Actin and actin-binding proteins. *Cold Spring Harb. Perspect. Biol.* **2016**, *8*, a018226. [CrossRef]

19. Nogales, E.; Wang, H.W. Structural intermediates in microtubule assembly and disassembly: How and why? *Curr. Opin. Cell Biol.* **2006**, *18*, 179–184. [CrossRef]

20. Gudimchuk, N.B.; McIntosh, J.R. Regulation of microtubule dynamics, mechanics and function through the growing tip. *Nat. Rev. Mol. Cell Biol.* **2021**, *22*, 777–795. [CrossRef]

21. Akıl, C.; Ali, S.; Tran, L.T.; Gaillard, J.; Li, W.; Hayashida, K.; Hirose, M.; Kato, T.; Oshima, A.; Fujishima, K.; et al. Structure and dynamics of *Odinarchaeota* tubulin and the implications for eukaryotic microtubule evolution. *Sci. Adv.* **2022**, *8*, eabm2225. [CrossRef] [PubMed]

22. Guru, A.; Saravanan, S.; Sharma, D.; Narasimha, M. The microtubule end-binding proteins EB1 and patronin modulate the spatiotemporal dynamics of myosin and pattern pulsed apical constriction. *Development* **2022**, *149*, dev199759. [CrossRef]

23. Martin, M.; Akhmanova, A. Coming into focus: Mechanisms of microtubule minus-end organization. *Trends Cell Biol.* **2018**, *28*, 574–588. [CrossRef] [PubMed]

24. Ferro, L.S.; Can, S.; Turner, M.A.; ElShenawy, M.M.; Yildiz, A. Kinesin and dynein use distinct mechanisms to bypass obstacles. *eLife* **2019**, *8*, e48629. [CrossRef] [PubMed]

25. Ivaska, J.; Pallari, H.M.; Nevo, J.; Eriksson, J.E. Novel functions of vimentin in cell adhesion, migration, and signaling. *Exp. Cell Res.* **2007**, *313*, 2050–2062. [CrossRef] [PubMed]

26. Bott, C.J.; Winckler, B. Intermediate filaments in developing neurons: Beyond structure. *Cytoskeleton* **2020**, *77*, 110–128. [CrossRef] [PubMed]

27. Moll, R.; Divo, M.; Langbein, L. The human keratins: Biology and pathology. *Histochem. Cell Biol.* **2008**, *129*, 705–733. [CrossRef]

28. Battaglia, R.A.; Delic, S.; Herrmann, H.; Snider, N.T. Vimentin on the move: New developments in cell migration. *F1000Research* **2018**, *7*, F1000 Faculty Rev-1796. [CrossRef]

29. Redmond, C.J.; Coulombe, P.A. Intermediate filaments as effectors of differentiation. *Curr. Opin. Cell Biol.* **2021**, *68*, 155–162. [CrossRef]

30. Pollard, T.D.; Borisy, G.G. Cellular motility driven by assembly and disassembly of actin filaments. *Cell* **2003**, *112*, 453–465. [CrossRef]

31. Dugina, V.B.; Shagieva, G.S.; Kopnin, P.B. Biological role of actin isoforms in mammalian cells. *Biochemistry* **2019**, *84*, 583–592. [CrossRef]

32. Rubenstein, P.A. The functional importance of multiple actin isoforms. *Bioessays* **1990**, *12*, 309–315. [CrossRef]

33. Carlier, M.F.; Shekhar, S. Global treadmilling coordinates actin turnover and controls the size of actin networks. *Nat. Rev. Mol. Cell Biol.* **2017**, *18*, 389–401. [CrossRef]

34. Izdebska, M.; Zielinska, W.; Halas-Wisniewska, M.; Grzanka, A. Involvement of actin and actin-binding proteins in carcinogenesis. *Cells* **2020**, *9*, 2245. [CrossRef]

35. Rotty, J.D.; Wu, C.; Bear, J.E. New insights into the regulation and cellular functions of the ARP2/3 complex. *Nat. Rev. Mol. Cell Biol.* **2013**, *14*, 7–12. [CrossRef] [PubMed]

36. Kage, F.; Winterhoff, M.; Dimchev, V.; Mueller, J.; Thalheim, T.; Freise, A.; Brühmann, S.; Kollasser, J.; Block, J.; Dimchev, G.; et al. FMNL formins boost lamellipodial force generation. *Nat. Commun.* **2017**, *8*, 14832. [CrossRef] [PubMed]

37. Davidson, P.M.; Cadot, B. Actin on and around the nucleus. *Trends Cell Biol.* **2021**, *31*, 211–223. [CrossRef]

38. Tann, J.Y.; Moore, A.W. MTOC organization and competition during neuron differentiation. *Results Probl. Cell Differ.* **2019**, *67*, 337–357. [PubMed]
39. Zehner, Z.E.; Paterson, B.M. Characterization of the chicken vimentin gene: Single copy gene producing multiple mRNAs. *Proc. Natl. Acad. Sci. USA* **1983**, *80*, 911–915. [CrossRef]
40. Lundin, V.F.; Leroux, M.R.; Stirling, P.C. Quality control of cytoskeletal proteins and human disease. *Trends Biochem. Sci.* **2010**, *35*, 288–297. [CrossRef]
41. Jaffe, A.B.; Hall, A. Rho GTPases: Biochemistry and biology. *Annu. Rev. Cell Dev. Biol.* **2005**, *21*, 247–269. [CrossRef] [PubMed]
42. Sakata-Yanagimoto, M.; Enami, T.; Yoshida, K.; Shiraishi, Y.; Ishii, R.; Miyake, Y.; Muto, H.; Tsuyama, N.; Sato-Otsubo, A.; Okuno, Y.; et al. Recurrent gain-of-function mutations of RHOA in diffuse-type gastric carcinoma. *Nat. Genet.* **2014**, *46*, 171–175. [CrossRef]
43. Lamouille, S.; Xu, J.; Derynck, R. Molecular mechanisms of epithelial-mesenchymal transition. *Nat. Rev. Mol. Cell Biol.* **2014**, *15*, 178–196. [CrossRef] [PubMed]
44. Ullmann, U.; In't Veld, P.; Gilles, C.; Sermon, K.; De, R.M.; Van, V.H.; Van, S.A.; Liebaers, I. Epithelial-mesenchymal transition process in human embryonic stem cells cultured in feeder-free conditions. *Mol. Hum. Reprod.* **2007**, *13*, 21–32. [CrossRef] [PubMed]
45. Mendez, M.G.; Kojima, S.; Goldman, R.D. Vimentin induces changes in cell shape, motility, and adhesion during the epithelial to mesenchymal transition. *FASEB J.* **2010**, *24*, 1838–1851. [CrossRef] [PubMed]
46. Hartland, E.L.; Ghosal, D.; Giogha, C. Manipulation of epithelial cell architecture by the bacterial pathogens *Listeria* and *Shigella*. *Curr. Opin. Cell Biol.* **2022**, *79*, 102131. [CrossRef] [PubMed]
47. Aktories, K.; Schwan, C.; Jank, T. *Clostridium difficile* toxin biology. *Annu. Rev. Microbiol.* **2017**, *71*, 281–307. [CrossRef]
48. Garzon, M.; Sosik, P.; Drastík, J.; Skalli, O. A self-controlled and self-healing model of bacterial cells. *Membranes* **2022**, *12*, 678. [CrossRef]
49. Pizarro-Cerdá, J.; Cossart, P. *Listeria monocytogenes*: Cell biology of invasion and intracellular growth. *Microbiol. Spectr.* **2018**, *6*, 6. [CrossRef]
50. Agaisse, H. Molecular and cellular mechanisms of *Shigella* flexneri dissemination. *Front. Cell. Infect. Microbiol.* **2016**, *11*, 29. [CrossRef]
51. Miao, C.; Zhao, S.; Etienne-Manneville, S.; Jiu, Y. The diverse actions of cytoskeletal vimentin in bacterial infection and host defense. *J. Cell Sci.* **2023**, *136*, jcs260509. [CrossRef]
52. Guo, Y.; Duan, M.; Wang, X.; Gao, J.; Guan, Z.; Zhang, M. Early events in rabies virus infection-Attachment, entry, and intracellular trafficking. *Virus Res.* **2019**, *263*, 217–225. [CrossRef] [PubMed]
53. Farhadi, L.; Ricketts, S.N.; Rust, M.J.; Das, M.; Robertson-Anderson, R.M.; Ross, J.L. Actin and microtubule crosslinkers tune mobility and control co-localization in a composite cytoskeletal network. *Soft Matter.* **2020**, *16*, 7191–7201. [CrossRef]
54. Cyrklaff, M.; Sanchez, C.P.; Kilian, N.; Bisseye, C.; Simpore, J.; Frischknecht, F.; Lanzer, M. Hemoglobins S and C interfere with actin remodeling in *Plasmodium falciparum*-infected erythrocytes. *Science* **2011**, *334*, 1283–1286. [CrossRef] [PubMed]
55. Nam, H.J.; Kang, J.K.; Kim, S.K.; Ahn, K.J.; Seok, H.; Park, S.J.; Chang, J.S.; Pothoulakis, C.; Lamont, J.T.; Kim, H. *Clostridium difficile* toxin A decreases acetylation of tubulin, leading to microtubule depolymerization through activation of histone deacetylase 6, and this mediates acute inflammation. *J. Biol. Chem.* **2010**, *285*, 32888–32896. [CrossRef] [PubMed]
56. Leroy, H.; Han, M.; Woottum, M.; Bracq, L.; Bouchet, J.; Xie, M.; Benichou, S. Virus-mediated cell-cell fusion. *Int. J. Mol. Sci.* **2020**, *21*, 9644. [CrossRef]
57. Zhang, Z.; Zheng, Y.; Niu, Z.; Zhang, B.; Wang, C.; Yao, X.; Peng, H.; Franca, D.N.; Wang, Y.; Zhu, Y.; et al. SARS-CoV-2 spike protein dictates syncytium-mediated lymphocyte elimination. *Cell Death Differ.* **2021**, *28*, 2765–2777. [CrossRef]
58. Kloc, M.; Uosef, A.; Wosik, J.; Kubiak, J.Z.; Ghobrial, R.M. Virus interactions with the actin cytoskeleton--what we know and do not know about SARS-CoV-2. *Arch. Virol.* **2022**, *167*, 737–749. [CrossRef] [PubMed]
59. Barbier, P.; Zejneli, O.; Martinho, M.; Lasorsa, A.; Belle, V.; Smet-Nocca, C.; Tsvetkov, P.O.; Devred, F.; Landrieu, I. Role of tau as a microtubule-associated protein: Structural and functional aspects. *Front. Aging Neurosci.* **2019**, *11*, 204. [CrossRef]
60. Alhogbani, T. Acute myocarditis associated with novel Middle East respiratory syndrome coronavirus. *Ann. Saudi Med.* **2016**, *36*, 78–80. [CrossRef]
61. Jackson, C.B.; Farzan, M.; Chen, B.; Choe, H. Mechanisms of SARS-CoV-2 entry into cells. *Nat. Rev. Mol. Cell Biol.* **2022**, *23*, 3–20. [CrossRef] [PubMed]
62. Cortese, M.; Lee, J.Y.; Cerikan, B.; Neufeldt, C.J.; Oorschot, V.M.J.; Köhrer, S.; Hennies, J.; Schieber, N.L.; Ronchi, P.; Mizzon, G.; et al. Integrative imaging reveals SARS-CoV-2-induced reshaping of subcellular morphologies. *Cell Host Microbe* **2020**, *28*, 853–866. [CrossRef] [PubMed]
63. Miranda-Saksena, M.; Denes, C.E.; Diefenbach, R.J.; Cunningham, A.L. Infection and transport of herpes simplex virus type 1 in neurons: Role of the cytoskeleton. *Viruses* **2018**, *10*, 92. [CrossRef] [PubMed]
64. Paluck, A.; Osan, J.; Hollingsworth, L.; Talukdar, S.N.; Saegh, A.A.; Mehedi, M. Role of ARP2/3 complex-driven actin polymerization in RSV infection. *Pathogens* **2021**, *11*, 26. [CrossRef]
65. Liu, X.; Nawaz, Z.; Guo, C.; Ali, S.; Naeem, M.A.; Jamil, T.; Ahmad, W.; Siddiq, M.U.; Ahmed, S.; Asif Idrees, M.; et al. Rabies virus exploits cytoskeleton network to cause early disease progression and cellular dysfunction. *Front. Vet. Sci.* **2022**, *9*, 889873. [CrossRef]

66. Piccinotti, S.; Kirchhausen, T.; Whelan, S.P. Uptake of rabies virus into epithelial cells by clathrin-mediated endocytosis depends upon actin. *J. Virol.* **2013**, *87*, 11637–11647. [CrossRef]
67. Lehmann, M.J.; Sherer, N.M.; Marks, C.B.; Pypaert, M.; Mothes, W. Actin- and myosin-driven movement of viruses along filopodia precedes their entry into cells. *J. Cell Biol.* **2005**, *170*, 317–325. [CrossRef]
68. Sheetz, M.P.; Turney, S.; Qian, H.; Elson, E.L. Nanometre-level analysis demonstrates that lipid flow does not drive membrane glycoprotein movements. *Nature* **1989**, *340*, 284–288. [CrossRef]
69. Svitkina, T.M.; Verkhovsky, A.B.; McQuade, K.M.; Borisy, G.G. Analysis of the actin-myosin II system in fish epidermal keratocytes: Mechanism of cell body translocation. *J. Cell Biol.* **1997**, *139*, 397–415. [CrossRef]
70. Cheng, Y.; Lou, J.X.; Liu, C.C.; Liu, Y.Y.; Chen, X.N.; Liang, X.D.; Zhang, J.; Yang, Q.; Go, Y.Y.; Zhou, B. Microfilaments and microtubules alternately coordinate the multi-step endosomal trafficking of classical swine fever virus. *J. Virol.* **2021**, *95*, e02436-20. [CrossRef]
71. Mylvaganam, S.; Freeman, S.A.; Grinstein, S. The cytoskeleton in phagocytosis and macropinocytosis. *Curr. Biol.* **2021**, *31*, R619–R632. [CrossRef] [PubMed]
72. Pastey, M.K.; Gower, T.L.; Spearman, P.W.; Crowe, J.E., Jr.; Graham, B.S. A RhoA-derived peptide inhibits syncytium formation induced by respiratory syncytial virus and parainfluenza virus type 3. *Nat. Med.* **2000**, *6*, 35–40. [CrossRef]
73. Hoppe, S.; Schelhaas, M.; Jaeger, V.; Liebig, T.; Petermann, P.; Knebel-Mörsdorf, D. Early herpes simplex virus type 1 infection is dependent on regulated Rac1/Cdc42 signalling in epithelial MDCKII cells. *J. Gen. Virol.* **2006**, *87*, 3483–3494. [CrossRef]
74. Cuartas-López, A.M.; Hernández-Cuellar, C.E.; Gallego-Gómez, J.C. Disentangling the role of PI3K/Akt, Rho GTPase and the actin cytoskeleton on dengue virus infection. *Virus Res.* **2018**, *256*, 153–165. [CrossRef]
75. Miller, M.S.; Hertel, L. Onset of human cytomegalovirus replication in fibroblasts requires the presence of an intact vimentin cytoskeleton. *J. Virol.* **2009**, *83*, 7015–7028. [CrossRef] [PubMed]
76. Sodeik, B. Mechanisms of viral transport in the cytoplasm. *Trends Microbiol.* **2000**, *8*, 465–472. [CrossRef] [PubMed]
77. Naghavi, M.H. HIV-1 capsid exploitation of the host microtubule cytoskeleton during early infection. *Retrovirology* **2021**, *18*, 19. [CrossRef] [PubMed]
78. Liu, M.; Schmidt, E.E.; Halford, W.P. ICP0 dismantles microtubule networks in herpes simplex virus-infected cells. *PLoS ONE* **2010**, *5*, e10975. [CrossRef]
79. Wang, J.; Fang, S.; Xiao, H.; Chen, B.; Tam, J.P.; Liu, D.X. Interaction of the coronavirus infectious bronchitis virus membrane protein with beta-actin and its implication in virion assembly and budding. *PLoS ONE* **2009**, *4*, e4908.
80. Richards, A.; Berth, S.H.; Brady, S.; Morfini, G. Engagement of neurotropic viruses in fast axonal transport: Mechanisms, potential role of host kinases and implications for neuronal dysfunction. *Front. Cell Neurosci.* **2021**, *15*, 684762. [CrossRef]
81. Shanda, S.K.; Wilson, D.W. UL36p is required for efficient transport of membrane-associated Herpes simplex virus type 1 along microtubules. *J. Virol.* **2008**, *82*, 7388–7394. [CrossRef]
82. Zaichick, S.V.; Bohannon, K.P.; Hughes, A.; Sollars, P.J.; Pickard, G.E.; Smith, G.A. The herpesvirus VP1/2 protein is an effector of dynein-mediated capsid transport and neuroinvasion. *Cell Host Microbe* **2013**, *13*, 193–203. [CrossRef]
83. Dodding, M.P.; Way, M. Coupling viruses to dynein and kinesin-1. *EMBO J.* **2011**, *30*, 3527–3539. [CrossRef]
84. Naghavi, M.H.; Walsh, D. Microtubule regulation and function during virus infection. *J. Virol.* **2017**, *91*, e00538-17. [CrossRef]
85. Pawlica, P.; Berthoux, L. Cytoplasmic dynein promotes HIV-1 uncoating. *Viruses* **2014**, *6*, 4195–4211. [CrossRef]
86. Yoder, A.; Guo, J.; Yu, D.; Cui, Z.; Zhang, X.E.; Wu, Y. Effects of microtubule modulators on HIV-1 infection of transformed and resting CD4 T cells. *J. Virol.* **2011**, *85*, 3020–3024. [CrossRef]
87. Neufeldt, C.J.; Cortese, M. Membrane architects: How positive-strand RNA viruses restructure the cell. *J. Gen. Virol.* **2022**, *103*, 10. [CrossRef]
88. Wen, Z.; Zhang, Y.; Lin, Z.; Shi, K.; Jiu, Y. Cytoskeleton--a crucial key in host cell for coronavirus infection. *J. Mol. Cell Biol.* **2020**, *12*, 968–979. [CrossRef] [PubMed]
89. Mallardo, M.; Schleich, S.; Krijnse Locker, J. Microtubule-dependent organization of vaccinia virus core-derived early mRNAs into distinct cytoplasmic structures. *Mol. Biol. Cell* **2001**, *12*, 3875–3891. [CrossRef] [PubMed]
90. Schepis, A.; Schramm, B.; de Haan, C.A.; Locker, J.K. Vaccinia virus-induced microtubule-dependent cellular rearrangements. *Traffic* **2006**, *7*, 308–323. [CrossRef] [PubMed]
91. Cibulka, J.; Fraiberk, M.; Forstova, J. Nuclear actin and lamins in viral infections. *Viruses* **2012**, *4*, 325–347. [CrossRef] [PubMed]
92. Forest, T.; Barnard, S.; Baines, J.D. Active intranuclear movement of herpesvirus capsids. *Nat. Cell Biol.* **2005**, *7*, 429–431. [CrossRef] [PubMed]
93. Jouvenet, N.; Monaghan, P.; Way, M.; Wileman, T. Transport of African swine fever virus from assembly sites to the plasma membrane is dependent on microtubules and conventional kinesin. *J. Virol.* **2004**, *78*, 7990–8001. [CrossRef] [PubMed]
94. Hyde, J.L.; Gillespie, L.K.; Mackenzie, J.M. Mouse norovirus 1 utilizes the cytoskeleton network to establish localization of the replication complex proximal to the microtubule organizing center. *J. Virol.* **2012**, *86*, 4110–4122. [CrossRef]
95. Appenzeller-Herzog, C.; Hauri, H.P. The ER-Golgi intermediate compartment (ERGIC): In search of its identity and function. *J. Cell Sci.* **2006**, *119*, 2173–2183. [CrossRef]
96. Yu, S.F.; Eastman, S.W.; Linial, M.L. Foamy virus capsid assembly occurs at a pericentriolar region through a cytoplasmic targeting/retention signal in Gag. *Traffic* **2006**, *7*, 966–977. [CrossRef]

97. Ng, M.L.; Lee, J.W.; Leong, M.L.; Ling, A.E.; Tan, H.C.; Ooi, E.E. Topographic changes in SARS coronavirus-infected cells at late stages of infection. *Emerg. Infect. Dis.* **2004**, *10*, 1907–1914. [CrossRef]
98. Bohn, W.; Rutter, G.; Hohenberg, H.; Mannweiler, K.; Nobis, P. Involvement of actin filaments in budding of measles virus: Studies on cytoskeletons of infected cells. *Virology* **1986**, *149*, 91–106. [CrossRef]
99. Milbradt, J.; Sonntag, E.; Wagner, S.; Strojan, H.; Wangen, C.; Lenac Rovis, T.; Lisnic, B.; Jonjic, S.; Sticht, H.; Britt, W.J.; et al. Human cytomegalovirus nuclear capsids associate with the core nuclear egress complex and the viral protein kinase pUL97. *Viruses* **2018**, *10*, 35. [CrossRef]
100. Iwami, S.; Takeuchi, J.S.; Nakaoka, S.; Mammano, F.; Clavel, F.; Inaba, H.; Kobayashi, T.; Misawa, N.; Aihara, K.; Koyanagi, Y.; et al. Cell-to-cell infection by HIV contributes over half of virus infection. *eLife* **2015**, *4*, e08150. [CrossRef]
101. Reh, L.; Magnus, C.; Schanz, M.; Weber, J.; Uhr, T.; Rusert, P.; Trkola, A. Capacity of broadly neutralizing antibodies to inhibit HIV-1 cell-cell transmission is strain- and epitope-dependent. *PLoS Pathog.* **2015**, *11*, e1004966. [CrossRef]
102. Zhong, P.; Agosto, L.M.; Ilinskaya, A.; Dorjbal, B.; Truong, R.; Derse, D.; Uchil, P.D.; Heidecker, G.; Mothes, W. Cell-to-cell transmission can overcome multiple donor and target cell barriers imposed on cell-free HIV. *PLoS ONE* **2013**, *8*, e53138. [CrossRef]
103. Kim, J.T.; Chang, E.; Sigal, A.; Baltimore, D. Dendritic cells efficiently transmit HIV to T cells in a tenofovir and raltegravir insensitive manner. *PLoS ONE* **2018**, *13*, e0189945. [CrossRef] [PubMed]
104. Rustom, A.; Saffrich, R.; Markovic, I.; Walther, P.; Gerdes, H.H. Nanotubular highways for intercellular organelle transport. *Science* **2004**, *303*, 1007–1010. [CrossRef]
105. Abounit, S.; Zurzolo, C. Wiring through tunneling nanotubes—From electrical signals to organelle transfer. *J. Cell Sci.* **2012**, *125*, 1089–1098. [CrossRef]
106. Zurzolo, C. Tunneling nanotubes: Reshaping connectivity. *Curr. Opin. Cell Biol.* **2021**, *71*, 139–147. [CrossRef] [PubMed]
107. Zhang, S.; Kazanietz, M.G.; Cooke, M. Rho GTPases and the emerging role of tunneling nanotubes in physiology and disease. *Am. J. Physiol. Cell Physiol.* **2020**, *319*, C877–C884. [CrossRef] [PubMed]
108. Tiwari, V.; Koganti, R.; Russell, G.; Sharma, A.; Shukla, D. Role of tunneling nanotubes in viral infection, neurodegenerative disease, and cancer. *Front. Immunol.* **2021**, *12*, 680891. [CrossRef]
109. Sewald, X.; Gonzalez, D.G.; Haberman, A.M.; Mothes, W. *In vivo* imaging of virological synapses. *Nat. Commun.* **2012**, *3*, 1320. [CrossRef]
110. Piguet, V.; Sattentau, Q. Dangerous liaisons at the virological synapse. *J. Clin. Investig.* **2004**, *114*, 605–610. [CrossRef]
111. Igakura, T.; Stinchcombe, J.C.; Goon, P.K.; Taylor, G.P.; Weber, J.N.; Griffiths, G.M.; Tanaka, Y.; Osame, M.; Bangham, C.R. Spread of HTLV-I between lymphocytes by virus-induced polarization of the cytoskeleton. *Science* **2003**, *299*, 1713–1716. [CrossRef]
112. Yashavantha Rao, H.C.; Jayabaskaran, C. The emergence of a novel coronavirus (SARS-CoV-2) disease and their neuroinvasive propensity may affect in COVID-19 patients. *J. Med. Virol.* **2020**, *92*, 786–790. [CrossRef]
113. Martin, N.; Welsch, S.; Jolly, C.; Briggs, J.A.; Vaux, D.; Sattentau, Q.J. Virological synapse-mediated spread of human immunodeficiency virus type 1 between T cells is sensitive to entry inhibition. *J. Virol.* **2010**, *84*, 3516–3527. [CrossRef]
114. Vasiliver-Shamis, G.; Cho, M.W.; Hioe, C.E.; Dustin, M.L. Human immunodeficiency virus type 1 envelope gp120-induced partial T-cell receptor signaling creates an F-actin-depleted zone in the virological synapse. *J. Virol.* **2009**, *83*, 11341–11355. [CrossRef]
115. Garcia, M.A.; Nelson, W.J.; Chavez, N. Cell-cell junctions organize structural and signaling networks. *Cold Spring Harb. Perspect. Biol.* **2018**, *10*, a029181. [CrossRef]
116. Heuser, S.; Hufbauer, M.; Marx, B.; Tok, A.; Majewski, S.; Pfister, H.; Akgül, B. The levels of epithelial anchor proteins β-catenin and zona occludens-1 are altered by E7 of human papillomaviruses 5 and 8. *J. Gen. Virol.* **2016**, *97*, 463–472. [CrossRef]
117. Labudová, M. Cell-to-cell transport in viral families: Faster than usual. *Acta Virol.* **2020**, *64*, 154–166. [CrossRef]
118. Sun, L.; Wu, J.; Du, F.; Chen, X.; Chen, Z.J. Cyclic GMP-AMP synthase is a cytosolic DNA sensor that activates the type I interferon pathway. *Science* **2013**, *339*, 786–791. [CrossRef] [PubMed]
119. Kawasaki, T.; Kawai, T.; Akira, S. Recognition of nucleic acids by pattern-recognition receptors and its relevance in autoimmunity. *Immunol. Rev.* **2011**, *243*, 61–73. [CrossRef] [PubMed]
120. Gack, M.U.; Nistal-Villán, E.; Inn, K.S.; García-Sastre, A.; Jung, J.U. Phosphorylation-mediated negative regulation of RIG-I antiviral activity. *J. Virol.* **2010**, *84*, 3220–3229. [CrossRef] [PubMed]
121. Wies, E.; Wang, M.K.; Maharaj, N.P.; Chen, K.; Zhou, S.; Finberg, R.W.; Gack, M.U. Dephosphorylation of the RNA sensors RIG-I and MDA5 by the phosphatase PP1 is essential for innate immune signaling. *Immunity* **2013**, *38*, 437–449. [CrossRef] [PubMed]
122. Wellington, A.; Emmons, S.; James, B.; Calley, J.; Grover, M.; Tolias, P.; Manseau, L. Spire contains actin binding domains and is related to ascidian posterior end mark-5. *Development* **1999**, *126*, 5267–5274. [CrossRef] [PubMed]
123. Torres, A.A.; Macilwee, S.L.; Rashid, A.; Cox, S.E.; Albarnaz, J.D.; Bonjardim, C.A.; Smith, G.L. The actin nucleator Spir-1 is a virus restriction factor that promotes innate immune signalling. *PLoS Pathog.* **2022**, *18*, e1010277. [CrossRef] [PubMed]
124. Fine, N.; Dimitriou, I.D.; Rottapel, R. Go with the flow: GEF-H1 mediated shear stress mechanotransduction in neutrophils. *Small GTPases* **2020**, *11*, 23–31. [CrossRef]
125. Chiang, H.S.; Zhao, Y.; Song, J.H.; Liu, S.; Wang, N.; Terhorst, C.; Sharpe, A.H.; Basavappa, M.; Jeffrey, K.L.; Reinecker, H.C. GEF-H1 controls microtubule-dependent sensing of nucleic acids for antiviral host defenses. *Nat. Immunol.* **2014**, *15*, 63–71. [CrossRef]
126. Fitzgerald, K.A.; McWhirter, S.M.; Faia, K.L.; Rowe, D.C.; Latz, E.; Golenbock, D.T.; Coyle, A.J.; Liao, S.M.; Maniatis, T. IKKepsilon and TBK1 are essential components of the IRF3 signaling pathway. *Nat. Immunol.* **2003**, *4*, 491–496. [CrossRef]

127. Stevens, C.; Henderson, P.; Nimmo, E.R.; Soares, D.C.; Dogan, B.; Simpson, K.W.; Barrett, J.C.; Wilson, D.C.; Satsangi, J. The intermediate filament protein, vimentin, is a regulator of NOD2 activity. *Gut* **2013**, *62*, 695–707. [CrossRef]
128. Hu, K.; Onintsoa Diarimalala, R.; Yao, C.; Li, H.; Wei, Y. EV-A71 mechanism of entry: Receptors/co-receptors, related pathways and inhibitors. *Viruses* **2023**, *15*, 785. [CrossRef]
129. Schäfer, G.; Graham, L.M.; Lang, D.M.; Blumenthal, M.J.; Bergant Marušič, M.; Katz, A.A. Vimentin modulates infectious internalization of human papillomavirus 16 pseudovirions. *J. Virol.* **2017**, *91*, e00307-17. [CrossRef]

Article

Enhancement of SARS-CoV-2 N Antigen-Specific T Cell Functionality by Modulating the Autophagy-Mediated Signal Pathway in Mice

Ziyu Wen [1,†], Yue Yuan [1,†], Yangguo Zhao [1], Haohang Wang [1], Zirong Han [1], Minchao Li [1], Jianhui Yuan [2,*] and Caijun Sun [1,3,*]

1. School of Public Health (Shenzhen), Shenzhen Campus of Sun Yat-Sen University, Shenzhen 518107, China; wenzy3@mail2.sysu.edu.cn (Z.W.); yuany263@mail2.sysu.edu.cn (Y.Y.); zhaoyg3@mail2.sysu.edu.cn (Y.Z.); wanghh7@mail2.sysu.edu.cn (H.W.); hanzr@mail2.sysu.edu.cn (Z.H.); liminchao@mail2.sysu.edu.cn (M.L.)
2. Nanshan District Center for Disease Control and Prevention, Shenzhen 518000, China
3. Key Laboratory of Tropical Disease Control (Sun Yat-Sen University), Ministry of Education, Guangzhou 510080, China
* Correspondence: jianhui_yuan@126.com (J.Y.); suncaijun@mail.sysu.edu.cn (C.S.)
† These authors contributed equally to this work.

Abstract: The frequent SARS-CoV-2 variants have caused a continual challenge, weakening the effectiveness of current vaccines, and thus it is of great importance to induce robust and conserved T cellular immunity for developing the next-generation vaccine against SARS-CoV-2 variants. In this study, we proposed a conception of enhancing the SARS-CoV-2 specific T cell functionality by fusing autophagosome-associated LC3b protein to the nucleocapsid (N) (N-LC3b). When compared to N protein alone, the N-LC3b protein was more effectively targeted to the autophagosome/lysosome/MHC II compartment signal pathway and thus elicited stronger CD4$^+$ and CD8$^+$ T cell immune responses in mice. Importantly, the frequency of N-specific polyfunctional CD4$^+$ and CD8$^+$ T cells, which can simultaneously secrete multiple cytokines (IFN-γ^+/IL-2$^+$/TNF-α^+), in the N-LC3b group was significantly higher than that in the N alone group. Moreover, there was a significantly improved T cell proliferation, especially for CD8$^+$ T cells in the N-LC3b group. In addition, the N-LC3b also induced a robust humoral immune response, characterized by the Th1-biased IgG2a subclass antibodies against the SARS-CoV-2 N protein. Overall, these findings demonstrated that our strategy could effectively induce a potential SARS-CoV-2 specific T cellular immunity with enhanced magnitude, polyfunctionality, and proliferation, and thus provided insights to develop a promising strategy for the design of a novel universal vaccine against SARS-CoV-2 variants and other emerging infectious diseases.

Keywords: SARS-CoV-2; N protein; autophagy; T cellular immunity

Citation: Wen, Z.; Yuan, Y.; Zhao, Y.; Wang, H.; Han, Z.; Li, M.; Yuan, J.; Sun, C. Enhancement of SARS-CoV-2 N Antigen-Specific T Cell Functionality by Modulating the Autophagy-Mediated Signal Pathway in Mice. *Viruses* **2023**, *15*, 1316. https://doi.org/10.3390/v15061316

Academic Editor: Elmostafa Bahraoui

Received: 3 May 2023
Revised: 28 May 2023
Accepted: 31 May 2023
Published: 2 June 2023

1. Introduction

The pandemic of coronavirus disease 2019 (COVID-19), caused by severe acute respiratory syndrome coronavirus 2 (SARS-CoV-2), has continued to threaten global public health [1]. The COVID-19 vaccine, as the most powerful weapon to control this pandemic, has been extensively developed, and at least 15 kinds of COVID-19 vaccines have been approved for clinical use by the World Health Organization (WHO), including inactivated vaccines, protein subunit vaccines, mRNA vaccines, and viral vector vaccines [2,3]. So far, these vaccines are mainly targeted to the S protein, which contains two subunits S1 and S2 that contribute to viral attachment, fusion, and entry, to induce neutralizing antibodies [4]. However, the frequent emergence of SARS-CoV-2 variants, such as Delta and Omicron, has greatly weakened vaccine effectiveness and caused breakthrough infections frequently due to the waned neutralizing antibody titers and the low frequency of virus-specific memory B cells [5–7]. Alternatively, it is of great importance to induce robust and conserved T cell-mediated immunity for developing the next-generation vaccine against SARS-CoV-2 variants.

Recent studies have revealed that the T cell immune responses play a critical role in controlling viral replication [6,8]. For example, a high frequency of SARS-CoV-2 specific T cells was identified in COVID-19 convalescent individuals [9,10]. Importantly, when compared to antibody responses, T cell immune responses are usually more conserved against viral variants [11–13]. Furthermore, the memory T cells usually have a long-term survival time [8]. For example, one study showed that the memory T cells can persist for more than 17 years in some convalescent individuals from SARS-CoV infection [14], and at least 20 months in some individuals who have recovered from SARS-CoV-2 infection [8]. Besides seropositive patients, the individuals with asymptomatic or mild disease courses of COVID-19 also had abundant memory T cell responses [15]. Thus, it is worth studying the possibility to develop a long-lasting universal COVID-19 vaccine by targeting broadly cross-reactive T cell epitopes.

The nucleocapsid (N) protein is conserved with approximately 90% amino acid homology between various SARS-CoV-2 variants, and it contains some cross-reactive T cell epitopes [16–18]. Therefore, it is expected to be a promising target for the universal COVID-19 vaccine. Previous studies demonstrated that the N-based vaccine generated partial protection against the SARS-CoV-2 challenge, and also enhanced the protection efficacy when combined with the S-antigen-based COVID-19 vaccine [19,20]. However, the immunogenicity of the natural N protein is relatively weak, and thus it can only elicit insufficient T cell immunity. Consequently, it is necessary to explore a novel strategy to improve the immunogenicity of the N antigen-based COVID-19 vaccine.

Autophagy, particularly macroautophagy, is a powerful tool that the host's cells use to defend against viral infections [21]. Autophagy contributes to the delivery and processing of endogenous antigens to MHC class II molecules by the cross-presentation mechanism [22,23]. The microtubule-associated protein 1 light chain 3 beta (LC3b) is one of the key components involved in macroautophagy and is usually dispersed throughout the cytoplasm in diffuse form (LC3-I). Upon the formation of the autophagosome, LC3-I is converted to the phosphatidylethanolamine-coupled LC3-II form, which can then become the punctate form at the autophagosome [24]. The recruitment of LC3-II is critical for the subsequent regulation of adaptive immune responses by the autophagosome/lysosome/MHC II compartment signal pathway. Actually, our previous work showed that when the Gag antigen of the simian immunodeficiency virus (SIV) was fused to the LC3b protein, it effectively enhanced the SIV antigen-specific T cell immunity with magnitude and polyfunctionality [25]. Based on this finding, we further verified this strategy to enhance the immunogenicity of the SARS-CoV-2 N antigen in the present study, and thus provided insights for the development of a T cell-based universal vaccine against SARS-CoV-2 variants.

2. Materials and Methods

2.1. Construction of DNA Vaccine

The sequences of the SARS-CoV-2 N gene and mouse LC3b gene were obtained from the National Center for Biotechnology Information (NCBI) and we conducted codon optimization for efficient expression in mammalian cells. The fusion gene, N-LC3b, was obtained by the overlap PCR method (Table S1). Specifically, the C terminal of the N gene was fused with the N terminal of the LC3b gene bridged with the GGGSGGG linker, and the Flag Tag was added to the C terminal of the fusion gene. Subsequently, the LC3b, the N, and the fused gene N-LC3b were cloned into the pVAX1 expression vector (Invitrogen, Carlsbad, CA, USA).

2.2. Western Blotting Analysis

293T cells (from human embryonic kidney cells) and Hela cells (from human cervical cancer cells) were cultured in complete Dulbecco's modified Eagle's medium (DMEM, Gibco, New York, NY, USA) containing 10% fetal bovine serum (FBS, Gibco, New York,

NY, USA) and 1% penicillin/streptomycin (Gibco, New York, NY, USA), at 37 °C in an atmosphere of 5% of CO_2. To verify the protein expression, 293T cells were seeded at a density of 5×10^5 cells in 6-well plates and transfected with the corresponding plasmids using Lipofectamine 2000 reagent (Invitrogen, Carlsbad, CA, USA) in OptiMEM serum-free medium for 24 h. To detect whether the decrease in N-LC3b was correlated with autophagy, Hela cells were transfected with pVAX-N, pVAX-N-LC3b, respectively, for 6 h and treated with rapamycin (RAPA) (100 nM) or chloroquine (CQ) (75 μM) for 24 h. Cells were lysed using RIPA cell lysis buffer (Beyotime, Shanghai, China) and SDS-PAGE was conducted. After the protein was transferred to the nitrocellulose filter (NC) membrane, anti-SARS-CoV-2 N antibody (CST, Boston, MA, USA), anti-LC3b antibodies (Sigma, Darmstadt, Germany), and anti-GAPDH antibody (Abmart, Shanghai, China) were incubated overnight at 4 °C. Next, the membrane was incubated with the horseradish peroxidase (HRP)-conjugated anti-rabbit IgG antibody or anti-mouse IgG antibody (Abclonal, Wuhan, China), and detected with a chemiluminescent HRP substrate (Tanon, Shanghai, China).

2.3. Confocal Microscopy

DC2.4 cells (from murine bone marrow-derived dendritic cells) were cultured in complete Roswell Park Memorial Institute 1640 medium (RPMI 1640, Gibco), containing 10% fetal bovine serum and 1% penicillin/streptomycin, at 37 °C in an atmosphere of 5% of CO_2. Hela cells or DC2.4 cells were seeded on microscope cover glasses and transfected with pVAX-N-LC3b or pVAX-N for 36 h, then treated with or without chloroquine (CQ) for 12 h. Cells were washed with phosphate-buffered saline (PBS) three times and fixed in 4% paraformaldehyde (Beyotime, Shanghai, China) for 1h at 4 °C. Next, cells were permeabilized and blocked with QuickBlock™ Blocking Buffer for Immunol Staining reagents (Beyotime, Shanghai, China) for 15 min at room temperature. Then, the cells were washed with PBST (PBS supplemented with Tween-20) and incubated with corresponding primary antibodies in 1% bovine serum albumin solution at a dilution of 1:200 overnight at 4 °C. After being washed with PBST, secondary antibodies were added and incubated for 2 h. Finally, the cells were stained with DAPI nucleic acid stain (1 mg/mL, Beyotime, Shanghai, China) for 15 min and imaged using a confocal microscope (Zeiss, Oberkochen, Germany).

2.4. Animal Immunization

Female, six-week-old, non-pathogenic BALB/c mice were raised in an SPF environment in this study. Twenty mice were randomly allocated to four groups, including the negative control group (PBS and pVAX-LC3b), and the experimental groups (pVAX-N and pVAX-N-LC3b). All mice were injected with 0.5% bupivacaine hydrochloride (Sigma, Darmstadt, Germany) in the gastrocnemius muscle of each hind leg three days before the initial immunization. On days 0 and 14, 50 μg of the corresponding DNA plasmid dissolved in 100 μL of sterile PBS was injected into the gastrocnemius muscle of both hind legs of the mice (50 μL for each leg). On day 28, mouse serum was collected for antibody detection by the enzyme-linked immunosorbent assay (ELISA). Mouse spleen lymphocytes were harvested and subjected to enzyme-linked immunosorbent spot (ELISPOT) assay, intracellular cytokine staining (ICS) assay, and carboxyfluorescein diacetate succinimidyl ester (CFSE)-based proliferation assay.

2.5. ELISA

SARS-CoV-2 nucleocapsid-specific antibodies were detected by ELISA as in our previously reported method [26,27]. In brief, 96-well EIA plates (Corning Inc, Corning, NY, USA) were coated with 100 ng per well of SARS-CoV-2 N protein antigens (Abclonal, Wuhan, China) in PBS overnight at 4 °C. Plates were washed with PBST three times and then blocked with 5% skimmed milk in PBST for 1 h at 37 °C. Mouse serum was diluted at 1:25, and then added to the washed plates and incubated for 2 h at 37 °C. After washing, plates were incubated with a 1:5000 dilution of HRP conjugated anti-mouse IgG secondary antibody (Abcam, Cambridge, UK), and HRP conjugated anti-mouse IgG1 secondary antibody (Abcam), HRP conjugated anti-mouse IgG2a secondary antibody (Proteintech, Wuhan, China), and HRP conjugated anti-mouse IgG2c secondary antibody (Abcam, Cambridge, UK), respectively, for 1 h at 37 °C. The color reaction was substrated with 3,3′,5,5′-Tetramethylbenzidine (TMB) and stopped with 1 M H_2SO_4. The value was read at a 450 nm wavelength using a Synergy HT Multi-Mode Plate Reader (BioTek, Winooski, VT, USA).

2.6. IFN-γ ELISPOT Assay

Interferon Gamma (IFN-γ) ELISPOT was performed using freshly isolated mouse splenic lymphocytes as previously described [28,29]. In brief, PVDF 96-well plates (Millipore, Billerica, MA) were pre-coated with mouse IFN-γ coating antibody (U-CyTech, Utrecht, The Netherlands) overnight at 4 °C. Mouse splenic lymphocytes were isolated using a density gradient medium (Dakewe Biotech, Shenzhen, China). The N peptide pools were 15 amino acids in length and overlapped by 11 amino acids covering the full-length SARS-CoV-2 N protein, and were synthesized by GenScript company, and then dissolved in dimethyl sulfoxide (DMSO). The 2.5×10^5 mouse splenocytes were plated into each well and stimulated with N peptide pool at a final concentration of 2 µg/mL for each peptide, while DMSO was performed as a mock simulation. After incubation for 24 h, the plates were incubated with biotinylated detection antibodies (U-CyTech, Utrecht, The Netherlands) and developed with alkaline phosphatase-conjugated streptavidin (U-CyTech, Utrecht, The Netherlands) and NBT/BCIP reagent (Pierce, Rockford, IL, USA). Finally, the spots were counted with an ELISPOT reader (Mabtech, Stockholm, Sweden).

2.7. ICS

ICS assay was performed according to our previous method [30]. Briefly, mouse splenic lymphocytes were seeded in the 96-well plates at 2×10^6 per well and stimulated with the SARS-CoV-2 N peptide pool (2 µg/mL for each peptide) at 37 °C for 1.5 h. Then, brefeldin A (BD, Franklin Lakes, NJ, USA) was added and incubated for 16 h at 37 °C. The cells were harvested and stained with anti-mouse CD3-FITC, anti-mouse CD4-BB700, and anti-mouse CD8-PE cy7 (BD, Franklin Lakes, NJ, USA) for 30 min, and protected from light at room temperature. Then, the cells were fixed and permeabilized with cytofix/cytoperm (BD, Franklin Lakes, NJ, USA) for 20 min, and protected from light at 4 °C. After being washed with perm/wash (BD, Franklin Lakes, NJ, USA), the cells were stained with anti-mouse IFN-γ-APC, anti-mouse IL-2-BV605, and anti-mouse TNF-α-PE (BD, Franklin Lakes, NJ, USA) for 1 h, and protected from light at 4 °C. Finally, the cells were washed with FACS washing buffer (PBS supplement with 2% heat-inactivated FBS) three times and resuspended in PBS. The data were obtained with Beckman CytExpert software and analyzed using FlowJo software (version 10.8.1).

2.8. CFSE-Based Proliferation Assay

Mouse splenic lymphocytes were resuspended at 1×10^6/mL in 0.1% FBS/PBS and incubated at 37 °C for 10 min with 0.2 µM CFSE (ThermoFisher, Waltham, MA, USA). An equal volume of ice-cold RPMI 1640 was added and an ice bath was run for 5 min to terminate the staining. After the addition of serum and washes with RPMI 1640, cells were resuspended at 1×10^6 cells/mL and plated into 48-well U-bottom plates at 500 µL volumes and simulated with a SARS-CoV-2 N peptide pool (2 µg/mL for each peptide). After being simulated for five days, cells were harvested and washed with FACS washing buffer and stained with anti-mouse CD3-PE, anti-mouse CD4-BB700, anti-mouse CD8-PE-Cy7, and fixable viability stain 780 (BD, Franklin Lakes, NJ, USA). Cells were washed and a flow cytometric analysis was conducted using Beckman CytExpert. The data were analyzed using FlowJo software (version 10.8.1).

2.9. Statistical Analysis

Graphical presentations and statistical analyses were performed using GraphPad Prism software version 8. Statistical significance was calculated using ANOVA with Holm–Sidak multiple comparisons tests. * $p < 0.05$; ** $p < 0.01$; *** $p < 0.001$; **** $p < 0.0001$.

3. Results

3.1. Functional Targeting of SARS-CoV-2 N-LC3b Fusion Antigen to Autophagosomes/ Lysosomes/MHC II Compartments Signal Pathway

We designed and constructed a series of recombinant DNA vectors carrying the mouse LC3b gene, SARS-CoV-2 N gene, and SARS-CoV-2 N-LC3b fusion gene, respectively (Figure 1A,B). Using anti-SARS-CoV-2 N antibodies to recognize the SARS-CoV-2 N-LC3b fusion proteins, we confirmed the appropriate protein expression of these constructs by Western blotting assay (Figure 1C).

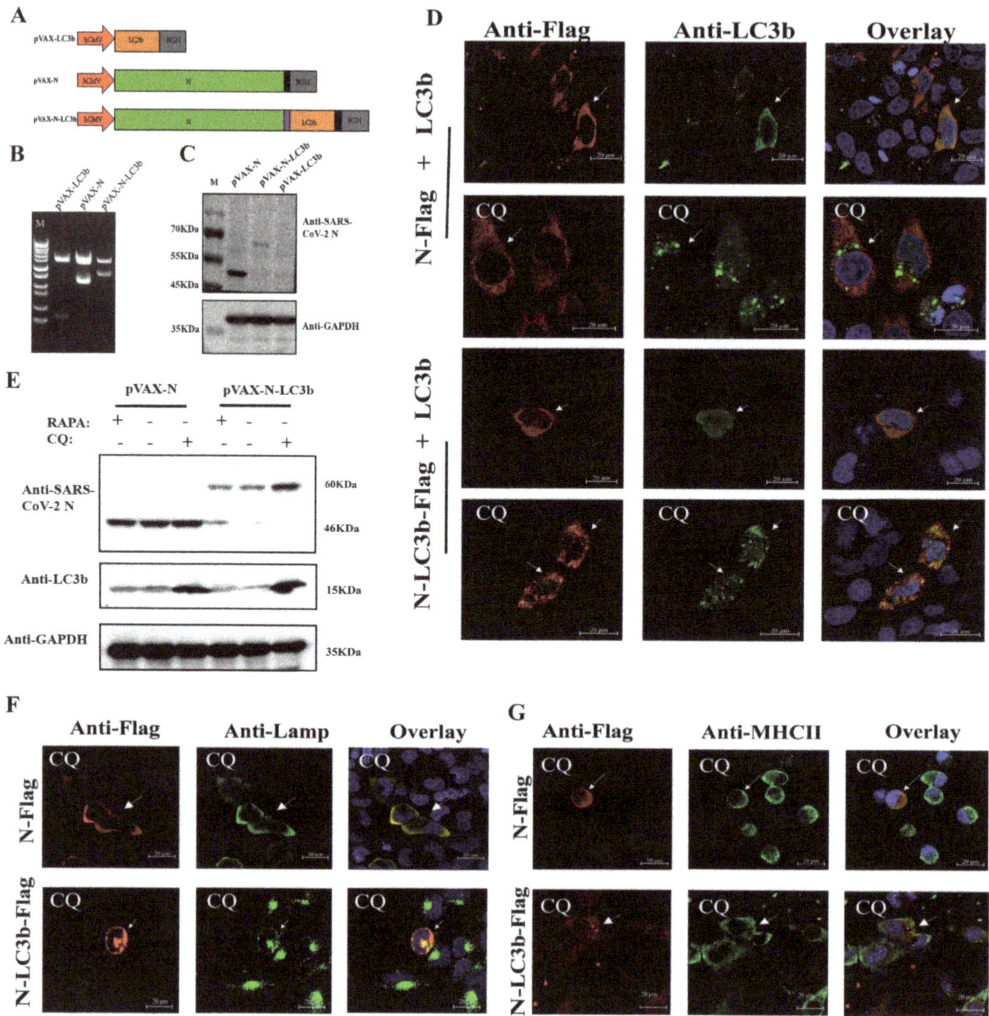

Figure 1. The SARS-CoV-2 N-LC3b fusion protein was effectively processed by autophagy pathway and presented to MHC II. (**A**) Schematic representation of constructs carrying various combinations

of mouse LC3b gene or SARS-CoV-2 under the CMV promoter denoted as pVAX-LC3b, pVAX-N, and pVAX-N-LC3b, respectively. (**B**) The plasmids were identified by HindIII, XbaI digestion. Lane1: DNA marker, 10,000, 8000, 6000, 5000, 4000, 3000, 2000, 1000, 750, 500, 250, 100 bp. Lane2: pVAX-LC3b; Lane3: pVAX-N; Lane4: pVAX-N-LC3b. (**C**) The protein expression was detected using anti-SARS-CoV-2 N antibodies and anti-GAPDH antibodies, respectively. The molecular weight of N protein: 46KDa; the molecular weight of N-LC3b fusion protein: 60KDa. (**D**) HeLa cells were co-transfected with pVAX-LC3b and pVAX-N or pVAX-N-LC3b plasmid with or without CQ (75 µM) treatment, and then stained with rabbit anti-LC3b IgG antibody and mouse anti-Flag IgG antibody. Subsequently, Cy3-labeled goat anti-mouse IgG (red fluorescence) and Alexa Fluor 488-labeled goat anti-rabbit IgG (green fluorescence) secondary antibodies were used; (**E**) Hela cells were transfected with pVAX-N, and pVAX-N-LC3b, respectively, and treated with RAPA (100 nM) or CQ. The protein expression was detected using anti-SARS-CoV-2 N antibodies and anti-mouse LC3b antibodies. GAPDH blots demonstrated that the overall protein was not affected after CQ and RAPA treatment. (**F**) HeLa cells and (**G**) DC2.4 cells were transfected with pVAX-N, pVAX-N-LC3b with CQ treatment, and then stained with mouse anti-LAMP-2 IgG antibody or mouse anti-MHC II IgG antibody and rabbit anti-Flag IgG antibody. Subsequently, Cy3-labeled goat anti-rabbit IgG (red fluorescence) and Alexa Fluor 488-labeled goat anti-mouse IgG (green fluorescence) secondary antibodies were used; the nucleus was stained with DAPI (blue fluorescence). The scale bar represents 20 µm. N: nucleocapsid protein. LC3b: microtubule-associated protein 1 light chain 3 beta. RAPA: rapamycin. CQ: chloroquine. The data represent three independent experiments.

We next investigated the subcellular localization of the SARS-CoV-2 N protein in the presence or absence of LC3b protein fusion using confocal microscopy. HeLa cells were co-transfected with pVAX-LC3b, pVAX-N-Flag, or pVAX-N-LC3b-Flag, and our results showed that the autophagosomes (green puncta) were co-localized with the N-LC3b fusion protein (red puncta) after being treated with CQ, which can block the fusion of autophagosomes with lysosomes, and thereby enable the visualization of the accumulation of autophagosomes (Figure 1D, bottom). However, the location of the N protein was significantly different from that of the N-LC3b fusion protein. The N protein alone was distributed homogenously throughout the cytoplasm whether CQ existed or not, and the autophagosomes (green puncta) were not co-localized with the N protein (red puncta) (Figure 1D, top). Moreover, the expression of the N-LC3b protein was significantly decreased after RAPA treatment, which can induce autophagic flux, and it obviously increased after CQ treatment, which might be attributed to CQ blocking the autophagy-mediated degradation of the N-LC3b protein, but the expression of the N protein alone was not changed after CQ treatment or RAPA treatment (Figure 1E). Subsequently, we also found that the N-LC3b protein was co-localized to LAMP II, which is a marker of endosomes/lysosomes (Figure 1F). To further verify whether the degradation of N-LC3b could be presented to the MHC II compartments, DC2.4 cells were transfected with pVAX-N and pVAX-N-LC3b, respectively, and then subjected to an immunofluorescence assay. The results showed that the N-LC3b protein was effectively co-localized to the MHC II compartments (Figure 1G). Taken together, these data demonstrated that the N-LC3b fusion protein can be functionally targeted to autophagosomes, processed by autophagy-mediated degradation in autolysosomes/lysosomes, and then presented to MHC II compartments for eliciting the subsequent adaptive immunity.

3.2. Enhancement of SARS-CoV-2 N Antigen-Specific T Cell Immune Responses by the N-LC3b Fusion Antigen

Then, we evaluated the immunogenicity of the N-LC3b fusion antigen in vivo. Mice were randomly allocated to four groups, including pVAX-Empty, pVAX-LC3b, pVAX-N, and pVAX-N-LC3b, and the antigen-specific antibodies and T cell immune responses were detected after immunization (Figure 2A). Our results demonstrated that the N antigen-specific T cell immune responses in the pVAX-N-LC3b group were greatly enhanced compared to the pVAX-N group. Of note, 1144 specific spot-forming cells (SFCs) per 10^6 spleen lymphocytes against SARS-CoV-2 N peptide pools were observed after the pVAX-N-LC3b immunization, which was 8.28 times higher than that in the pVAX-N group (Figure 2B). Meanwhile, the total frequencies of IFN-γ^+ or TNF-α^+ CD8$^+$ T cells, responding with N-specific peptide pools in the pVAX-N-LC3b group, were significantly higher than those in the pVAX-N group (Figure 2C). Furthermore, the induction of polyfunctional T cells secreting multiple cytokines was also analyzed in this study (Figure 2D), and the results showed that there was a higher frequency of polyfunctional CD4$^+$ T cells (Figure 2E) and CD8$^+$ T cells (Figure 2F) in the pVAX-N-LC3b group than that in the pVAX-N group. Overall, these results demonstrated that the N-LC3b fusion antigen elicited a potent antigen-specific CD4$^+$ T and CD8$^+$ T cellular immunity, with enhanced magnitude and polyfunctionality.

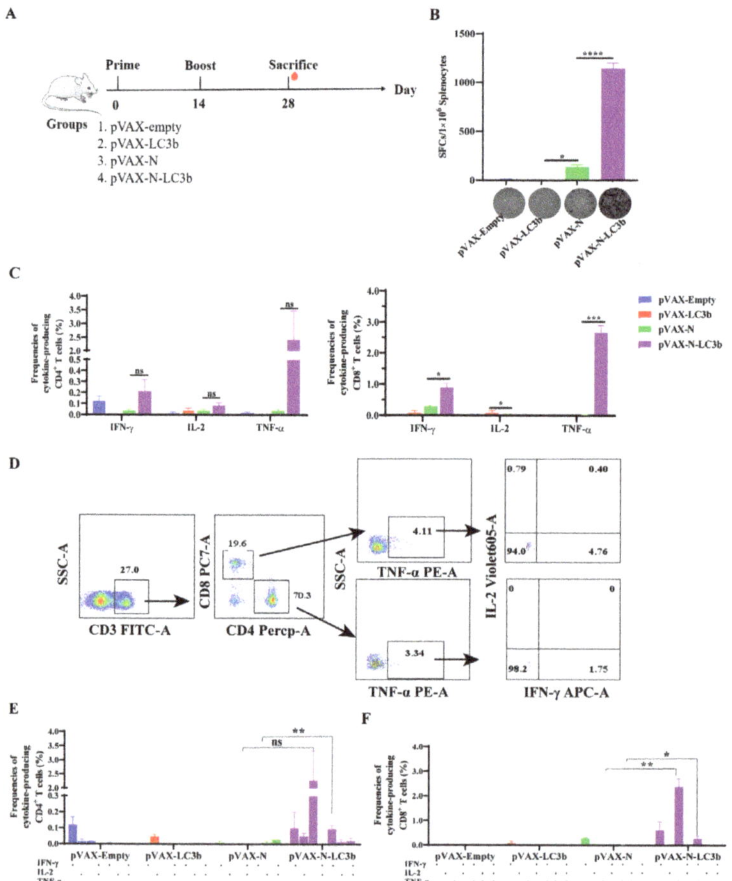

Figure 2. The SARS-CoV-2 N antigen-specific T cell immune responses in response to the N-LC3b fusion antigen. (**A**) The schedule of immunization experiment in mice. (**B**) The column showed the SARS-CoV-2 N-specific SFCs per 1×10^6 spleen lymphocytes as measured by IFN-γ ELISPOT assay. The bottom of the graph showed the original picture for our ELISPOT assay (2.5×10^5 cells per well). (**C**) The frequencies of total IFN-γ, TNF-α, or IL-2 cytokine-positive CD4$^+$ T cells (**left**) and CD8$^+$ T cells (**right**). (**D**) The graph represented the gating strategy of polyfunctional T cells analysis. Multiple cytokines-positive CD4$^+$ T cells (**E**) and CD8$^+$ T cells (**F**) were analyzed by multicolor flow cytometry. Two independent experiments for animal immunization were repeated. The data were shown as the mean ± SD for each group ($n = 5$). ns: no significance. * $p < 0.05$, ** $p < 0.01$, *** $p < 0.001$, **** $p < 0.0001$, ns: no significance.

3.3. Enhancement of SARS-CoV-2 N Antigen-Specific T Cell Proliferation by the N-LC3b Fusion Antigen

Recent studies demonstrated that SARS-CoV-2 specific CD4⁺ T and CD8⁺ T cells in COVID-19 convalescent individuals had strong ex vivo proliferation capacities, implying that the induction of T lymphocyte proliferation should be an important immunological parameter to evaluate an effective COVID-19 vaccine candidate. We, therefore, evaluated the SARS-CoV-2 antigen-induced proliferation capacity in response to the N-LC3b fusion antigen immunization. The fresh splenocytes from the immunized mice were labeled with CFSE and stimulated for 5 days with N-specific peptide pools, and then analyzed using flow cytometry. The percentage of CFSE low cells of both CD3⁺ CD4⁺ T cells (Figure 3A) and CD3⁺ CD8⁺ T cells (Figure 3B) in the pVAX-N-LC3b group exhibited a significant increase compared to that in the pVAX-N group. Thus, the N-LC3b fusion antigen effectively induced an improved proliferation capacity of the antigen-specific CD3⁺ CD4⁺ T and CD3⁺ CD8⁺ T lymphocytes, which is important for long-lasting immune protection.

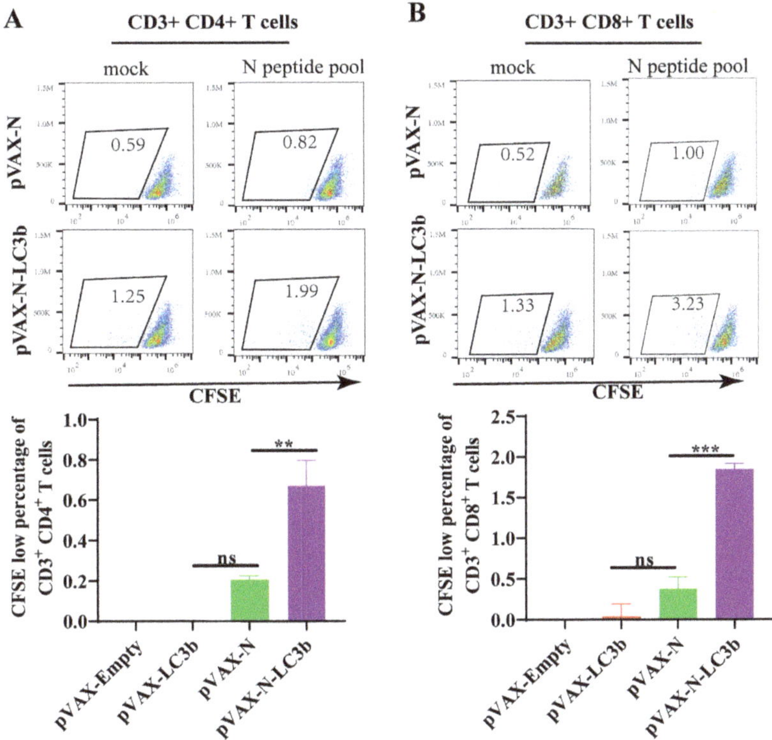

Figure 3. SARS-CoV-2 N antigen-specific T cell proliferation detected by CFSE-based staining. The scatter plot showed the proliferation capacity of CD3⁺ CD4⁺ T cells (**A**) and CD3⁺ CD8⁺ T cells (**B**) in response to SARS-CoV-2 N peptides stimulation or DMSO stimulation (mock). The CFSE-low cells (lower fluorescence intensity) represent these daughter cells after cell proliferation. Columns represent the percentage of N-specific CFSE-low cells in response to stimulation (subtracted by the mock). These data were shown as the mean ± SD for each group ($n = 5$). ** $p < 0.01$, *** $p < 0.001$, ns: no significance.

3.4. Induction of Th1-Biased Immunity by the N-LC3b Fusion Antigen

Furthermore, we also detected the humoral immune responses elicited by our strategy. Our result showed that both the N antigen and the N-LC3b fusion antigen could effectively induce the IgG antibody immune responses (Figure 4A). Moreover, we analyzed the subclass of these IgG antibodies. In mice, IgG1 is a marker of a Th2-biased immune response, while IgG2a and IgG2c antibodies represent a Th1-biased immune response. We found that the pVAX-N immunization mainly induced the IgG1, IgG2a, and Ig2c antibodies, of which the IgG1 antibody was the most induced. However, the pVAX-N-LC3b preferentially induced the IgG2a antibody (Figure 4B–D). The ratio of IgG1/IgG2a showed that the N-LC3b fusion antigen induced a Th1-biased immune response in mice (Figure 4E).

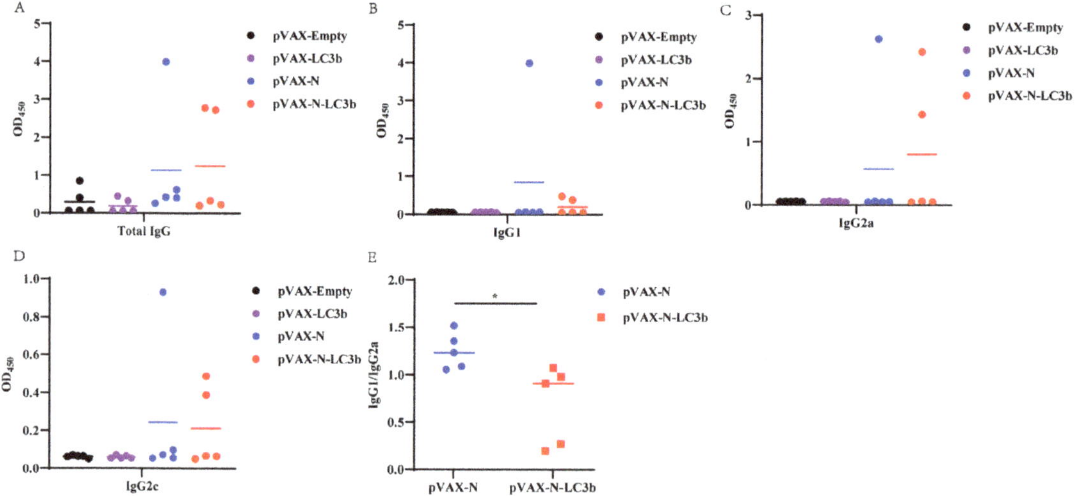

Figure 4. Induction of Th1-biased immunity by the N-LC3b fusion antigen. Sera (at diluted 1:25) of the immunized mice on day 28 were analyzed for the SARS-CoV-2 N specific total IgG (**A**), IgG1 (**B**), IgG2a (**C**), or IgG2c (**D**) subclass antibodies. (**E**) The ratio of IgG1/IgG2a. The data were shown as the mean ± SD for each group ($n = 5$). * $p < 0.05$.

4. Discussion

The accumulating data have shown that T cells play a pivotal role in fighting SARS-CoV-2 infection, and, most likely, in forming immunological memory following recovery from COVID-19 [31]. Some recovered COVID-19 patients had robust SARS-CoV-2 specific T cell immune responses but no obvious SARS-CoV-2 neutralizing antibody was detected, implying the importance of T cell immunity in controlling COVID-19 progression [15]. More importantly, T cell-mediated immunity can effectively play a critical role against viral variants, which can rapidly escape immune recognition by neutralizing antibodies [32]. T cells differentiate into effector cells by specifically recognizing the MHC-antigen peptide complex presented by antigen-presenting cells. Classically, the endogenous antigen peptides binding with MHC class I molecules are recognized by CD8[+] T cells, while the exogenous antigen peptides degraded by the lysosomal pathway binding with MHC class II molecules are recognized by CD4[+] T cells. In addition, antigen-presenting cells might also process exogenous antigens to CD8[+] T cells via the MHC class I pathway but not CD4[+] T cells, also known as cross-presentation. Interestingly, macroautophagy can have a cross-presentation function to deliver some endogenous proteins to MHC class II molecules and thus enhance the adaptive T cell immune responses [33,34]. Both our work and the other group's study showed that the antigen design by fusing autophagy-associated protein LC3b effectively elicited a robust immune response against influenza, *Mycobacterium* tuberculosis (*Mtb*), and simian immunodeficiency virus (SIV) infections [34,35]. Consistent with these findings, we herein found that when compared to N antigen alone, the N-LC3b antigen was more effectively targeted to the autophagosome/lysosome/MHC II compartment signal pathway, and effectively elicited the stronger polyfunctional CD4[+] and CD8[+] T cell immune responses in mice.

Another notable observation is that the pVAX-N immunization induced a Th2-biased immune response, while the pVAX-N-LC3b immunization induced a Th1-biased immune response. An effective COVID-19 vaccine should induce the "right" antibodies and T cell responses because the "wrong" antibodies might increase the risk of immunopathologies. For example, some case reports indicated that patients with high SARS-CoV-2 IgG levels were more likely to develop severe disease [36]. Moreover, the potential risks of antibody-dependent enhancement (ADE) and enhanced respiratory disease (ERD) has been observed in animal models when immunized with some COVID-19 vaccines [37], which mainly induced the Th2-biased immune responses [38]. Although these immunized mice had significantly lower viral load after the SARS-CoV-2 challenge, they also had an eosinophilic infiltrate into the lungs, which was accompanied by ERD-related immunopathology in lung tissue [39]. Interestingly, our strategy can regulate the bias of the Th1/Th2 immune responses, as revealed by the inverted ratio of the IgG1/IgG2a subclass induced by the N-LC3b fusion antigen when compared to N antigen alone. In addition, the cytokine profile of splenocytes from the pVAX-N-LC3b immunized mice also supported a preferable Th1-biased immune response.

Our strategy induced the promising immunogenicity of the novel COVID-19 vaccine candidate in mice, but there are some limitations. For instance, we only used the DNA-based vaccine to verify this hypothesis as a model; other strategies including an mRNA-based vaccine and viral-vector-based vaccine might be more suitable to deliver this designed antigen to elicit a more robust immune response. In addition, the protective efficacy against SARS-CoV-2 infections was not evaluated in this study because of our limited resources for the high-level biosafety laboratory. In conclusion, we developed a simple and effective strategy to induce a potential SARS-CoV-2 specific T cellular immunity with enhanced magnitude, polyfunctionality, and proliferation by targeting the autophagy-mediated signal pathway, and our findings warranted a further study on protection efficacy in an animal infection model as a novel T cell-based COVID-19 vaccine candidate. In addition, this N-LC3b fusion protein is a promising component in the prime/boost vaccination strategy to elicit a balanced Th1/Th2 immunity against SARS-CoV-2 infection.

Supplementary Materials: The following supporting information can be downloaded at https://www.mdpi.com/article/10.3390/v15061316/s1, Table S1: Primer Sequences.

Author Contributions: CS conceived and designed the experiments. Z.W., Y.Y., Y.Z., H.W., Z.H. and M.L. performed the experiments. Z.W. and Y.Y. analyzed the data. J.Y. contributed the materials and reagents. Z.W., Y.Y. and C.S. wrote the manuscript. All authors have read and agreed to the published version of the manuscript.

Funding: This research was funded by the National Key R&D Program of China (2022YFE0203100, 2021YFC2300103), the National Natural Science Foundation of China (82271786, 81971927), Science and Technology Planning Project of Guangdong Province, China (2021B1212040017), Shenzhen Science and Technology Program (JSGG20200225152008136), and Sanming Project of Medicine in Shenzhen Nanshan (No. SZSM202103008).

Institutional Review Board Statement: Animal experiments were carried out in accordance with the recommendations in the Institutional and National Guidelines for Animal Care and Use. The protocol was approved by the Committee on the Ethics of Animal Experiments of Sun Yat-sen University, Guangzhou, China (SYSU-IACUC-2022-001856). All procedures were performed under ether anesthesia and suffering was minimized.

Informed Consent Statement: Not applicable.

Data Availability Statement: Not applicable.

Acknowledgments: We appreciate all other members in our group for their helpful advice and discussion on improving this project.

Conflicts of Interest: The authors declare that they have no conflicts of interest. The funders had no role in the study design, data collection and analysis, decision to publish, or the preparation of the manuscript.

References

1. WHO Organization. WHO Coronavirus (COVID-19) Dashboard. Available online: https://covid19.who.int/ (accessed on 21 February 2023).
2. McDonald, I.; Murray, S.M.; Reynolds, C.J.; Altmann, D.M.; Boyton, R.J. Comparative systematic review and meta-analysis of reactogenicity, immunogenicity and efficacy of vaccines against SARS-CoV-2. *NPJ Vaccines* **2021**, *6*, 74. [CrossRef] [PubMed]
3. Kim, J.H.; Marks, F.; Clemens, J.D. Looking beyond COVID-19 vaccine phase 3 trials. *Nat. Med.* **2021**, *27*, 205–211. [CrossRef] [PubMed]
4. Du, L.; He, Y.; Zhou, Y.; Liu, S.; Zheng, B.J.; Jiang, S. The spike protein of SARS-CoV—A target for vaccine and therapeutic development. *Nat. Rev. Microbiol.* **2009**, *7*, 226–236. [CrossRef] [PubMed]
5. Hacisuleyman, E.; Hale, C.; Saito, Y.; Blachere, N.E.; Bergh, M.; Conlon, E.G.; Schaefer-Babajew, D.J.; DaSilva, J.; Muecksch, F.; Gaebler, C.; et al. Vaccine Breakthrough Infections with SARS-CoV-2 Variants. *N. Engl. J. Med.* **2021**, *384*, 2212–2218. [CrossRef] [PubMed]
6. Keeton, R.; Tincho, M.B.; Ngomti, A.; Baguma, R.; Benede, N.; Suzuki, A.; Khan, K.; Cele, S.; Bernstein, M.; Karim, F.; et al. T cell responses to SARS-CoV-2 spike cross-recognize Omicron. *Nature* **2022**, *603*, 488–492. [CrossRef]
7. McEwen, A.E.; Cohen, S.; Bryson-Cahn, C.; Liu, C.; Pergam, S.A.; Lynch, J.; Schippers, A.; Strand, K.; Whimbey, E.; Mani, N.S.; et al. Variants of Concern Are Overrepresented Among Postvaccination Breakthrough Infections of Severe Acute Respiratory Syndrome Coronavirus 2 (SARS-CoV-2) in Washington State. *Clin. Infect. Dis.* **2022**, *74*, 1089–1092. [CrossRef]
8. Primorac, D.; Brlek, P.; Matišić, V.; Molnar, V.; Vrdoljak, K.; Zadro, R.; Parčina, M. Cellular Immunity-The Key to Long-Term Protection in Individuals Recovered from SARS-CoV-2 and after Vaccination. *Vaccines* **2022**, *10*, 442. [CrossRef]
9. Peng, Y.; Mentzer, A.J.; Liu, G.; Yao, X.; Yin, Z.; Dong, D.; Dejnirattisai, W.; Rostron, T.; Supasa, P.; Liu, C.; et al. Broad and strong memory CD4(+) and CD8(+) T cells induced by SARS-CoV-2 in UK convalescent individuals following COVID-19. *Nat. Immunol.* **2020**, *21*, 1336–1345. [CrossRef]
10. Cassaniti, I.; Percivalle, E.; Bergami, F.; Piralla, A.; Comolli, G.; Bruno, R.; Vecchia, M.; Sambo, M.; Colaneri, M.; Zuccaro, V.; et al. SARS-CoV-2 specific T-cell immunity in COVID-19 convalescent patients and unexposed controls measured by ex vivo ELISpot assay. *Clin. Microbiol. Infect.* **2021**, *27*, 1029–1034. [CrossRef]
11. Tarke, A.; Coelho, C.H.; Zhang, Z.; Dan, J.M.; Yu, E.D.; Methot, N.; Bloom, N.I.; Goodwin, B.; Phillips, E.; Mallal, S.; et al. SARS-CoV-2 vaccination induces immunological T cell memory able to cross-recognize variants from Alpha to Omicron. *Cell* **2022**, *185*, 847–859.e811. [CrossRef]
12. Li, M.; Zeng, J.; Li, R.; Wen, Z.; Cai, Y.; Wallin, J.; Shu, Y.; Du, X.; Sun, C. Rational Design of a Pan-Coronavirus Vaccine Based on Conserved CTL Epitopes. *Viruses* **2021**, *13*, 333. [CrossRef] [PubMed]
13. De Marco, L.; D'Orso, S.; Pirronello, M.; Verdiani, A.; Termine, A.; Fabrizio, C.; Capone, A.; Sabatini, A.; Guerrera, G.; Placido, R.; et al. Assessment of T-cell Reactivity to the SARS-CoV-2 Omicron Variant by Immunized Individuals. *JAMA Netw. Open* **2022**, *5*, e2210871. [CrossRef]
14. Le Bert, N.; Tan, A.T.; Kunasegaran, K.; Tham, C.Y.L.; Hafezi, M.; Chia, A.; Chng, M.H.Y.; Lin, M.; Tan, N.; Linster, M.; et al. SARS-CoV-2-specific T cell immunity in cases of COVID-19 and SARS, and uninfected controls. *Nature* **2020**, *584*, 457–462. [CrossRef] [PubMed]
15. Sekine, T.; Perez-Potti, A.; Rivera-Ballesteros, O.; Strålin, K.; Gorin, J.B.; Olsson, A.; Llewellyn-Lacey, S.; Kamal, H.; Bogdanovic, G.; Muschiol, S.; et al. Robust T Cell Immunity in Convalescent Individuals with Asymptomatic or Mild COVID-19. *Cell* **2020**, *183*, 158–168.e14. [CrossRef]
16. Oliveira, S.C.; de Magalhães, M.T.Q.; Homan, E.J. Immunoinformatic Analysis of SARS-CoV-2 Nucleocapsid Protein and Identification of COVID-19 Vaccine Targets. *Front. Immunol.* **2020**, *11*, 587615. [CrossRef]
17. Bai, Z.; Cao, Y.; Liu, W.; Li, J. The SARS-CoV-2 Nucleocapsid Protein and Its Role in Viral Structure, Biological Functions, and a Potential Target for Drug or Vaccine Mitigation. *Viruses* **2021**, *13*, 1115. [CrossRef] [PubMed]
18. Holmes, K.V.; Enjuanes, L. Virology. The SARS coronavirus: A postgenomic era. *Science* **2003**, *300*, 1377–1378. [CrossRef]
19. Matchett, W.E.; Joag, V.; Stolley, J.M.; Shepherd, F.K.; Quarnstrom, C.F.; Mickelson, C.K.; Wijeyesinghe, S.; Soerens, A.G.; Becker, S.; Thiede, J.M.; et al. Nucleocapsid vaccine elicits spike-independent SARS-CoV-2 protective immunity. *bioRxiv* **2021**. [CrossRef]
20. Hasanpourghadi, M.; Novikov, M.; Ambrose, R.; Chekaoui, A.; Newman, D.; Ding, J.; Giles-Davis, W.; Xiang, Z.; Zhou, X.Y.; Liu, Q.; et al. Heterologous chimpanzee adenovirus vector immunizations for SARS-CoV-2 spike and nucleocapsid protect hamsters against COVID-19. *Microbes Infect.* **2022**, *25*, 105082. [CrossRef]
21. Choi, Y.; Bowman, J.W.; Jung, J.U. Autophagy during viral infection—A double-edged sword. *Nat. Rev. Microbiol.* **2018**, *16*, 341–354. [CrossRef]
22. Shibutani, S.T.; Saitoh, T.; Nowag, H.; Münz, C.; Yoshimori, T. Autophagy and autophagy-related proteins in the immune system. *Nat. Immunol.* **2015**, *16*, 1014–1024. [CrossRef] [PubMed]
23. Dasari, V.; Rehan, S.; Tey, S.K.; Smyth, M.J.; Smith, C.; Khanna, R. Autophagy and proteasome interconnect to coordinate cross-presentation through MHC class I pathway in B cells. *Immunol. Cell Biol.* **2016**, *94*, 964–974. [CrossRef] [PubMed]
24. Mizushima, N.; Yoshimori, T.; Levine, B. Methods in mammalian autophagy research. *Cell* **2010**, *140*, 313–326. [CrossRef] [PubMed]

25. Jin, Y.; Sun, C.; Feng, L.; Li, P.; Xiao, L.; Ren, Y.; Wang, D.; Li, C.; Chen, L. Regulation of SIV antigen-specific CD4+ T cellular immunity via autophagosome-mediated MHC II molecule-targeting antigen presentation in mice. *PLoS ONE* **2014**, *9*, e93143. [CrossRef]
26. Li, M.; Chen, J.; Liu, Y.; Zhao, J.; Li, Y.; Hu, Y.; Chen, Y.Q.; Sun, L.; Shu, Y.; Feng, F.; et al. Rational design of AAVrh10-vectored ACE2 functional domain to broadly block the cell entry of SARS-CoV-2 variants. *Antiviral. Res.* **2022**, *205*, 105383. [CrossRef]
27. Luo, H.; Jia, T.; Chen, J.; Zeng, S.; Qiu, Z.; Wu, S.; Li, X.; Lei, Y.; Wang, X.; Wu, W.; et al. The Characterization of Disease Severity Associated IgG Subclasses Response in COVID-19 Patients. *Front. Immunol.* **2021**, *12*, 632814. [CrossRef]
28. Sun, C.; Zhang, L.; Zhang, M.; Liu, Y.; Zhong, M.; Ma, X.; Chen, L. Induction of balance and breadth in the immune response is beneficial for the control of SIVmac239 replication in rhesus monkeys. *J. Infect.* **2010**, *60*, 371–381. [CrossRef]
29. Wen, Z.; Fang, C.; Liu, X.; Liu, Y.; Li, M.; Yuan, Y.; Han, Z.; Wang, C.; Zhang, T.; Sun, C. A recombinant Mycobacterium smegmatis-based surface display system for developing the T cell-based COVID-19 vaccine. *Hum. Vaccines Immunother.* **2023**, *19*, 2171233. [CrossRef]
30. Sun, C.; Feng, L.; Zhang, Y.; Xiao, L.; Pan, W.; Li, C.; Zhang, L.; Chen, L. Circumventing antivector immunity by using adenovirus-infected blood cells for repeated application of adenovirus-vectored vaccines: Proof of concept in rhesus macaques. *J. Virol.* **2012**, *86*, 11031–11042. [CrossRef]
31. Diao, B.; Wang, C.; Tan, Y.; Chen, X.; Liu, Y.; Ning, L.; Chen, L.; Li, M.; Liu, Y.; Wang, G.; et al. Reduction and Functional Exhaustion of T Cells in Patients With Coronavirus Disease 2019 (COVID-19). *Front. Immunol.* **2020**, *11*, 827. [CrossRef]
32. Aurisicchio, L.; Brambilla, N.; Cazzaniga, M.E.; Bonfanti, P.; Milleri, S.; Ascierto, P.A.; Capici, S.; Vitalini, C.; Girolami, F.; Giacovelli, G.; et al. A first-in-human trial on the safety and immunogenicity of COVID-eVax, a cellular response-skewed DNA vaccine against COVID-19. *Mol. Ther.* **2022**, *31*, 788–800. [CrossRef] [PubMed]
33. Ireland, J.M.; Unanue, E.R. Autophagy in antigen-presenting cells results in presentation of citrullinated peptides to CD4 T cells. *J. Exp. Med.* **2011**, *208*, 2625–2632. [CrossRef] [PubMed]
34. Schmid, D.; Pypaert, M.; Münz, C. Antigen-loading compartments for major histocompatibility complex class II molecules continuously receive input from autophagosomes. *Immunity* **2007**, *26*, 79–92. [CrossRef] [PubMed]
35. Mannar, D.; Saville, J.W.; Zhu, X.; Srivastava, S.S.; Berezuk, A.M.; Tuttle, K.S.; Marquez, A.C.; Sekirov, I.; Subramaniam, S. SARS-CoV-2 Omicron variant: Antibody evasion and cryo-EM structure of spike protein-ACE2 complex. *Science* **2022**, *375*, 760–764. [CrossRef]
36. Okba, N.M.A.; Müller, M.A.; Li, W.; Wang, C.; GeurtsvanKessel, C.H.; Corman, V.M.; Lamers, M.M.; Sikkema, R.S.; de Bruin, E.; Chandler, F.D.; et al. Severe Acute Respiratory Syndrome Coronavirus 2-Specific Antibody Responses in Coronavirus Disease Patients. *Emerg. Infect. Dis.* **2020**, *26*, 1478–1488. [CrossRef]
37. Wang, Q.; Zhang, L.; Kuwahara, K.; Li, L.; Liu, Z.; Li, T.; Zhu, H.; Liu, J.; Xu, Y.; Xie, J.; et al. Immunodominant SARS Coronavirus Epitopes in Humans Elicited both Enhancing and Neutralizing Effects on Infection in Non-human Primates. *ACS Infect. Dis.* **2016**, *2*, 361–376. [CrossRef]
38. Luan, N.; Li, T.; Wang, Y.; Cao, H.; Yin, X.; Lin, K.; Liu, C. Th2-Oriented Immune Serum After SARS-CoV-2 Vaccination Does Not Enhance Infection In Vitro. *Front. Immunol.* **2022**, *13*, 882856. [CrossRef]
39. Liu, L.; Wei, Q.; Lin, Q.; Fang, J.; Wang, H.; Kwok, H.; Tang, H.; Nishiura, K.; Peng, J.; Tan, Z.; et al. Anti-spike IgG causes severe acute lung injury by skewing macrophage responses during acute SARS-CoV infection. *JCI Insight* **2019**, *4*, e123158. [CrossRef]

Review

Autoantibodies to Interferons in Infectious Diseases

Eugenia Quiros-Roldan [1], Alessandra Sottini [2], Simona Giulia Signorini [2], Federico Serana [2], Giorgio Tiecco [1] and Luisa Imberti [3,*]

[1] Department of Infectious and Tropical Diseases, ASST Spedali Civili, Brescia and University of Brescia, 25123 Brescia, Italy; eugeniaquiros@yahoo.it (E.Q.-R.); g.tiecco@unibs.it (G.T.)

[2] Clinical Chemistry Laboratory, ASST Spedali Civili of Brescia, 25123 Brescia, Italy; alessandra.sottini@asst-spedalicivili.it (A.S.); simona.signorini@asst-spedalicivili.it (S.G.S.); federico.serana@asst-spedalicivili.it (F.S.)

[3] Section of Microbiology, University of Brescia, P. le Spedali Civili, 1, 25123 Brescia, Italy

* Correspondence: limberti@yahoo.it

Abstract: Anti-cytokine autoantibodies and, in particular, anti-type I interferons are increasingly described in association with immunodeficient, autoimmune, and immune-dysregulated conditions. Their presence in otherwise healthy individuals may result in a phenotype characterized by a predisposition to infections with several agents. For instance, anti-type I interferon autoantibodies are implicated in Coronavirus Disease 19 (COVID-19) pathogenesis and found preferentially in patients with critical disease. However, autoantibodies were also described in the serum of patients with viral, bacterial, and fungal infections not associated with COVID-19. In this review, we provide an overview of anti-cytokine autoantibodies identified to date and their clinical associations; we also discuss whether they can act as enemies or friends, i.e., are capable of acting in a beneficial or harmful way, and if they may be linked to gender or immunosenescence. Understanding the mechanisms underlying the production of autoantibodies could improve the approach to treating some infections, focusing not only on pathogens, but also on the possibility of a low degree of autoimmunity in patients.

Keywords: antiviral immunity; autoantibodies; COVID-19; cytokines; interferons

Citation: Quiros-Roldan, E.; Sottini, A.; Signorini, S.G.; Serana, F.; Tiecco, G.; Imberti, L. Autoantibodies to Interferons in Infectious Diseases. *Viruses* **2023**, *15*, 1215. https://doi.org/10.3390/v15051215

Academic Editors: Caijun Sun and Feng Ma

Received: 6 May 2023
Revised: 18 May 2023
Accepted: 19 May 2023
Published: 22 May 2023

1. Introduction

In the 20th century, latent infections in asymptomatic individuals were yet to be well studied and further elucidated. This issue led to the recognition that most infectious agents are lethal for only a small percentage of infected individuals. There is no doubt that an infectious agent is necessary to trigger clinical disease; however, it is just as indisputable that the bacteria, fungi, viruses, and parasites are far from being the only responsible cause of severe disease or death. Therefore, one of the main questions is as follows: what are the characteristics for which a patient suffering from an infectious disease could be life threatening? The enormous interindividual clinical variability observed in most infections appears to be due to genetic and immunological determinants [1]. Apart from genetic background, the great diversity of the human immune system, which is forged based on age, gender-related behaviors, diet, environmental exposure, and microbiome, is a powerful defense against opportunistic pathogens; however, at the same time, it can be the substrate on which immune-associated diseases develop [2].

One of the most important immunological mediators in the natural defense against infectious agents is the group of interferons (IFNs). They are potent cell growth regulators with immunomodulatory activity, but are best known for their antiviral activity [3].

A common phenomenon reported in many globally relevant infections is autoimmunity, with infections and other highly inflammatory diseases being associated with the presence of autoantibodies (aAbs). Therefore, aAbs can be broadly divided into "common" types that are found in apparently healthy individuals and that, through binding a variety of microbial components, provide the first line of defense against infections [4], and "pathogenic" types, which contribute to various immune-mediated diseases.

The large number of studies that previously focused on aAbs targeting cellular antigens, such as dsDNA and lipids, but also immune molecules, such as cytokines, underscores the importance that autoimmunity can play during infections. The role played by aAbs during specific infections is beginning to be an emerging topic of interest, especially after the discovery that anti-type I IFNs (IFN-I) aAbs play a fundamental role in the evolution of severe acute respiratory syndrome coronavirus 2 (SARS-CoV-2) infection [5].

The aim of our paper is to review the functional properties of aAbs produced against cytokines, looking at IFNs in particular, which are important anti-infective components of the immune system, in order to enhance understanding of their role in old and emerging infectious diseases.

2. Production of aAbs

2.1. Types of aAbs

The production of antibodies (Abs) is a vital way in which the adaptive immune system works either to recognize and neutralize or eliminate antigens and pathogens. Although thought to be absent in healthy individuals, due to the immune tolerance mechanism [6], the common Abs that react with self-molecules are found in healthy subjects [7]. They also fulfill the definition of aAb since they are self-reactive, rather than self-specific, and characterized by a broad reactivity directed against well-conserved public epitopes [8].

A type of common aAbs can be generated when molecules of infectious agents share similarity with foreign and self-peptides. Through a mechanism defined as "molecular mimicry" [9], these proteins may activate self-reactive T or B cells, thus producing cross-reactive Abs. For instance, the Abs produced against a viral phosphoprotein of measles or type 1 Herpes simplex viruses (HSV) cross-react with an intermediate filament protein of human cells [10]. Similarly, the significant sequence homology between Coxsackie virus P2-C protein and glutamate decarboxylase in humans may trigger type 1 diabetes [11], while ankylosing spondylitis, systemic lupus erythematosus (SLE), and Lyme disease may be induced by antigens such as pulD from *Klebsiella* sp., OSP-A from *Borrelia* sp., and nuclear antigen-1 from Epstein–Barr virus [12,13]. More recently, others potential molecular mimicry candidates were identified using bioinformatics techniques [14,15].

Another class of aAbs includes those defined as "natural" because they are present in the blood without any evident antigenic stimulation. Accordingly, this type of aAbs was even identified in mice raised under germ-free conditions [16]. Unlike adaptive Abs, these aAbs are synthesized by CD20+CD27+CD43+CD70−B1 lymphocytes and marginal-zone B cells [17,18] and do not undergo affinity maturation via antigen stimulation or extensive somatic mutation [19]. Their functions are not clear, though they may play a role in the maintenance of immune homeostasis, regulation of the immune response, resistance to infections, and transport and functional modulation of biologically active molecules [19,20].

The presence of aAbs rarely induces improvement in the disease, except in the cases of anti-cytokine aAbs associated with mild autoimmune diseases or specific anti-cancer aAbs that can be found in cancer patients with better survival [21–25]. On the contrary, they often cause adverse effects, playing a major role in several infectious, autoimmune, cardiovascular, neurological, and neurodegenerative diseases, as well as in metabolic dysfunction and cancers [26–33].

Anti-cytokine aAbs were initially described in an increasing number of primary immunodeficiencies with autoimmune features, especially autoimmune polyendocrine syndrome type I (APS-1), which is a disease of defective T cell-mediated central tolerance. However, these aAbs were proposed as an emerging alternative pathological mechanism leading to impaired immune response and susceptibility to infections.

2.2. Anti-Cytokine aAbs

In recent years, life-threatening diseases caused by anti-cytokine aAbs received widespread attention. Anti-cytokine aAbs found in healthy individuals [34,35] often are not strongly inhibitory, not necessarily associated with a respective neutralizing activity, and typically detected at lower binding titers [36–39]. Therefore, they seem to be a physiologic mechanism used to control the immune response [22]. In contrast, pathogenic anti-cytokine aAbs, which are usually polyclonal IgGs, may affect cytokine biology through diminishing or augmenting signaling or altering their half-lives in the circulation [22,38,40–43]. Anti-cytokines aAbs were found in patients with SLE, Sjogren's syndrome, and rheumatoid arthritis [43]. In addition, diseases due to aAbs targeting specific cytokines or cytokine pathways, which are classified in a unique category termed "phenocopies of primary immunodeficiency", comprise acquired immunodeficiency characterized by the presence of some anti-cytokine aAbs, notably to IFN-γ, interleukin (IL)-6, IL-17, IL-22, and granulocyte macrophage colony stimulating factor (GM-CSF). These diseases were found in particular, but not exclusively, in adult patients who showed phenotypic manifestations similar to those that occur due to pathogenic variants in genes encoding for the specific cytokines, their receptors, or molecules mediating cytokine signal transduction [44]. In addition, aAbs against pro-inflammatory cytokines are also found in multiple sclerosis that affect young adults [22]; however, their biological role is not yet clarified.

Anti-cytokine aAbs received great attention in recent years, especially in explaining the enormous phenotypic variability in infections, as well as the different incidence and inter-individual response variability. The development of several infectious diseases are described to be associated with the presence of aAbs targeting a number of cytokines [45]. Anti-IL-2 aAbs were found in patients with human immunodeficiency virus (HIV) infection [46]. Anti-GM-CSF aAbs were detected in patients with cryptococcosis, nocardiosis, non-tuberculous mycobacteria (NTM), and histoplasmosis [47,48], while anti-IL-6 aAbs were associated with severe bacterial infections, such as *Escherichia coli*, *Streptococcus intermedius*, and *Staphylococcus aureus* [49], including staphylococcal sepsis [50]. IL-17A, IL-17F, and IL-22 are considered important in mucosal immunity, principally in chronic mucocutaneous candidiasis [51].

Therefore, the presence of anti-cytokine aAbs can have severe consequences and cause highly varied manifestations.

Very recent anti-cytokine aAbs were found in >50% of critically ill patients with non-SARS-CoV-2 infections, i.e., caused by other viral and fungal pathogens, as well as known or suspected bacterial pathogens [52]. These aAbs were far more common in infected versus uninfected patients and, importantly, were seen not only in multiple respiratory viral infections, but also in the non-respiratory bacterial infections observed in patients admitted to the intensive care unit (ICU). Moreover, while most of these aAbs were present at the onset of infections, some can emerge over time and persist for at least 28 days after infection.

3. Role of IFNs and Anti-IFN aAbs in Infectious Diseases

3.1. IFNs in Infectious Diseases

There are three families of IFNs (Table 1): IFN-I (mainly IFN-α, IFN-β, and IFN-ω), IFN type II (IFN-II; IFN-γ), and IFN type III (IFN-III; IFN-λ).

Table 1. Types of IFNs.

	Other Designation	Official Gene Definition	Chromosome	Protein	Receptor
Type I IFNs					
IFN-alpha	IFN-α1, -α2, -α4, -α5, -α6, -α7, -α8, -α10, -α13, -α14, -α16, -α17, -α21	IFNA1, IFNA2, IFNA4, IFNA5, IFNA6, IFNA7, IFNA8, IFNA10, IFNA11P, IFNA12P, IFNA13, IFNA14, IFNA16, IFNA17, IFNA20P, IFNA21, IFNA22P	9p21.3	19 kDa, 20 glycosylated, 165–166 a, 188–189 aa (human)	IFNAR1; IFNAR2
IFN-beta	IFN-β	IFNB1		20 kDa, 22 kDa glycosilated, 166 aa, 187 aa (human)	
IFN-omega	IFN-ω	IFNW1		22 kDa glycosilated, 187 aa, 195 aa (human)	
IFN-epsilon	IFN-ε	IFNE		24.4 kDa, 187 aa, 208 aa (human)	
IFN-kappa	IFN-κ	IFNK		19 kDa, 182 aa 207 aa (human)	
IFN-zeta	IFN-ζ (limitin)			21.7 kDa glycosilated, 182 aa	
IFN-tau	IFN-τ	IFNT		19–24 kDa, 172 aa	
IFN-nu	IFN-v	IFNNP1		NA	
Type II IFN IFN-gamma	IFN-γ	IFNG	12q15	17 kDa, 115–175 aa, 166 aa (human)	IFNGR1; IFNGR2
Type III IFNs					
IFN-lamda	IFN-λ1 (IL-29)	IFNL1	19q13.2	21 kDa, 23–35 glycosilated, 200 aa (human)	IFNLR1; IL-10R2
	IFN-λ2 (IL-28A)	IFNL2		22 kDa, 24 glycosilated, 200 aa (human)	
	IFN-λ3 (IL-28B)	IFNL3		21 kDa, 24 glycosilated, 196 aa (human)	
	IFN-λ4	IFNL4		179 aa	

Greatest amount of information was taken from Negishi et al. [53] and GeneCards®: The Human Gene Database [54].

In humans, they constitute the first line of defense in response to invading pathogens. Indeed, IFN-I signal transduction pathways were previously identified as a critical factor limiting cytomegalovirus (CMV) infection and replication [55]. IFN-I contributes to the control of latent HSV infections, especially those caused by Alphaherpesvirinae HSV and varicella zoster virus (VZV). Similarly, it is crucial for HIV-1 infection of monocytes and macrophages [56,57], and its deficiency is related to significant impairment of the immune response during productive HIV-1 infection and infection latency [58]. Finally, dysregulation of the IFN-I signaling pathway by *Mycobacterium tuberculosis* (MT) also leads to exacerbation of HIV-1 infection in macrophages [59]. These data underline the importance of IFN-I responses towards HIV-1 infection [60] and, probably, in co-infections of HIV-1-infected patients with CMV and MT. However, the effects of IFN-I on the outcome of different infections are very complex, with both protective and detriment effects according to type of micro-organism and types of IFNs involved [61]. One of most representative examples is MT infection, for which IFN-I signaling demonstrated both pathogenic and protective roles [62].

IFN-II, as a key player in driving cellular immunity, can orchestrate numerous protective functions to heighten immune responses during several infectious diseases. Indeed, it has important immunomodulatory effects because it increases antigen processing and presentation, enables leukocyte trafficking, induces an anti-viral state, boosts the antimicrobial functions, and affects cellular proliferation and apoptosis [63]. IFN-II is important in endowing protection against bacterial infections, such as *Chlamydia* [64–66], *Staphylococcus aureus* [67], MT [68], NTM [69], *Salmonella* [70], and *Listeria* [71]. In addition, the protective benefits of IFN-II can be observed in the context of viral infections because its production via Natural Killer (NK) cells can successfully limit Hepatitis C virus proliferation in HIV-1-infected patients [72], while IFN-II treatment enhances survival of neurons infected with VZV [73]. Finally, IFN-II plays a pivotal role in host resistance to parasite invasions, which happen during *Leishmania* and *Toxoplasma* infections [63].

The most recently found member of IFNs, IFN-III was originally thought to act in parallel to IFN-I to activate compartmentalized antiviral responses [74]. Subsequent studies provided increasing evidence for distinct roles for each IFN family [75], and it seems that IFN-III can be also a critical instructor of antifungal neutrophil responses [76] and main players in protecting intestinal cells against enteric virus infections [77].

3.2. aAbs against IFNs in Non-Infectious Diseases

During the initial stages of an infection, there is a balance between mechanisms that promote or inhibit micro-organism invasion. Usually, the inhibiting mechanisms are able to clear the infection; however, sometimes, some bacteria or viruses can emerge and evade the host interferon response. The IFN response that helps host cells fight off invading pathogens occurs in two phases: an initial intracellular phase, in which infected cells produce IFNs, and an intercellular phase, in which infection-induced IFNs are secreted into the extracellular environment. Secreted IFNs bind to IFN receptors on surrounding cells, leading to the synthesis of proteins and more IFNs and resulting in a rapid clearance of pathogens [78]. When chronic, recurrent, hard-to-control, or unusually serious infections of common pathogens compared with the normal population occur, a potential deficiency in the IFN defense can be considered. Regardless of the type of IFN, the production of high titers of anti-IFN aAbs in serum interrupts the activation of the downstream response pathway through blocking the combination between IFNs and their receptor, resulting in increased infection rates [45]. Therefore, anti-IFN aAbs occur in previously healthy people who develop chronic, recurring, and difficult-to-control infections [45]. Although aAbs neutralizing the activity of IFN-I were detected for the first time about four decades ago in a 77-year-old woman with disseminated Herpes zoster virus [79], their presence was considered clinically silent in the general population until the Coronavirus Disease 19 (COVID-19) pandemic caused by SARS-CoV-2.

There are multiple methods for detecting aAbs against, including preferentially neutralizing assay [80], radioimmunoprecipitation and real time PCR [81], time-resolved immunofluorometric assay [82], radioimmunoassay [83], magnetic-beads-based assay [84], cell-based autoantibody assay (CBAA) [85], microarray-based assay [86], and homemade [5,87] and commercially available enzyme-linked immunosorbent assays (ELISA). These methods were developed and improved in recent years and are used principally to detect anti-IFN-I aAbs for diagnostic purposes.

Anti-IFN aAbs were initially found in patients treated with IFN-α or IFN-β [88], before being found in patients with chronic graft-versus-host disease following allogeneic bone marrow transplantation [89], myasthenia gravis [90], or thymoma [91] as well as in some women with SLE [43]. They were also previously detected in most patients with APS-1 [92]; in some patients with combined immunodeficiency due to hypomorphic recombination activating genes (RAG)-1 or RAG-2, Omenn's syndrome, leaky severe combined immunodeficiency, T-cell lymphopenia, and ataxia-telangiectasia [93]; in men with X-linked enteropathic polyendocrine immunodysregulation and forkhead box P3 (FOXP3) mutations [94]; and in women with incontinentia pigmenti and heterozygous "null" mutations in X-linked NEMO syndrome [5].

The biological mechanism behind anti-IFN-II aAbs' role in diseases remains unclear. Several studies showed that they are able to block the binding of IFN-γ to its receptor, which inhibits the early signal transduction, as well as the downstream biological consequences of IFN-II binding, which include the upregulation of tumor necrosis factor (TNF)-α and IL-12 production [95]. Accordingly, neutralizing anti-IFN-II aAbs are the cause of adult-onset immunodeficiency (AOID) and the associated Sweet syndrome [96], which are both characterized by the increased risk of opportunistic infections (OIs), as reported in the next chapter. The anti-IFN-II aAbs present in some healthy subjects did not show neutralizing activity or effects on IL-12 production [97].

To our knowledge, anti-IFN-III aAbs have so far only been studied in patients with rheumatologic diseases; however, in these patients, they did not exhibit neutralizing activity [43].

3.3. aAbs against IFNs in Infectious Diseases

IFNs have immunoregulatory functions during infection and immune responses, and, accordingly, defective activity of IFNs significantly contributes to infections' severity [1]. In particular, since IFN-I play a role in tightening barriers at mucosal interfaces, Abs that neutralize its activity can be linked with a negative outcome during respiratory infections [98].

It should be emphasized that aAbs directed against human IFN-α were first observed in a patient with VZV in 1981 [99]; only 20 years later, they were described in relation to the emergence of anti-IFN-I aAbs in HCV-infected patients during IFN-α treatment. Moreover, their presence was predictive of breakthrough despite an initial response to treatment, suggesting the pathogenic role of these aAbs in the loss efficacy of IFN-α therapy and HCV reactivation [100,101]. After a further 20 years, scientific interest moved to the role of anti-IFN aAbs in the course of respiratory infections. Indeed, recent data reported that neutralizing anti-IFN-I aAbs (against IFN-α2 alone or with IFN-ω) can be involved in critical influenza pneumonia, as they are present in about 5% of cases of life-threatening respiratory infections. After adjustments for age and sex, the presence of high concentrations of both IFN-α2- and IFN-ω-neutralizing aAbs were found to induce the highest risk of critical influenza pneumonia in patients <70 years old [102]. Similarly, these aAbs seem to play a role in complicating the severe respiratory syndrome, with a high fatality rate caused by Middle East respiratory coronavirus (MERS CoV) recorded. The 93.3% of MERS CoV-infected patients with anti-IFN-I aAbs (IFN-α2, IFN-β, and/or IFN-ω) were critically ill and needed to be admitted to the ICU, compared to just 66% of patients without aAbs. However, the presence of anti-IFN-I aAbs was not associated with different clinical outcomes or responses to treatment with IFN-β1b or antiviral drugs [103].

On the contrary, a low incidence (1.1%) of anti-IFN-I aAbs was found in a cohort of critically ill patients with acute respiratory failure. The study, however, included patients with both infectious (rhinovirus, influenza, parainfluenza, and seasonal coronavirus infections) and non-infectious etiologies [104]. The importance of anti-IFN-I aAbs during infections with severe pulmonary involvement was confirmed based on the fact that no increased prevalence of neutralizing anti-IFN-I aAbs were found in patients with idiopathic pulmonary fibrosis compared with general population [105].

IFN-I seems have a role during Flaviviruses infections, including those caused by yellow fever virus (YFV), dengue, West Nile, and Zika viruses [98], though no studies were performed that searched for anti-IFN-I aAbs during these emerging infections. However, anti-IFN-I aAbs were detected in one third of subjects with severe adverse events following vaccination for YFV, albeit only for vaccinations that used an attenuated alive virus. High titers of circulating aAbs against at least 14 of the several IFN-I were found. The authors demonstrated their IFN-I-neutralizing activity in vitro, blocking the protective effect of IFN-α2 against YFV vaccine strains [106]. Of interest is the potential link between the presence of anti-IFN-I aAbs and possible reactivation of latent virus infections, particularly HSV (CMV, HSV-1/2, or both) in critically ill COVID-19 patients [107].

Finally, our group found an elevated prevalence (11.6%) of anti-IFN-I aAbs in HIV-1-infected patients with OIs (neutralizing activity was not determined); however, no statistically significant differences were found for viro/immunological characteristics (CD4 and CD8 cell counts and viral load) between patients with and without anti-IFN-I aAbs (submitted). While the anti-IFN-II aAbs were first identified in a group of HIV-infected patients [40], those with neutralizing activity were first described in 2004 in the context of selective susceptibility to NTM infection [108]. Accordingly, neutralizing anti-IFN-II aAbs were detected in 88% of patients with multiple OIs in Asia [109] and in 62% of subjects with disseminated NTM and without HIV infection [110]. Nevertheless, the association between their presence and intracellular infections, especially in the contest of the onset of immunodeficiency, was clearly established years later. Indeed, it is well known that anti-IFN-II aAbs play a critical role in the pathogenesis of AOID, which is also referred to as AIDS-like syndrome [111], in which infectious diseases caused by opportunistic pathogens were firstly reported in adults without known immunodeficiency. Although genetic factors (HLA-DQB1*05:01 andHLA-DQB1*05:02) are associated with a very high risk of critical AOID and environmental exposure contributes to AOID, the autoimmunity caused by anti-IFN-II aAbs is critical for the syndrome's pathogenesis. It is also known that anti-IFN-II aAbs titers were strongly associated with the severity of infections, which was related to their neutralizing activity [112].

Recent findings confirmed the presence of IFN-II aAbs in the course of disseminated NTM disease, non-typhoid *Salmonella*, *Cryptococcus*, and VZV infections, particularly in Asian populations with AOID [113]. Interestingly, NTM is the most common pathogenic micro-organism in patients with AOID showing high titers of anti-IFN-II aAbs [113]. High titers of highly neutralizing anti-IFN-II aAbs were also reported by several groups in sporadic cases of disease caused by low-virulence *Mycobacteria* and MT, as well as after Bacillus Calmette–Guérin vaccination [44]. An association between anti-IFN-II aAbs and infections caused by other opportunistic pathogens, including those caused by other bacteria, e.g., *Burkholderia*, fungi, e.g., *Penicillium*, *Histoplasma*, and *Candida* sp., and viruses, in particular VZV and CMV, were also described in other studies [114,115].

To our knowledge, the presence of anti-IFN-III aAbs were described in patients with non-SARS-CoV-2 respiratory infections, where they seemed to display a neutralizing activity [52].

3.4. aAbs against IFN-I in COVID-19

The COVID-19 pandemic began at the end of January 2020, causing almost 7 million deaths worldwide and more than 680 million infections [116]. SARS-CoV-2 can cause infections of very different severities, ranging from asymptomatic forms to extremely serious cases that require hospitalization in ICU and cause death [117]. As IFNs represent the first line of defense during the early phase of viral infection, the levels of these cytokines were described as relevant in determining the outcome of SARS-CoV-2 infection. In particular, low levels of IFNs in the lungs or peripheral blood renders SARS-CoV-2 capable of evading innate recognition [118–120]. IFN-I levels may change due to many factors, i.e., age [121], gender [122], genetic defects in IFN-related encoding genes [123], and the presence of neutralizing anti-IFN-I aAbs [5]. Up to 10% of elderly patients were positive for anti-IFN-I aAbs (neutralizing IFN-α2, β and/or ω); this percentage may increase with COVID-19 severity [5,124,125]. The presence of anti-IFN-I aAbs mainly correlates to severity of COVID-19 in males over 65 years old [87,126]. Many compelling studies confirmed the link between anti-IFN-I aAbs and COVID-19 severity (Table 2).

Table 2. Number and percentage of positive samples for anti-IFN-I aAbs.

COVID-19 Gravity	Numbers	Country	IFN-I aAbs Tested	Neutralizing Activity against IFNs	Total	Percentage	Publication Year	Reference
Recovered	19	Colombia	IFN-α	No	5	26.3	2021	[127]
Severe	18		IFN-α	No	3	16.7		
All	172	USA	IFN-α and IFN-ω	Yes	9	5.2	2021	[128]
All	210	The Netherlands	IFN-α and IFN-ω	No	35	17	2021	[129]
All	35		IFN-α and IFN-ω	Yes	6	17		

Table 2. *Cont.*

COVID-19 Gravity	Numbers	Country	IFN-I aAbs Tested	Neutralizing Activity against IFNs	Total	Percentage	Publication Year	Reference
Severe	47	Spain	IFN-α and IFN-ω	10 ng/mL	5	10.6	2021	[130]
Critical	16		IFN-α and IFN-ω	10 ng/mL	3	18.7		
Severe	47		IFN-β	10 ng/mL	0	0		
Critical	16		IFN-β	10 ng/mL	0	0		
Convalescent	116	USA	IFN-α	No	4	3.0	2021	[131]
Convalescent	116		IFN-α and IFN-ω	10 ng/mL	2	1.5		
Critical	26	France	IFN-α and IFN-ω	Yes	8	30.7	2021	[132]
Severe	44	Italy	IFN-α, IFN-ω and IFN-β	Yes	2	4.5	2021	[133]
Critical	135			Yes	23	17		
Severe	84	France	IFN-α	No	21	25	2021	[134]
			IFN-α and IFN-ω	Yes	15	18		
Severe	623	Consortium	IFN-α and IFN-ω	10 ng/mL	22	3.53	2021	[87]
Critical	3136		IFN-α and IFN-ω	10 ng/mL	307	9.8		
Severe	522		IFN-α and IFN-ω	100 pg/mL	34	6.5		
Critical	3595		IFN-α and IFN-ω	100 pg/mL	489	13.6		
Severe	187		IFN-β	10 ng/mL	0	0		
Critical	1773		IFN-β	10 ng/mL	23	1.3		
All	8	USA	IFN-α and IFN-ω	No	1	12.8	2021	[135]
N/A	51	USA and Germany	IFN-α	No	23	45	2021	[136]
Severe	102	Several	IFN-α	10 ng/mL	6	6	2021	[137]
Critical	26		IFN-α	10 ng/mL	5	19		
Critical	275	Spain	IFN-α and IFN-ω	no	49	17.8	2021	[138]
Critical	275			10 ng/mL	26	9.5		
Severe	49	USA	IFN-α and IFN-ω	Yes	4	8.2	2021	[139]
Critical	86	Russia	IFN-α and IFN-ω	No	9	10.5	2021	[140]
Critical	47	France	IFN-α and IFN-ω	Yes	2	4.2	2022	[141]
Critical	139	France	IFN-α and IFN-ω	no	107	77	2022	[142]
			IFN-α and IFN-ω	10 ng and 100 pg/mL	11	7.9		
			IFN-β	10 ng/mL	0	0		
Deceased	11		IFN-α, IFN-ω	10 ng/mL	6	55		
Severe	70	Russia	IFN-α	no	13	18	2022	[143]
Severe	97	The Netherlands	IFN-α	Yes	7	7	2022	[144]
Fatal	38			Yes	5	13		
Severe	52	Belgium	IFN-α	?	8	15.3	2022	[145]
All	360	Italy	IFN-α	No	27	7.5	2022	[146]
			IFN-α	Yes	13	3.6		
			IFN-β	No	37	10.3		
			IFN-β	Yes	1	0.3		
Critical	237	Germany	IFN-α and IFN-ω	Yes	18	7.5	2022	[147]
Severe	235	Japan	IFN-α and IFN-ω	10 ng/mL	5	2.1	2022	[148]
Critical	170		IFN-α and IFN-ω	10 ng/mL	10	5.9		
Severe	235		IFN-α and IFN-ω	100 pg/mL	6	2.6		
Critical	170		IFN-α and IFN-ω	100 pg/mL	18	10.6		
Critical	103	Switzerland	IFN-α and IFN-ω	yes	11	10.7	2022	[107]
			IFN-β	yes	0	0		
Severe/	16	France	IFN-α	10^2 pg/mL	4	25	2022	[149]
critical in	13		IFN-ω	10^2 pg/mL	4	31		
SLE	12		IFN-β	10^4 pg/mL	2	17		
Critical	925	France	IFN-α and IFN-ω	Yes	96	10.3	2022	[124]

SLE: Systemic lupus erythematosus.

In a recent meta-analysis including more than 7700 patients, the positive rate of anti-IFN-I aAbs was found to be 5% (95% CI, 3–8%); however, this rate reached 10% (95% CI, 7–14%) when analysis was restricted to patients with severe infections [150].

The evidence that anti-IFN-I aAbs are capable of altering the course of COVID-19 through perturbing the immune response to SARS-CoV-2 and tissue homeostasis was also provided through data obtained regarding these aAbs with mouse surrogates, which led to increased disease severity in a mouse model of SARS-CoV-2 infection [128].

Nonetheless, different data are provided by a recent investigation that demonstrated the presence of autoreactive polyclonal B-cell activation and aAbs production, but did not demonstrate a correlation between anti-IFN-I aAbs levels and COVID-19 severity [151]. These conflicting results could be explained via different assays used to detect aAbs with high or low affinity, different patient populations tested, or the different contribution to disease outcomes of pre-existing or infection-induced neutralizing anti-IFN-I aAbs.

We also studied whether the presence of anti-IFN-I aAbs could have a role in SARS-CoV-2 breakthrough infections in vaccinated patients. Breakthroughs were reported worldwide, with most of them being asymptomatic or mild cases; the few breakthrough critical cases were mainly described in immune-depressed patients. A total of 20% of these breakthroughs can occur in patients with normal antibody response to the vaccine who also carry aAbs neutralizing IFN-α2 and/or IFN-ω [87].

Finally, the presence of anti-IFN-I aAbs was previously proposed as a possible driver of post-acute COVID-19, which is also known as "long COVID" [152–154], in which a persistent immune response seems be the inducing mechanism [155]. However, anti-IFN-α2 aAbs were uncommon in long-COVID patients [156].

Until now, the relationship between anti-IFN-II aAbs and COVID-19 severity was little explored. Only a pilot study in Taiwan described the presence of anti-IFN-II aAbs with neutralizing activity in the 18% of COVID-19 patients with severe/critical illness. The prevalence was statistically higher compared with non-severe COVID-19 patients or healthy controls. Moreover, median titers of anti-IFN-II aAbs were higher in severe/critical patients than in patients with mild/moderate disease or healthy controls [157].

Anti-IFN-III aAbs were recently detected via the Molecular Indexing of Proteins by Self-Assembly technology in patients with life-threatening COVID-19, though they were not detected in plasma samples of healthy subjects or convalescent plasma from non-hospitalized individuals with COVID-19 [158].

4. Significance of the Production of aAbs against IFNs

4.1. Is There a Gender Bias for aAbs Production?

Recent results demonstrated that aAbs production among healthy subjects did not show a gender bias because the median numbers and the weighted prevalence of 77 common aAbs were similar between males and females [159]. This result stands in contrast to the evidences that autoimmune diseases disproportionally affect females compared with males. While the risk of contracting autoimmune diseases is up to four times greater in women than in men, the mechanism for this sex bias is still obscure. Several hypotheses were previously proposed, including that women have an evolutionarily conserved tendency toward an enhanced activation of B cells resulting in higher levels of antibody production, which may be responsible for the increased incidence of antibody-driven autoimmune diseases [160]. This suggestion agreed with the fact that autoimmune diseases in females are associated with antibody-mediated pathology, whereas in males they are preferentially associated with acute inflammation [161].

Gender differences in circulating anti-cytokines aAbs were observed in atherosclerosis, with an upregulation of anti-TNF-α, anti-IL-1α, and anti-IL-1β IgG levels are more likely to occur in female than in male patients [162]. Similarly, among patients with acquired pulmonary alveolar proteinosis, which is an ultrarare autoimmune disease characterized by accumulation of excess surfactant in the alveoli, leading to pulmonary insufficiency, men are predominantly affected (male:female ratio of 65:1), and high levels of aAbs that neutralize GM-CSF signaling were detected [163]. In addition, 10 of 14 patients with severe disseminated NTB and no other evidence of immunodeficiency who produced anti-IFN-II aAbs were females [164]. Finally, only 2.6% of females with life-threatening COVID-19 showed anti-IFN-I aAbs, compared to 12.5% of males [5]. In particular, neutralizing anti-IFN-I aAbs were detected in 94% of males with critical COVID-19 pneumonia. The recently published meta-analysis [150] confirmed that the generally (not calculated according to disease severity) higher prevalence of neutralizing aAbs in males (5%) than in females (2%). Different data were obtained regarding 130 critically ill COVID-19 Swiss patients, in which 11.3% of males and 13.0% of females showed detectable anti-IFN-α2 aAbs, while 7.5% of males and 8.7% of females presented anti-IFN-ω aAbs in their plasma, which were not present in plasma of 130 healthy donors [107]. Moreover, in our laboratory, we tested the presence of anti-IFN-I aAbs in 349 critically ill male and female COVID-19 patients (Table 3; previously unpublished data). We found a percentage of anti-IFN-I aAbs-positive patients comparable to that of previously published studies (see Table 2); however, unlike Busnadiego et al. [107], we did not observe a gender difference in our patients. This result can be due to the fact that females may carry more non-neutralizing anti-IFN-I aAbs and that the female age was slightly higher than that of males. Furthermore, females also appeared to have a more critical disease, since 50% (vs. 35% of males) of them died, although this difference was not significant.

Table 3. Characteristics of male and female patients tested for presence of anti-IFN-I-aAbs.

	Males	**Females**	*p*-**Value**
Critical patients, number (%)	219 (52)	130 (45)	0.092
Age (years), mean ($\pm$SD)	72 $\pm$ 13	75 $\pm$ 12	0.023
Age (years) range	40–98	38–99	-
Anti-IFN-I aAbs, number (%)	28 (13)	16 (12)	1.000
Age (years), mean ($\pm$SD)	75 $\pm$ 10	78 $\pm$ 14	0.332
Deceased, number (%)	10 (36)	8 (50)	0.525
ICU	14 (50)	3 (19)	0.057
Days of hospitalization, mean ($\pm$SD)	27 $\pm$ 20	23 $\pm$ 16	0.477
Vaccinated for SARS-CoV-2, number (%)	14 (50)	10 (63)	0.534
Cardiovascular diseases	13 (46)	5 (31)	0.361
Hypertension	13 (46)	8 (50)	1.000
Dyslipidemia	8 (29)	0 (0)	0.036
Diabetes	11 (39)	2 (13)	0.089
Solid tumor	5 (18)	4 (25)	0.702
Neurologic diseases	6 (21)	6 (38)	0.303
COVID-19 complications	25 (89)	14 (88)	1.000
Acute respiratory distress syndrome	25 (89)	14 (88)	1.000

Analyses of aAbs against IFN-α2 and IFN-ω were performed using ELISA method [5,87], with few modifications. Means were compared via *t*-test, while proportions were compared via Fisher's exact test (differences with $p < 0.05$ are considered significant and shown in bold font).

Notably, anti-IL-6 aAbs were predominantly elevated in asymptomatic COVID-19 females [165]. However, the reported data, together with those included in another study [166], revealed that despite classic autoimmune diseases being more prevalent in females, a paradoxical male predominance of autoimmune activation illness is present in the setting of severe COVID-19.

4.2. Are Infection-Related aAbs Associated to Immunosenescence?

Age-associated changes in the immune system heavily increase the risk of bacterial and viral infections in the elderly [167]; however, the relationship between aAbs production, aging, and infectious diseases has not yet been formally demonstrated. Moreover, reasons for the generation of aAbs are not exactly clarified, although several hypotheses were proposed. Potential hypotheses are as follows: (a) tolerance defects and inflammation; (b) modification of antigen expression; (c) changes in exposure or presentation of antigens;

(d) cellular death mechanisms; (e) combination of genetic and environmental factors (e.g., simultaneous exposure to microorganisms with certain toxins and hazardous chemicals); and (f) infections with viral proteins with sequences similar to a human protein (in around 20 autoimmune diseases, aAbs are generated due to cross-reactivity to infectious agent proteins) [168,169]. The thymic involution, which is a naturally occurring part of the aging process that leads to reduced thymus activity, also increased the likelihood of high autoimmune incidence [170]. Interestingly, the rate of thymic involution can be regulated via numerous growth hormones and sex steroids, as well as via metabolic activity, and involution appears to occur more rapidly in males than females [170], thus further supporting a gender bias in aAbs induction (see previous section). Therefore, those factors that reduce thymus activity, including some cancers, age, sex or certain other diseases/disorders and lifestyles, which are associated with immunosenescence, could also increase individual risk of developing aAbs.

In the last few years, increased levels of aAbs were also associated with the accumulation of Age-Associated B Cells (ABCs) [171], which are one of the immune changes that characterize the immunosenescence [172]. Indeed, ABCs were observed to secrete aAbs [173], which were first described more than 50 years ago [174,175]. It is also of note that ABC-like cells producing aAbs were identified during HCV infection [176], and that ABCs expand in the presence of viral infections not associated with autoimmune disease, such as murine CMV [177], influenza virus [178], HIV-1 [179], and SARS-CoV-2 [180]. In this last infection, the expansion of ABCs may be responsible for the increased levels of anti-IFN-I aAbs, which are associated with the higher risk of critical COVID-19 infection in the elderly population [5,87].

However, the age-associated increase in these specific aAbs is not restricted to patients with ongoing infections, being also observed in apparently healthy populations. Indeed, the prevalence of anti-IFN-I aAbs neutralizing 100 ng/mL of IFN-α and/or IFN-ω increased from 1.1% in individuals under the age of 70 years old to more than 4.4% in those over the age of 70 years old, and up to 7.1% for those with aged between 80 and 85 years old [5,123,137]. The reasons why this prevalence decreases in patients aged >85 are not clear; one cause can be the fact that most of individuals died before the COVID-19 pandemic from other illnesses aggravated due to the presence of the aAbs, with only aAbs-negative ones remaining only [123]. Another plausible explanation is that "long-lived" individuals have passed a certain threshold of age selection and, therefore, their physiological parameters might differ noticeably from those of the general population. This possibility is supported by the increased percentages of naive T cells in the CD4+ T-cell subset, higher prevalence of low frequency clonotypes, and slightly higher T-cell receptor diversity observed in healthy individuals with an average age of 82 years old compared to younger subjects [181].

However, all currently reported information reported is in contrast with trends recently demonstrated in healthy individuals, in whom the number of aAbs increased with age from infancy to adolescence, when they reached a plateau [159]. This specific observation suggests that while the response to infectious agents (and may be vaccines) might contribute to the production of aAbs through molecular mimicry, this mechanism does not appear to continue to accumulate aAbs throughout life.

4.3. Are aAbs Detected during Infections Dangerous or Protective?

The ability of anti-IFN-I aAbs to neutralize soluble IFN-I from binding to their receptor on the surface of cells was proposed as the most straightforward mechanism through which these aAbs could promote virus replication and subsequent disease [182]. However, the significance of aAbs production and their function was not definitively established, since they may mediate diverse immunological functions depending on their specific interaction with the target cytokine. Circulating cytokine/aAbs immune complexes are probably in equilibrium with their free cytokine and free aAbs in concentrations that vary based on the levels of cytokine that need to be neutralized. In addition, their concentration, epitope specificity, avidity, isotype, and subclass may influence the capacity of these molecules to neutralize their related cytokine. It was also suggested that aAbs produced against cytokines may play a role in the physiological regulation of their biological activities via neutralizing the targets or prolonging the half-life [183].

Anti-cytokine aAbs with a moderate affinity for self-antigens provide a first line of defense against infections, probably have housekeeping functions, and contribute to the homeostasis of the immune system [4]. Furthermore, they may play a direct pathogenic role in susceptibility to infection, rather than arising as an immune response to the pathogenic micro-organism, because, in APS-1, they may be detected prior to the development of the associated infectious disease [184]. It is important to take into account that the biological significance of anti-cytokine aAbs must be evaluated in the context of disease, because they may play a role in modulating disease activity in autoimmune conditions and may also increase susceptibility to infections in certain immune-deficient patients [183]. In addition, the function of anti-cytokines aAbs during infections could theoretically be very different: they could be beneficial, harmful, or have both effects. The final case is supported by a large body of evidence indicating that aAbs induced during malarial infection are associated with disease severity and clinical outcomes, but are also capable of mediating protection against *Plasmodium* sp. [185]. It was also proposed that aAbs against inflammatory cytokines might protect against untoward inflammation [4] and be generated when such a response is of benefit for the host. On the contrary, the new aAbs, including anti-cytokine aAbs, which were recently detected in severe COVID-19, may directly cause harm, such as blood clotting, blood vessel inflammation, and tissue damage [136], and seemed to be associated with long-COVID symptoms [186], even if these data are controversial [155]. These aAbs perturb immune function and impair virological control via inhibiting immunoreceptor signaling and altering peripheral immune cell composition [128].

Finally, it is important to remember the possibility that the production of aAbs can be limited in time and can stop when the triggering stimulus fails. For instance, while anti-IFN-I aAbs remained stable in patients with AIRE deficiency and thymic malignancies, anti-IFN-I aAbs with neutralizing activity peaked soon after COVID-19 onset and declined to undetectable levels during convalescence [187]. Therefore, depending on context, these aAbs may serve beneficial "housekeeping" functions through removing surplus danger signals from circulation or, conversely, induce disease emergence.

The detection of aAbs can be useful for early diagnosis and prognosis, as observed in cardiovascular diseases and cancers, because they may be detected well in advance of clinical manifestations, enabling earlier identification of patients that may benefit from effective treatments with a targeted approach [188,189]. Therefore, aAbs found at an early stage of COVID-19 and non-COVID-19 infections seem to predict the disease severity and possible long-term effects, thus potentially facilitating more effective therapy [52,136]. Accordingly, anti-cytokine aAbs (e.g., antibodies against IFN-α, IFN-ϵ, IL-6, IL-22, GM-CSF, and TNF-α) were proposed for COVID-19 and non-COVID-19 treatment, although with different results, since some improved clinical outcomes, while others had no benefit [190]. For instance, subcutaneous IFN-β treatment of hospitalized patients did not seem to improve COVID-19 clinical outcomes [130]. Interestingly, in a phase II clinical trial, a single dose of pegylated IFN-III as an early antiviral treatment for COVID-19 did not inhibit or increase B-cell antibody responses measured in plasma, instead accelerating viral clearance [191]. Therefore, it was proposed that IFN-III could potentially be a superior choice of treatment compared to other IFN-I in SARS-CoV-2-infected patients [192].

Recently, the presence of aAbs against specific chemokines in Italian and Swiss subjects was found of help in the identification of convalescent individuals with favorable acute and long-COVID disease courses. Anti-chemokine monoclonal antibodies derived from these individuals block leukocyte migration and, thus, may be beneficial through modulation of the inflammatory response [153].

5. Conclusions

The extent of the COVID-19 pandemic, the large availability of biological samples enriched by clinical information, and the allocation of a large amount of research funds and dedicated staff rapidly advanced our knowledge of the balance between direct SARS-CoV-2-induced damage and inflammatory responses triggered by the virus and how this balance contributes to the broad spectrum of disease severity. The studies carried out subsequently made possible to detect a high prevalence of anti-IFN-I aAbs in both COVID-19 and the serum of patients with other non-SARS-CoV-2 viral, bacterial, and fungal infections.

aAbs can have different roles in these diseases and can act both as enemies or friends, i.e., capable of acting in a beneficial or harmful way. Establishing the levels and stability of aAbs could be useful to discriminate between the two possibilities, as it would be help us to understand the characteristics of the other components of the immune system of subjects in whom aAbs are identified. Indeed, their discovery in elderly subjects indicates that they may be linked to other defects related to immunosenescence or inflammaging.

Several questions remain to be answered regardin the significance of the presence of anti-IFNs aAbs in healthy individuals and infected patients. For instance, it must be elucidated why naturally-occurring anti-IFN-I aAbs were mainly described as binding the IFN-β subtypes and/or IFN-ω, while aAbs against IFN-β or other IFN types, such as IFN-III, appeared to be much rarer [5,87,158]. Furthermore, it must be determined whether there might be a normal physiological role for low or transient levels of anti-IFN-I aAb. Therefore, future studies on different infections and involving a larger number of patients are desirable to define the impact, long-term duration, and clinical implications of the production of such aAbs. Understanding the immunological mechanisms underlying the production of aAbs could improve the approach to some infections, focusing not only on pathogens, but also on a possible low degree of autoimmunity in these patients.

Author Contributions: Each author has made substantial contributions to the conception or design of the manuscript and has approved the submission. Conceptualization: E.Q.-R. and L.I.; Writing—Original Draft Preparation: E.Q.-R. and L.I.; Methodology: A.S. and S.G.S.; Data curation: F.S.; Writing—Review and Editing, E.Q.-R., A.S., S.G.S., F.S., G.T. and L.I. All authors have read and agreed to the published version of the manuscript.

Funding: This research received no external funding.

Institutional Review Board Statement: Not applicable.

Informed Consent Statement: Not applicable.

Data Availability Statement: Not applicable.

Acknowledgments: We thank Michail S. Lionakis for providing sera from APS-1 patients, which were used as controls in the anti-IFN-I aAbs assay, and Paul Bastard for providing assistance in test development.

Conflicts of Interest: The authors declare no conflict of interest.

References

1. Casanova, J.L.; Abel, L. Mechanisms of viral inflammation and disease in humans. *Science* **2021**, *374*, 1080–1086. [CrossRef]
2. Liston, A.; Humblet-Baron, S.; Duffy, D.; Goris, A. Human immune diversity: From evolution to modernity. *Nat. Immunol.* **2021**, *22*, 1479–1489. [CrossRef] [PubMed]
3. Katze, M.G.; He, Y.; Gale, M., Jr. Viruses and interferon: A fight for supremacy. *Nat. Rev. Immunol.* **2002**, *2*, 675–687. [CrossRef] [PubMed]
4. Elkon, K.; Casali, P. Nature and functions of autoantibodies. *Nat. Clin. Pract. Rheumatol.* **2008**, *4*, 491–498. [CrossRef] [PubMed]
5. Bastard, P.; Rosen, L.B.; Zhang, Q.; Michailidis, E.; Hoffmann, H.H.; Zhang, Y.; Dorgham, K.; Philippot, Q.; Rosain, J.; Béziat, V.; et al. Autoantibodies against type I IFNs in patients with life-threatening COVID-19. *Science* **2020**, *370*, eabd4585. [CrossRef]
6. Nemazee, D. Mechanisms of central tolerance for B cells. *Nat. Rev. Immunol.* **2017**, *17*, 281–294. [CrossRef]
7. Nagele, E.P.; Han, M.; Acharya, N.K.; DeMarshall, C.; Kosciuk, M.C.; Nagele, R.G. Natural IgG autoantibodies are abundant and ubiquitous in human sera, and their number is influenced by age, gender, and disease. *PLoS ONE* **2013**, *8*, e60726. [CrossRef]

8. Oppezzo, P.; Dighiero, G. Autoanticorps, tolérance et auto-immunité [Autoantibodies, tolerance and autoimmunity]. *Pathol. Biol.* **2003**, *51*, 297–304. (In French) [CrossRef]

9. Olson, J.K.; Croxford, J.L.; Calenoff, M.A.; Dal Canto, M.C.; Miller, S.D. A virus-induced molecular mimicry model of multiple sclerosis. *J. Clin. Investig.* **2001**, *108*, 311–318. [CrossRef]

10. Fujinami, R.S.; Oldstone, M.B.; Wroblewska, Z.; Frankel, M.E.; Koprowski, H. Molecular mimicry in virus infection: Crossreaction of measles virus phosphoprotein or of herpes simplex virus protein with human intermediate filaments. *Proc. Natl. Acad. Sci. USA* **1983**, *80*, 2346–2350. [CrossRef]

11. Atkinson, M.A.; Bowman, M.A.; Campbell, L.; Darrow, B.L.; Kaufman, D.L.; Maclaren, N.K. Cellular immunity to a determinant common to glutamate decarboxylase and coxsackie virus in insulin-dependent diabetes. *J. Clin. Investig.* **1994**, *94*, 2125–2129. [CrossRef] [PubMed]

12. Fielder, M.; Pirt, S.J.; Tarpey, I.; Wilson, C.; Cunningham, P.; Ettelaie, C.; Binder, A.; Bansal, S.; Ebringer, A. Molecular mimicry and ankylosing spondylitis: Possible role of a novel sequence in pullulanase of Klebsiella pneumoniae. *FEBS Lett.* **1995**, *369*, 243–248. [CrossRef] [PubMed]

13. McClain, M.T.; Heinlen, L.D.; Dennis, G.J.; Roebuck, J.; Harley, J.B.; James, J.A. Early events in lupus humoral autoimmunity suggest initiation through molecular mimicry. *Nat. Med.* **2005**, *11*, 85–89. [CrossRef] [PubMed]

14. Doxey, A.C.; McConkey, B.J. Prediction of molecular mimicry candidates in human pathogenic bacteria. *Virulence* **2013**, *4*, 453–466. [CrossRef]

15. Venigalla, S.S.K.; Premakumar, S.; Janakiraman, V. A possible role for autoimmunity through molecular mimicry in alphavirus mediated arthritis. *Sci. Rep.* **2020**, *10*, 938–950. [CrossRef]

16. Haury, M.; Sundblad, A.; Grandien, A.; Barreau, C.; Coutinho, A.; Nobrega, A. The repertoire of serum IgM in normal mice is largely independent of external antigenic contact. *Eur. J. Immunol.* **1997**, *27*, 1557–1563. [CrossRef]

17. Palma, J.; Tokarz-Deptuła, B.; Deptuła, J.; Deptuła, W. Natural antibodies—Facts known and unknown. *Cent. Eur. J. Immunol.* **2018**, *43*, 466–475. [CrossRef]

18. Griffin, D.O.; Holodick, N.E.; Rothstein, T.L. Human B1 cells in umbilical cord and adult peripheral blood express the novel phenotype CD20+ CD27+ CD43+ CD70−. *J. Exp. Med.* **2011**, *208*, 67–80. [CrossRef]

19. Coutinho, A.; Kazatchkine, M.D.; Avrameas, S. Natural autoantibodies. *Curr. Opin. Immunol.* **1995**, *7*, 812–818. [CrossRef]

20. Shoenfeld, Y.; Toubi, E. Protective autoantibodies: Role in homeostasis, clinical importance, and therapeutic potential. *Arthritis Rheumatol.* **2005**, *52*, 2599–2606. [CrossRef]

21. Watanabe, M.; Uchida, K.; Nakagaki, K.; Trapnell, B.C.; Nakata, K. High avidity cytokine autoantibodies in health and disease: Pathogenesis and mechanisms. *Cytokine Growth Factor Rev.* **2010**, *21*, 263–273. [CrossRef] [PubMed]

22. Cappellano, G.; Orilieri, E.; Woldetsadik, A.D.; Boggio, E.; Soluri, M.F.; Comi, C.; Sblattero, D.; Chiocchetti, A.; Dianzani, U. Anti-cytokine autoantibodies in autoimmune diseases. *Am. J. Clin. Exp. Immunol.* **2012**, *1*, 136–146. [PubMed]

23. Tabuchi, Y.; Shimoda, M.; Kagara, N.; Naoi, Y.; Tanei, T.; Shimomura, A.; Shimazu, K.; Kim, S.J.; Noguchi, S. Protective effect of naturally occurring anti-HER2 autoantibodies on breast cancer. *Breast Cancer Res. Treat.* **2016**, *157*, 55–63. [CrossRef]

24. Gillissen, M.A.; de Jong, G.; Kedde, M.; Yasuda, E.; Levie, S.E.; Moiset, G.; Hensbergen, P.J.; Bakker, A.Q.; Wagner, K.; Villaudy, J.; et al. Patient-derived antibody recognizes a unique CD43 epitope expressed on all AML and has antileukemia activity in mice. *Blood Adv.* **2017**, *1*, 1551–1564. [CrossRef] [PubMed]

25. von Mensdorff-Pouilly, S.; Verstraeten, A.A.; Kenemans, P.; Snijdewint, F.G.; Kok, A.; Van Kamp, G.J.; Paul, M.A.; Van Diest, P.J.; Meijer, S.; Hilgers, J. Survival in early breast cancer patients is favorably influenced by a natural humoral immune response to polymorphic epithelial mucin. *J. Clin. Oncol.* **2000**, *18*, 574–583. [CrossRef]

26. Britschgi, M.; Olin, C.E.; Johns, H.T.; Takeda-Uchimura, Y.; LeMieux, M.C.; Rufibach, K.; Rajadas, J.; Zhang, H.; Tomooka, B.; Robinson, W.H.; et al. Neuroprotective natural antibodies to assemblies of amyloidogenic peptides decrease with normal aging and advancing Alzheimer's disease. *Proc. Natl. Acad. Sci. USA* **2009**, *106*, 12145–12150. [CrossRef]

27. Ercolini, A.M.; Miller, S.D. The role of infections in autoimmune disease. *Clin. Exp. Immunol.* **2009**, *155*, 1–15. [CrossRef]

28. Kazarian, M.; Laird-Offringa, I.A. Small-cell lung cancer-associated autoantibodies: Potential applications to cancer diagnosis, early detection, and therapy. *Mol. Cancer* **2011**, *10*, 33–52. [CrossRef]

29. Menconi, F.; Marcocci, C.; Marinò, M. Diagnosis and classification of Graves' disease. *Autoimmun. Rev.* **2014**, *13*, 398–402. [CrossRef]

30. De Virgilio, A.; Greco, A.; Fabbrini, G.; Inghilleri, M.; Rizzo, M.I.; Gallo, A.; Conte, M.; Rosato, C.; Ciniglio Appiani, M.; de Vincentiis, M. Parkinson's disease: Autoimmunity and neuroinflammation. *Autoimmun. Rev.* **2016**, *15*, 1005–1011. [CrossRef]

31. Leslie, R.D.; Palmer, J.; Schloot, N.C.; Lernmark, A. Diabetes at the crossroads: Relevance of disease classification to pathophysiology and treatment. *Diabetologia* **2016**, *59*, 13–20. [CrossRef] [PubMed]

32. Ludwig, R.J.; Vanhoorelbeke, K.; Leypoldt, F.; Kaya, Z.; Bieber, K.; McLachlan, S.M.; Komorowski, L.; Luo, J.; Cabral-Marques, O.; Hammers, C.M.; et al. Mechanisms of autoantibody-induced pathology. *Front. Immunol.* **2017**, *8*, 603–645. [CrossRef] [PubMed]

33. Meier, L.A.; Binstadt, B.A. The contribution of autoantibodies to inflammatory cardiovascular pathology. *Front. Immunol.* **2018**, *9*, 911–925. [CrossRef] [PubMed]

34. Ansari, R.; Rosen, L.B.; Lisco, A.; Gilden, D.; Holland, S.M.; Zerbe, C.S.; Bonomo, R.A.; Cohen, J.I. Primary and acquired immunodeficiencies associated with severe varicella-zoster virus infections. *Clin. Infect. Dis.* **2021**, *73*, e2705–e2712. [CrossRef]

35. Ikeda, Y.; Toda, G.; Hashimoto, N.; Umeda, N.; Miyake, K.; Yamanaka, M.; Kurokowa, K. Naturally occurring anti-interferon-alpha 2a antibodies in patients with acute viral hepatitis. *Clin. Exp. Immunol.* **1991**, *85*, 80–84. [CrossRef]
36. Hansen, M.B.; Svenson, M.; Bendtzen, K. Human anti-interleukin 1 alpha antibodies. *Immunol. Lett.* **1991**, *30*, 133–139. [CrossRef]
37. Bayat, A.; Burbelo, P.D.; Browne, S.K.; Quinlivan, M.; Martinez, B.; Holland, S.M.; Buvanendran, A.; Kroin, J.S.; Mannes, A.J.; Breuer, J.; et al. Anti-cytokine autoantibodies in postherpetic neuralgia. *J. Transl. Med.* **2015**, *13*, 333–341. [CrossRef]
38. Kärner, J.; Pihlap, M.; Ranki, A.; Krohn, K.; Trebusak Podkrajsek, K.; Bratanic, N.; Battelino, T.; Willcox, N.; Peterson, P.; Kisand, K. IL-6-specific autoantibodies among APECED and thymoma patients. *Immun. Inflamm. Dis.* **2016**, *4*, 235–243. [CrossRef]
39. von Stemann, J.H.; Rigas, A.S.; Thørner, L.W.; Rasmussen, D.G.K.; Pedersen, O.B.; Rostgaard, K.; Erikstrup, C.; Ullum, H.; Hansen, M.B. Prevalence and correlation of cytokine-specific autoantibodies with epidemiological factors and C-reactive protein in 8,972 healthy individuals: Results from the Danish Blood Donor Study. *PLoS ONE* **2017**, *12*, e0179981. [CrossRef]
40. Caruso, A.; Foresti, I.; Gribaudo, G.; Bonfanti, C.; Pollara, P.; Dolei, A.; Landolfo, S.; Turano, A. Anti-interferon-gamma antibodies in sera from HIV infected patients. *J. Biol. Regul. Homeost. Agents* **1989**, *3*, 8–12.
41. Sunder-Plassmann, G.; Sedlacek, P.L.; Sunder-Plassmann, R.; Derfler, K.; Swoboda, K.; Fabrizii, V.; Hirschl, M.M.; Balcke, P. Anti-interleukin-1 alpha autoantibodies in hemodialysis patients. *Kidney Int.* **1991**, *40*, 787–791. [CrossRef]
42. Takasaki, J.; Ogawa, Y. Anti-interleukin-8 autoantibody in the tracheobronchial aspirate of infants with chronic lung disease. *Pediatr. Int.* **2001**, *43*, 48–52. [CrossRef] [PubMed]
43. Gupta, S.; Tatouli, I.P.; Rosen, L.B.; Hasni, S.; Alevizos, I.; Manna, Z.G.; Rivera, J.; Jiang, C.; Siegel, R.M.; Holland, S.M.; et al. Distinct functions of autoantibodies against interferon in systemic lupus erythematosus: A comprehensive analysis of anticytokine autoantibodies in common rheumatic diseases. *Arthritis Rheumatol.* **2016**, *68*, 1677–1687. [CrossRef] [PubMed]
44. Ku, C.L.; Chi, C.Y.; von Bernuth, H.; Doffinger, R. Autoantibodies against cytokines: Phenocopies of primary immunodeficiencies? *Hum. Genet.* **2020**, *139*, 783–794. [CrossRef] [PubMed]
45. Chen, L.F.; Yang, C.D.; Cheng, X.B. Anti-interferon autoantibodies in adult-onset immunodeficiency syndrome and severe COVID-19 infection. *Front. Immunol.* **2021**, *12*, 788368. [CrossRef] [PubMed]
46. Bost, K.L.; Hahn, B.H.; Saag, M.S.; Shaw, G.M.; Weigent, D.A.; Blalock, J.E. Individuals infected with HIV possess antibodies against IL-2. *Immunology* **1988**, *65*, 611–615. [PubMed]
47. Salvator, H.; Cheng, A.; Rosen, L.B.; Williamson, P.R.; Bennett, J.E.; Kashyap, A.; Ding, L.; Kwon-Chung, K.J.; Namkoong, H.; Zerbe, C.S.; et al. Neutralizing GM-CSF autoantibodies in pulmonary alveolar proteinosis, cryptococcal meningitis and severe nocardiosis. *Respir. Res.* **2022**, *23*, 280–289. [CrossRef]
48. Kuo, C.Y.; Wang, S.Y.; Shih, H.P.; Tu, K.H.; Huang, W.C.; Ding, J.Y.; Lin, C.H.; Yeh, C.F.; Ho, M.W.; Chang, S.C.; et al. Disseminated cryptococcosis due to anti-granulocyte-macrophage colony-stimulating factor autoantibodies in the absence of pulmonary alveolar proteinosis. *J. Clin. Immunol.* **2017**, *37*, 143–152. [CrossRef]
49. Nanki, T.; Onoue, I.; Nagasaka, K.; Takayasu, A.; Ebisawa, M.; Hosoya, T.; Shirai, T.; Sugihara, T.; Hirata, S.; Kubota, T.; et al. Suppression of elevations in serum C reactive protein levels by anti-IL-6 autoantibodies in two patients with severe bacterial infections. *Ann. Rheum. Dis.* **2013**, *72*, 1100–1102. [CrossRef]
50. Bloomfield, M.; Parackova, Z.; Cabelova, T.; Pospisilova, I.; Kabicek, P.; Houstkova, H.; Sediva, A. Anti-IL6 autoantibodies in an infant with CRP-less septic shock. *Front. Immunol.* **2019**, *10*, 2629–2635. [CrossRef]
51. Ling, Y.; Puel, A. IL-17 and infections. *Actas Dermosifiliogr.* **2014**, *105*, 34–40. [CrossRef] [PubMed]
52. Feng, A.; Yang, E.Y.; Moore, A.R.; Dhingra, S.; Chang, S.E.; Yin, X.; Pi, R.; Mack, E.K.; Völkel, S.; Geßner, R.; et al. Autoantibodies are highly prevalent in non-SARS-CoV-2 respiratory infections and critical illness. *JCI Insight* **2023**, *8*, e163150. [CrossRef] [PubMed]
53. Negishi, H.; Taniguchi, T.; Yanai, H. The interferon (IFN) class of cytokines and the IFN regulatory factor (IRF) transcription factor family. *Cold Spring Harb. Perspect. Biol.* **2018**, *10*, a028423. [CrossRef] [PubMed]
54. GeneCards®: The Human Gene Database. Available online: https://www.genecards.org/Search/Keyword?queryString=interferon (accessed on 26 April 2023).
55. Dell'Oste, V.; Biolatti, M.; Galitska, G.; Griffante, G.; Gugliesi, F.; Pasquero, S.; Zingoni, A.; Cerboni, C.; De Andrea, M. Tuning the orchestra: HCMV vs. innate immunity. *Front. Microbiol.* **2020**, *11*, 661–681. [CrossRef] [PubMed]
56. Meylan, P.R.; Guatelli, J.C.; Munis, J.R.; Richman, D.D.; Kornbluth, R.S. Mechanisms for the inhibition of HIV replication by interferons-alpha, -beta, and -gamma in primary human macrophages. *Virology* **1993**, *193*, 138–148. [CrossRef]
57. Baca-Regen, L.; Heinzinger, N.; Stevenson, M.; Gendelman, H.E. Alpha interferon-induced antiretroviral activities: Restriction of viral nucleic acid synthesis and progeny virion production in human immunodeficiency virus type 1-infected monocytes. *J. Virol.* **1994**, *68*, 7559–7565. [CrossRef]
58. Ranganath, N.; Sandstrom, T.S.; Fadel, S.; Côté, S.C.; Angel, J.B. Type I interferon responses are impaired in latently HIV infected cells. *Retrovirology* **2016**, *13*, 66–84. [CrossRef]
59. Dupont, M.; Rousset, S.; Manh, T.V.; Monard, S.C.; Pingris, K.; Souriant, S.; Vahlas, Z.; Velez, T.; Poincloux, R.; Maridonneau-Parini, I.; et al. Dysregulation of the IFN-I signaling pathway by Mycobacterium tuberculosis leads to exacerbation of HIV-1 infection of macrophages. *J. Leukoc. Biol.* **2022**, *112*, 1329–1342. [CrossRef]
60. Sandler, N.G.; Bosinger, S.E.; Estes, J.D.; Zhu, R.T.; Tharp, G.K.; Boritz, E.; Levin, D.; Wijeyesinghe, S.; Makamdop, K.N.; del Prete, G.Q.; et al. Type I interferon responses in rhesus macaques prevent SIV infection and slow disease progression. *Nature* **2014**, *511*, 601–605. [CrossRef]

61. Kovarik, P.; Castiglia, V.; Ivin, M.; Ebner, F. Type I interferons in bacterial infections: A balancing act. *Front. Immunol.* **2016**, *7*, 652–670. [CrossRef]
62. Mundra, A.; Yegiazaryan, A.; Karsian, H.; Alsaigh, D.; Bonavida, V.; Frame, M.; May, N.; Gargaloyan, A.; Abnousian, A.; Venketaraman, V. Pathogenicity of type I interferons in mycobacterium tuberculosis. *Int. J. Mol. Sci.* **2023**, *24*, 3919. [CrossRef] [PubMed]
63. Kak, G.; Raza, M.; Tiwari, B.K. Interferon-gamma (IFN-γ): Exploring its implications in infectious diseases. *Biomol. Concepts* **2018**, *9*, 64–79. [CrossRef] [PubMed]
64. Rothfuchs, A.G.; Trumstedt, C.; Wigzell, H.; Rottenberg, M.E. Intracellular bacterial infection-induced IFN-gamma is critically but not solely dependent on Toll-like receptor 4-myeloid differentiation factor 88-IFN-alpha beta-STAT1 signaling. *J. Immunol.* **2004**, *172*, 6345–6353. [CrossRef] [PubMed]
65. Green, A.M.; Difazio, R.; Flynn, J.L. IFN-γ from CD4 T cells is essential for host survival and enhances CD8 T cell function during Mycobacterium tuberculosis infection. *J. Immunol.* **2013**, *190*, 270–277. [CrossRef]
66. Naglak, E.K.; Morrison, S.G.; Morrison, R.P. IFNγ is required for optimal antibody-mediated immunity against genital chlamydia infection. *Infect. Immun.* **2016**, *84*, 3232–3242. [CrossRef]
67. Beekhuizen, H.; van de Gevel, J.S. Gamma interferon confers resistance to infection with Staphylococcus aureus in human vascular endothelial cells by cooperative proinflammatory and enhanced intrinsic antibacterial activities. *Infect. Immun.* **2007**, *75*, 5615–5626. [CrossRef]
68. Flynn, J.L.; Chan, J.; Triebold, K.J.; Dalton, D.K.; Stewart, T.A.; Bloom, B.R. An essential role for interferon gamma in resistance to Mycobacterium tuberculosis infection. *J. Exp. Med.* **1993**, *178*, 2249–2254. [CrossRef]
69. Krisnawati, D.I.; Liu, Y.C.; Lee, Y.J.; Wang, Y.T.; Chen, C.L.; Tseng, P.C.; Lin, C.F. Functional neutralization of anti-IFN-γ autoantibody in patients with nontuberculous mycobacteria infection. *Sci. Rep.* **2019**, *9*, 5682–5691. [CrossRef]
70. Bao, S.; Beagley, K.W.; France, M.P.; Shen, J.; Husband, A.J. Interferon-gamma plays a critical role in intestinal immunity against Salmonella typhimurium infection. *Immunology* **2000**, *99*, 464–472. [CrossRef]
71. Zenewicz, L.A.; Shen, H. Innate and adaptive immune responses to Listeria monocytogenes: A short overview. *Microbes Infect.* **2007**, *9*, 1208–1215. [CrossRef]
72. Kokordelis, P.; Krämer, B.; Körner, C.; Boesecke, C.; Voigt, E.; Ingiliz, P.; Glässner, A.; Eisenhardt, M.; Wolter, F.; Kaczmarek, D.; et al. An effective interferon-gamma-mediated inhibition of hepatitis C virus replication by natural killer cells is associated with spontaneous clearance of acute hepatitis C in human immunodeficiency virus-positive patients. *Hepatology* **2014**, *59*, 814–827. [CrossRef] [PubMed]
73. Baird, N.L.; Bowlin, J.L.; Hotz, T.J.; Cohrs, R.J.; Gilden, D. Interferon gamma prolongs survival of varicella-zoster virus-infected human neurons In vitro. *J. Virol.* **2015**, *89*, 7425–7427. [CrossRef] [PubMed]
74. Prokunina-Olsson, L.; Muchmore, B.; Tang, W.; Pfeiffer, R.M.; Park, H.; Dickensheets, H.; Hergott, D.; Porter-Gill, P.; Mumy, A.; Kohaar, I.; et al. A variant upstream of IFNL3 (IL28B) creating a new interferon gene IFNL4 is associated with impaired clearance of hepatitis C virus. *Nat. Genet.* **2013**, *45*, 164–171. [CrossRef]
75. Kotenko, S.V.; Durbin, J.E. Contribution of type III interferons to antiviral immunity: Location, location, location. *J. Biol. Chem.* **2017**, *292*, 7295–7303. [CrossRef]
76. Espinosa, V.; Dutta, O.; McElrath, C.; Du, P.; Chang, Y.J.; Cicciarelli, B.; Pitler, A.; Whitehead, I.; Obar, J.J.; Durbin, J.E.; et al. Type III interferon is a critical regulator of innate antifungal immunity. *Sci. Immunol.* **2017**, *2*, eaan5357. [CrossRef] [PubMed]
77. Stanifer, M.L.; Guo, C.; Doldan, P.; Boulant, S. Importance of type I and III interferons at respiratory and intestinal barrier surfaces. *Front. Immunol.* **2020**, *11*, 608645. [CrossRef]
78. Devasthanam, A.S. Mechanisms underlying the inhibition of interferon signaling by viruses. *Virulence* **2014**, *5*, 270–277. [CrossRef]
79. Pozzetto, B.; Mogensen, K.E.; Tovey, M.G.; Gresser, I. Characteristics of autoantibodies to human interferon in a patient with varicella-zoster disease. *J. Infect. Dis.* **1984**, *150*, 707–713. [CrossRef]
80. Meager, A.; Visvalingam, K.; Peterson, P.; Möll, K.; Murumägi, A.; Krohn, K.; Eskelin, P.; Perheentupa, J.; Husebye, E.; Kadota, Y.; et al. Anti-interferon autoantibodies in autoimmune polyendocrinopathy syndrome type 1. *PLoS Med.* **2006**, *3*, e289. [CrossRef]
81. Capra, R.; Sottini, A.; Cordioli, C.; Serana, F.; Chiarini, M.; Caimi, L.; Padovani, A.; Bergamaschi, R.; Imberti, L. IFNbeta bioavailability in multiple sclerosis patients: MxA versus antibody-detecting assays. *J. Neuroimmunol.* **2007**, *189*, 102–110. [CrossRef]
82. Zhang, L.; Barker, J.M.; Babu, S.; Su, M.; Stenerson, M.; Cheng, M.; Shum, A.; Zamir, E.; Badolato, R.; Law, A.; et al. A robust immunoassay for anti-interferon autoantibodies that is highly specific for patients with autoimmune polyglandular syndrome type 1. *Clin. Immunol.* **2007**, *125*, 131–137. [CrossRef] [PubMed]
83. Oftedal, B.E.; Wolff, A.S.; Bratland, E.; Kämpe, O.; Perheentupa, J.; Myhre, A.G.; Meager, A.; Purushothaman, R.; Ten, S.; Husebye, E.S. Radioimmunoassay for autoantibodies against interferon omega; its use in the diagnosis of autoimmune polyendocrine syndrome type I. *Clin. Immunol.* **2008**, *129*, 163–169. [CrossRef] [PubMed]
84. Ding, L.; Mo, A.; Jutivorakool, K.; Pancholi, M.; Holland, S.M.; Browne, S.K. Determination of human anticytokine autoantibody profiles using a particle-based approach. *J. Clin. Immunol.* **2012**, *32*, 238–245. [CrossRef] [PubMed]
85. Breivik, L.; Oftedal, B.E.; Bøe Wolff, A.S.; Bratland, E.; Orlova, E.M.; Husebye, E.S. A novel cell-based assay for measuring neutralizing autoantibodies against type I interferons in patients with autoimmune polyendocrine syndrome type 1. *Clin. Immunol.* **2014**, *153*, 220–227. [CrossRef] [PubMed]

86. Rosenberg, J.M.; Price, J.V.; Barcenas-Morales, G.; Ceron-Gutierrez, L.; Davies, S.; Kumararatne, D.S.; Döffinger, R.; Utz, P.J. Protein microarrays identify disease-specific anti-cytokine autoantibody profiles in the landscape of immunodeficiency. *J. Allergy Clin. Immunol.* **2016**, *137*, 204–213.e3. [CrossRef]

87. Bastard, P.; Gervais, A.; Le Voyer, T.; Rosain, J.; Philippot, Q.; Manry, J.; Michailidis, E.; Hoffmann, H.H.; Eto, S.; Garcia-Prat, M.; et al. Autoantibodies neutralizing type I IFNs are present in ~4% of uninfected individuals over 70 years old and account for ~20% of COVID-19 deaths. *Sci. Immunol.* **2021**, *6*, eabl4340. [CrossRef]

88. Vallbracht, A.; Treuner, J.; Flehmig, B.; Joester, K.E.; Niethammer, D. Interferon-neutralizing antibodies in a patient treated with human fibroblast interferon. *Nature* **1981**, *289*, 496–497. [CrossRef]

89. Prümmer, O.; Bunjes, D.; Wiesneth, M.; Arnold, R.; Porzsolt, F.; Heimpel, H. High-titre interferon-alpha antibodies in a patient with chronic graft-versus-host disease after allogeneic bone marrow transplantation. *Bone Marrow Transplant.* **1994**, *14*, 483–486.

90. Meager, A.; Wadhwa, M.; Dilger, P.; Bird, C.; Thorpe, R.; Newsom-Davis, J.; Willcox, N. Anti-cytokine autoantibodies in autoimmunity: Preponderance of neutralizing autoantibodies against interferon-alpha, interferon-omega and interleukin-12 in patients with thymoma and/or myasthenia gravis. *Clin. Exp. Immunol.* **2003**, *132*, 128–136. [CrossRef]

91. Shiono, H.; Wong, Y.L.; Matthews, I.; Liu, J.L.; Zhang, W.; Sims, G.; Meager, A.; Beeson, D.; Vincent, A.; Willcox, N. Spontaneous production of anti-IFN-alpha and anti-IL-12 autoantibodies by thymoma cells from myasthenia gravis patients suggests autoimmunization in the tumor. *Int. Immunol.* **2003**, *15*, 903–913. [CrossRef]

92. Levin, M. Anti-interferon auto-antibodies in autoimmune polyendocrinopathy syndrome type 1. *PLoS Med.* **2006**, *3*, e292. [CrossRef] [PubMed]

93. Walter, J.E.; Rosen, L.B.; Csomos, K.; Rosenberg, J.M.; Mathew, D.; Keszei, M.; Ujhazi, B.; Chen, K.; Lee, Y.N.; Tirosh, I.; et al. Broad-spectrum antibodies against self-antigens and cytokines in RAG deficiency. *J. Clin. Investig.* **2015**, *125*, 4135–4148. [CrossRef] [PubMed]

94. Rosenberg, J.M.; Maccari, M.E.; Barzaghi, F.; Allenspach, E.J.; Pignata, C.; Weber, G.; Torgerson, T.R.; Utz, P.J.; Bacchetta, R. Neutralizing anti-cytokine autoantibodies against interferon-α in immunodysregulation polyendocrinopathy enteropathy X-linked. *Front. Immunol.* **2018**, *9*, 544–552. [CrossRef] [PubMed]

95. Chawansuntati, K.; Rattanathammethee, K.; Wipasa, J. Minireview: Insights into anti-interferon-γ autoantibodies. *Exp. Biol. Med.* **2021**, *246*, 790–795. [CrossRef]

96. Kiratikanon, S.; Phinyo, P.; Rujiwetpongstorn, R.; Patumanond, J.; Tungphaisal, V.; Mahanupab, P.; Chaiwarith, R.; Tovanabutra, N.; Chiewchanvit, S.; Chuamanochan, M. Adult-onset immunodeficiency due to anti-interferon-gamma autoantibody-associated Sweet syndrome: A distinctive entity. *J. Dermatol.* **2022**, *49*, 133–141. [CrossRef]

97. Lin, C.H.; Chi, C.Y.; Shih, H.P.; Ding, J.Y.; Lo, C.C.; Wang, S.Y.; Kuo, C.Y.; Yeh, C.F.; Tu, K.H.; Liu, S.H.; et al. Identification of a major epitope by anti-interferon-γ autoantibodies in patients with mycobacterial disease. *Nat. Med.* **2016**, *22*, 994–1001. [CrossRef]

98. Acosta, P.L.; Byrne, A.B.; Hijano, D.R.; Talarico, L.B. Human type I interferon antiviral effects in respiratory and reemerging viral infections. *J. Immunol. Res.* **2020**, *2020*, 1372494. [CrossRef]

99. Mogensen, K.E.; Daubas, P.; Gresser, I.; Sereni, D.; Varet, B. Patient with circulating antibodies to alpha-interferon. *Lancet* **1981**, *2*, 1227–1228. [CrossRef]

100. Leroy, V.; Baud, M.; de Traversay, C.; Maynard-Muet, M.; Lebon, P.; Zarski, J.P. Role of anti-interferon antibodies in breakthrough occurrence during alpha 2a and 2b therapy in patients with chronic hepatitis C. *J. Hepatol.* **1998**, *28*, 375–381. [CrossRef]

101. Hoffmann, R.M.; Berg, T.; Teuber, G.; Prümmer, O.; Leifeld, L.; Jung, M.C.; Spengler, U.; Zeuzem, S.; Hopf, U.; Pape, G.R. Interferon-antibodies and the breakthrough phenomenon during ribavirin/interferon-alpha combination therapy and interferon-alpha monotherapy of patients with chronic hepatitis C. *Z. Gastroenterol.* **1999**, *37*, 715–723.

102. Zhang, Q.; Pizzorno, A.; Miorin, L.; Bastard, P.; Gervais, A.; Le Voyer, T.; Bizien, L.; Manry, J.; Rosain, J.; Philippot, Q.; et al. Autoantibodies against type I IFNs in patients with critical influenza pneumonia. *J. Exp. Med.* **2022**, *219*, e20220514. [CrossRef] [PubMed]

103. Alotaibi, F.; Alharbi, N.K.; Rosen, L.B.; Asiri, A.Y.; Assiri, A.M.; Balkhy, H.H.; Al Jeraisy, M.; Mandourah, Y.; AlJohani, S.; Al Harbi, S.; et al. Type I interferon autoantibodies in hospitalized patients with Middle East respiratory syndrome and association with outcomes and treatment effect of interferon beta-1b in MIRACLE clinical trial. *Influenza Other Respir. Viruses* **2023**, *17*, e13116. [CrossRef] [PubMed]

104. Ghale, R.; Spottiswoode, N.; Anderson, M.S.; Mitchell, A.; Wang, G.; Calfee, C.S.; DeRisi, J.L.; Langelier, C.R. Prevalence of type-1 interferon autoantibodies in adults with non-COVID-19 acute respiratory failure. *Respir. Res.* **2022**, *23*, 354–358. [CrossRef] [PubMed]

105. Philippot, Q.; Bastard, P.; Puel, A.; Casanova, J.L.; Cobat, A.; Laouénan, C.; Tardivon, C.; Crestani, B.; Borie, R. No increased prevalence of autoantibodies neutralizing type I IFNs in idiopathic pulmonary fibrosis patients. *Respir. Res.* **2023**, *24*, 87–90. [CrossRef]

106. Bastard, P.; Michailidis, E.; Hoffmann, H.H.; Chbihi, M.; Le Voyer, T.; Rosain, J.; Philippot, Q.; Seeleuthner, Y.; Gervais, A.; Materna, M.; et al. Auto-antibodies to type I IFNs can underlie adverse reactions to yellow fever live attenuated vaccine. *J. Exp. Med.* **2021**, *218*, e20202486. [CrossRef]

107. Busnadiego, I.; Abela, I.A.; Frey, P.M.; Hofmaenner, D.A.; Scheier, T.C.; Schuepbach, R.A.; Buehler, P.K.; Brugger, S.D.; Hale, B.G. Critically ill COVID-19 patients with neutralizing autoantibodies against type I interferons have increased risk of herpesvirus disease. *PLoS Biol.* **2022**, *20*, e3001709. [CrossRef]

108. Höflich, C.; Sabat, R.; Rosseau, S.; Temmesfeld, B.; Slevogt, H.; Döcke, W.D.; Grütz, G.; Meisel, C.; Halle, E.; Göbel, U.B.; et al. Naturally occurring anti-IFN-gamma autoantibody and severe infections with Mycobacterium cheloneae and Burkholderia cocovenenans. *Blood* **2004**, *103*, 673–675. [CrossRef]

109. Browne, S.K.; Burbelo, P.D.; Chetchotisakd, P.; Suputtamongkol, Y.; Kiertiburanakul, S.; Shaw, P.A.; Kirk, J.L.; Jutivorakool, K.; Zaman, R.; Ding, L.; et al. Adult-onset immunodeficiency in Thailand and Taiwan. *N. Engl. J. Med.* **2012**, *367*, 725–734. [CrossRef]

110. Aoki, A.; Sakagami, T.; Yoshizawa, K.; Shima, K.; Toyama, M.; Tanabe, Y.; Moro, H.; Aoki, N.; Watanabe, S.; Koya, T.; et al. Clinical significance of Interferon-γ neutralizing autoantibodies against disseminated nontuberculous mycobacterial disease. *Clin. Infect. Dis.* **2018**, *66*, 1239–1245. [CrossRef]

111. Döffinger, R.; Helbert, M.R.; Barcenas-Morales, G.; Yang, K.; Dupuis, S.; Ceron-Gutierrez, L.; Espitia-Pinzon, C.; Barnes, N.; Bothamley, G.; Casanova, J.L.; et al. Autoantibodies to interferon-gamma in a patient with selective susceptibility to mycobacterial infection and organ-specific autoimmunity. *Clin. Infect. Dis.* **2004**, *38*, e10-4. [CrossRef]

112. Hong, G.H.; Ortega-Villa, A.M.; Hunsberger, S.; Chetchotisakd, P.; Anunnatsiri, S.; Mootsikapun, P.; Rosen, L.B.; Zerbe, C.S.; Holland, S.M. Natural history and evolution of anti-Interferon-γ autoantibody-associated immunodeficiency syndrome in Thailand and the United States. *Clin. Infect. Dis.* **2020**, *71*, 53–62. [CrossRef] [PubMed]

113. Chen, P.K.; Liao, T.L.; Chang, S.H.; Yeo, K.J.; Chou, C.H.; Chen, D.Y. High-titer anti-interferon-γ neutralizing autoantibodies linked to opportunistic infections in patients with adult-onset still's disease. *Front. Med. (Lausanne)* **2023**, *9*, 1097514. [CrossRef] [PubMed]

114. Barcenas-Morales, G.; Cortes-Acevedo, P.; Doffinger, R. Anticytokine autoantibodies leading to infection: Early recognition, diagnosis and treatment options. *Curr. Opin. Infect. Dis.* **2019**, *32*, 330–336. [CrossRef] [PubMed]

115. Taur, P.D.; Gowri, V.; Pandrowala, A.A.; Iyengar, V.V.; Chougule, A.; Golwala, Z.; Chandak, S.; Agarwal, R.; Keni, P.; Dighe, N.; et al. Clinical and molecular findings in mendelian susceptibility to mycobacterial diseases: Experience from India. *Front. Immunol.* **2021**, *12*, 631298. [CrossRef] [PubMed]

116. Worldmeter, COVID-19 Coronavirus Pandemic. Available online: https://www.worldometers.info/coronavirus/ (accessed on 26 April 2023).

117. Mehta, O.P.; Bhandari, P.; Raut, A.; Kacimi, S.E.O.; Huy, N.T. Coronavirus Disease (COVID-19): Comprehensive review of clinical presentation. *Front. Public Health* **2021**, *8*, 582932. [CrossRef]

118. Blanco-Melo, D.; Nilsson-Payant, B.E.; Liu, W.C.; Uhl, S.; Hoagland, D.; Møller, R.; Jordan, T.X.; Oishi, K.; Panis, M.; Sachs, D.; et al. Imbalanced host response to SARS-CoV-2 drives development of COVID-19. *Cell* **2020**, *181*, 1036–1045.e9. [CrossRef]

119. Hadjadj, J.; Yatim, N.; Barnabei, L.; Corneau, A.; Boussier, J.; Smith, N.; Péré, H.; Charbit, B.; Bondet, V.; Chenevier-Gobeaux, C.; et al. Impaired type I interferon activity and inflammatory responses in severe COVID-19 patients. *Science* **2020**, *369*, 718–724. [CrossRef]

120. Galani, I.E.; Rovina, N.; Lampropoulou, V.; Triantafyllia, V.; Manioudaki, M.; Pavlos, E.; Koukaki, E.; Fragkou, P.C.; Panou, V.; Rapti, V.; et al. Untuned antiviral immunity in COVID-19 revealed by temporal type I/III interferon patterns and flu comparison. *Nat. Immunol.* **2021**, *22*, 32–40. [CrossRef]

121. Pierangeli, A.; Gentile, M.; Oliveto, G.; Frasca, F.; Sorrentino, L.; Matera, L.; Nenna, R.; Viscido, A.; Fracella, M.; Petrarca, L.; et al. Comparison by age of the local interferon response to SARS-CoV-2 suggests a role for IFN-ε and -ω. *Front. Immunol.* **2022**, *13*, 873232. [CrossRef]

122. Spiering, A.E.; de Vries, T.J. Why females do better: The X chromosomal TLR7 gene-dose effect in COVID-19. *Front. Immunol.* **2021**, *12*, 756262. [CrossRef]

123. Puel, A.; Bastard, P.; Bustamante, J.; Casanova, J.L. Human autoantibodies underlying infectious diseases. *J. Exp. Med.* **2022**, *219*, e20211387. [CrossRef] [PubMed]

124. Arrestier, R.; Bastard, P.; Belmondo, T.; Voiriot, G.; Urbina, T.; Luyt, C.E.; Gervais, A.; Bizien, L.; Segaux, L.; Ben Ahmed, M.; et al. Auto-antibodies against type I IFNs in > 10% of critically ill COVID-19 patients: A prospective multicentre study. *Ann. Intensive Care* **2022**, *12*, 121–132. [CrossRef] [PubMed]

125. Bastard, P.; Vazquez, S.; Liu, J.; Laurie, M.T.; Wang, C.Y.; Gervais, A.; Le Voyer, T.; Bizien, L.; Zamecnik, C.; Philippot, Q.; et al. Vaccine breakthrough hypoxemic COVID-19 pneumonia in patients with auto-Abs neutralizing type I IFNs. *Sci. Immunol.* **2022**, eabp8966. [CrossRef]

126. Manry, J.; Bastard, P.; Gervais, A.; Le Voyer, T.; Rosain, J.; Philippot, Q.; Michailidis, E.; Hoffmann, H.H.; Eto, S.; Garcia-Prat, M.; et al. The risk of COVID-19 death is much greater and age dependent with type I IFN autoantibodies. *Proc. Natl. Acad. Sci. USA* **2022**, *119*, e2200413119. [CrossRef]

127. Acosta-Ampudia, Y.; Monsalve, D.M.; Rojas, M.; Rodríguez, Y.; Gallo, J.E.; Salazar-Uribe, J.C.; Santander, M.J.; Cala, M.P.; Zapata, W.; Zapata, M.I.; et al. COVID-19 convalescent plasma composition and immunological effects in severe patients. *J. Autoimmun.* **2021**, *118*, 102598. [CrossRef]

128. Wang, E.Y.; Mao, T.; Klein, J.; Dai, Y.; Huck, J.D.; Jaycox, J.R.; Liu, F.; Zhou, T.; Israelow, B.; Wong, P.; et al. Diverse functional autoantibodies in patients with COVID-19. *Nature* **2021**, *595*, 283–288. [CrossRef] [PubMed]

129. Koning, R.; Bastard, P.; Casanova, J.L.; Brouwer, M.C.; van de Beek, D.; Amsterdam, U.M.C. COVID-19 Biobank Investigators. Autoantibodies against type I interferons are associated with multi-organ failure in COVID-19 patients. *Intensive Care Med.* **2021**, *47*, 704–706. [CrossRef] [PubMed]

130. Troya, J.; Bastard, P.; Planas-Serra, L.; Ryan, P.; Ruiz, M.; de Carranza, M.; Torres, J.; Martínez, A.; Abel, L.; Casanova, J.L.; et al. Neutralizing autoantibodies to type I IFNs in >10% of patients with severe COVID-19 pneumonia hospitalized in Madrid, Spain. *J. Clin. Immunol.* **2021**, *41*, 914–922. [CrossRef]

131. Vazquez, S.E.; Bastard, P.; Kelly, K.; Gervais, A.; Norris, P.J.; Dumont, L.J.; Casanova, J.L.; Anderson, M.S.; DeRisi, J.L. Neutralizing autoantibodies to type I interferons in COVID-19 convalescent donor plasma. *J. Clin. Immunol.* **2021**, *41*, 1169–1171. [CrossRef]

132. Lopez, J.; Mommert, M.; Mouton, W.; Pizzorno, A.; Brengel-Pesce, K.; Mezidi, M.; Villard, M.; Lina, B.; Richard, J.C.; Fassier, J.B.; et al. Early nasal type I IFN immunity against SARS-CoV-2 is compromised in patients with autoantibodies against type I IFNs. *J. Exp. Med.* **2021**, *218*, e20211211. [CrossRef]

133. Abers, M.S.; Rosen, L.B.; Delmonte, O.M.; Shaw, E.; Bastard, P.; Imberti, L.; Quaresima, V.; Biondi, A.; Bonfanti, P.; Casta-gnoli, R.; et al. Neutralizing type-I interferon autoantibodies are associated with delayed viral clearance and intensive care unit admission in patients with COVID-19. *Immunol. Cell Biol.* **2021**, *99*, 917–921. [CrossRef]

134. Goncalves, D.; Mezidi, M.; Bastard, P.; Perret, M.; Saker, K.; Fabien, N.; Pescarmona, R.; Lombard, C.; Walzer, T.; Casanova, J.L.; et al. Antibodies against type I interferon: Detection and association with severe clinical outcome in COVID-19 patients. *Clin. Transl. Immunol.* **2021**, *10*, e1327. [CrossRef]

135. Ziegler, C.G.K.; Miao, V.N.; Owings, A.H.; Navia, A.W.; Tang, Y.; Bromley, J.D.; Lotfy, P.; Sloan, M.; Laird, H.; Williams, H.B.; et al. Impaired local intrinsic immunity to SARS-CoV-2 infection in severe COVID-19. *Cell* **2021**, *184*, 4713–4733.e22. [CrossRef]

136. Chang, S.E.; Feng, A.; Meng, W.; Apostolidis, S.A.; Mack, E.; Artandi, M.; Barman, L.; Bennett, K.; Chakraborty, S.; Chang, I.; et al. New-onset IgG autoantibodies in hospitalized patients with COVID-19. *Nat. Commun.* **2021**, *12*, 5417. [CrossRef]

137. van der Wijst, M.G.P.; Vazquez, S.E.; Hartoularos, G.C.; Bastard, P.; Grant, T.; Bueno, R.; Lee, D.S.; Greenland, J.R.; Sun, Y.; Perez, R.; et al. Type I interferon autoantibodies are associated with systemic immune alterations in patients with COVID-19. *Sci. Transl. Med.* **2021**, *13*, eabh2624. [CrossRef]

138. Solanich, X.; Rigo-Bonnin, R.; Gumucio, V.D.; Bastard, P.; Rosain, J.; Philippot, Q.; Perez-Fernandez, X.L.; Fuset-Cabanes, M.P.; Gordillo-Benitez, M.Á.; Suarez-Cuartin, G.; et al. Pre-existing autoantibodies neutralizing high concentrations of type I interferons in almost 10% of COVID-19 patients admitted to intensive care in Barcelona. *J. Clin. Immunol.* **2021**, *41*, 1733–1744. [CrossRef]

139. Yee, D.; Tso, M.; Shaw, E.; Rosen, L.B.; Samuels, E.; Bastard, P.; Casanova, J.L.; Holland, S.M.; Su, H.C.; Richard, S.A.; et al. Type I interferon autoantibodies are detected in those with critical COVID-19, including a young female patient. *Open Forum Infect. Dis.* **2021**, *8*, S325–S326. [CrossRef]

140. Savvateeva, E.; Filippova, M.; Valuev-Elliston, V.; Nuralieva, N.; Yukina, M.; Troshina, E.; Baklaushev, V.; Ivanov, A.; Gryadunov, D. Microarray-based detection of antibodies against SARS-CoV-2 proteins, common respiratory viruses and type I interferons. *Viruses* **2021**, *13*, 2553. [CrossRef]

141. Carapito, R.; Li, R.; Helms, J.; Carapito, C.; Gujja, S.; Rolli, V.; Guimaraes, R.; Malagon-Lopez, J.; Spinnhirny, P.; Lederle, A.; et al. Identification of driver genes for critical forms of COVID-19 in a deeply phenotyped young patient cohort. *Sci. Transl. Med.* **2022**, *14*, eabj7521. [CrossRef]

142. Chauvineau-Grenier, A.; Bastard, P.; Servajean, A.; Gervais, A.; Rosain, J.; Jouanguy, E.; Cobat, A.; Casanova, J.L.; Rossi, B. Autoantibodies neutralizing type I interferons in 20% of COVID-19 deaths in a French hospital. *J. Clin. Immunol.* **2022**, *42*, 459–470. [CrossRef]

143. Petrikov, S.S.; Borovkova, N.V.; Popugaev, K.A.; Storozheva, M.V.; Kvasnikov, A.M.; Godkov, M.A. Anti-interferon alpha autoantibodies and their significance in COVID-19. *Russ. J. Inf. Immun.* **2022**, *12*, 279–287. [CrossRef]

144. Raadsen, M.P.; Gharbharan, A.; Jordans, C.C.E.; Mykytyn, A.Z.; Lamers, M.M.; van den Doel, P.B.; Endeman, H.; van den Akker, J.P.C.; GeurtsvanKessel, C.H.; Koopmans, M.P.G.; et al. Interferon-α2 auto-antibodies in convalescent plasma therapy for COVID-19. *J. Clin. Immunol.* **2022**, *42*, 232–239. [CrossRef]

145. Steels, S.; Van Elslande, J.; Leuven COVID-Study Group; De Munter, P.; Bossuyt, X. Transient increase of pre-existing anti-IFN-α2 antibodies induced by SARS-CoV-2 infection. *J. Clin. Immunol.* **2022**, *42*, 742–745. [CrossRef]

146. Frasca, F.; Scordio, M.; Santinelli, L.; Gabriele, L.; Gandini, O.; Criniti, A.; Pierangeli, A.; Angeloni, A.; Mastroianni, C.M.; d'Ettorre, G.; et al. Anti-IFN-α/-ω neutralizing antibodies from COVID-19 patients correlate with downregulation of IFN response and laboratory biomarkers of disease severity. *Eur. J. Immunol.* **2022**, *52*, 1120–1128. [CrossRef]

147. Akbil, B.; Meyer, T.; Stubbemann, P.; Thibeault, C.; Staudacher, O.; Niemeyer, D.; Jansen, J.; Mühlemann, B.; Doehn, J.; Tabeling, C.; et al. Early and rapid identification of COVID-19 patients with neutralizing type I interferon auto-antibodies. *J. Clin. Immunol.* **2022**, *42*, 1111–1129. [CrossRef]

148. Eto, S.; Nukui, Y.; Tsumura, M.; Nakagama, Y.; Kashimada, K.; Mizoguchi, Y.; Utsumi, T.; Taniguchi, M.; Sakura, F.; Noma, K.; et al. Neutralizing type I interferon autoantibodies in Japanese patients with severe COVID-19. *J. Clin. Immunol.* **2022**, *42*, 1360–1370. [CrossRef]

149. Mathian, A.; Breillat, P.; Dorgham, K.; Bastard, P.; Charre, C.; Lhote, R.; Quentric, P.; Moyon, Q.; Mariaggi, A.A.; Mouries-Martin, S.; et al. Lower disease activity but higher risk of severe COVID-19 and herpes zoster in patients with systemic lupus erythematosus with pre-existing autoantibodies neutralising IFN-α. *Ann. Rheum. Dis.* **2022**, *81*, 1695–1703. [CrossRef]

150. Wang, X.; Tang, Q.; Li, H.; Jiang, H.; Xu, J.; Bergquist, R.; Qin, Z. Autoantibodies against type I interferons in COVID-19 infection: A systematic review and meta-analysis. *Int. J. Infect. Dis.* **2023**, *130*, 147–152. [CrossRef]

151. Jiang, W.; Johnson, D.; Adekunle, R.; Heather, H.; Xu, W.; Cong, X.; Wu, X.; Fan, H.; Andersson, L.M.; Robertson, J.; et al. COVID-19 is associated with bystander polyclonal autoreactive B cell activation as reflected by a broad autoantibody production, but none is linked to disease severity. *J. Med. Virol.* **2023**, *95*, e28134. [CrossRef]

152. Center for Disease Control and Prevention, Long COVID or Post-COVID Conditions. Available online: https://www.cdc.gov/coronavirus/2019-ncov/long-term-effects/index.html (accessed on 26 April 2023).

153. Muri, J.; Cecchinato, V.; Cavalli, A.; Shanbhag, A.A.; Matkovic, M.; Biggiogero, M.; Maida, P.A.; Moritz, J.; Toscano, C.; Ghovehoud, E.; et al. Autoantibodies against chemokines post-SARS-CoV-2 infection correlate with disease course. *Nat. Immunol.* **2023**, *24*, 604–611. [CrossRef]

154. Lavi, Y.; Vojdani, A.; Halpert, G.; Sharif, K.; Ostrinski, Y.; Zyskind, I.; Lattin, M.T.; Zimmerman, J.; Silverberg, J.I.; Rosenberg, A.Z.; et al. Dysregulated levels of circulating autoantibodies against neuronal and nervous system autoantigens in COVID-19 patients. *Diagnostics* **2023**, *13*, 687. [CrossRef]

155. Mantovani, A.; Morrone, M.C.; Patrono, C.; Santoro, M.G.; Schiaffino, S.; Remuzzi, G.; Bussolati, G.; COVID-19 Commission of the Accademia Nazionale dei Lincei. Long Covid: Where we stand and challenges ahead. *Cell Death Differ.* **2022**, *29*, 1891–1900. [CrossRef] [PubMed]

156. Peluso, M.J.; Mitchell, A.; Wang, C.Y.; Takahashi, S.; Hoh, R.; Tai, V.; Durstenfeld, M.S.; Hsue, P.Y.; Kelly, J.D.; Martin, J.N. Low prevalence of interferon α autoantibodies in people experiencing symptoms of post-Coronavirus Disease 2019 (COVID-19) conditions, or long COVID. *J. Infect Dis.* **2023**, *227*, 246–250. [CrossRef] [PubMed]

157. Chen, P.K.; Yeo, K.J.; Chang, S.H.; Liao, T.L.; Chou, C.H.; Lan, J.L.; Chang, C.K.; Chen, D.Y. The detectable anti-interferon-γ autoantibodies in COVID-19 patients may be associated with disease severity. *Virol. J.* **2023**, *20*, 33. [CrossRef]

158. Credle, J.J.; Gunn, J.; Sangkhapreecha, P.; Monaco, D.R.; Zheng, X.A.; Tsai, H.J.; Wilbon, A.; Morgenlander, W.R.; Rastegar, A.; Dong, Y. Unbiased discovery of autoantibodies associated with severe COVID-19 via genome-scale self-assembled DNA-barcoded protein libraries. *Nat. Biomed. Eng.* **2022**, *6*, 992–1003. [CrossRef]

159. Shome, M.; Chung, Y.; Chavan, R.; Park, J.G.; Qiu, J.; LaBaer, J. Serum autoantibodyome reveals that healthy individuals share common autoantibodies. *Cell Rep.* **2022**, *39*, 110873. [CrossRef]

160. Kronzer, V.L.; Bridges, S.L., Jr.; Davis, J.M., 3rd. Why women have more autoimmune diseases than men: An evolutionary perspective. *Evol. Appl.* **2020**, *14*, 629–633. [CrossRef]

161. Fairweather, D.; Frisancho-Kiss, S.; Rose, N.R. Sex differences in autoimmune disease from a pathological perspective. *Am. J. Pathol.* **2008**, *173*, 600–609. [CrossRef]

162. Wang, P.; Zhao, H.; Wang, Z.; Zhang, X. Circulating natural antibodies to inflammatory cytokines are potential biomarkers for atherosclerosis. *J. Inflamm.* **2018**, *15*, 22–38. [CrossRef]

163. Ataya, A.; Knight, V.; Carey, B.C.; Lee, E.; Tarling, E.J.; Wang, T. The role of GM-CSF autoantibodies in infection and autoimmune pulmonary alveolar proteinosis: A concise review. *Front. Immunol.* **2021**, *12*, 752856. [CrossRef]

164. Browne, S.K.; Holland, S.M. Anticytokine autoantibodies in infectious diseases: Pathogenesis and mechanisms. *Lancet Infect. Dis.* **2010**, *10*, 875–885. [CrossRef]

165. Liu, Y.; Ebinger, J.E.; Mostafa, R.; Budde, P.; Gajewski, J.; Walker, B.; Joung, S.; Wu, M.; Bräutigam, M.; Hesping, F.; et al. Paradoxical sex-specific patterns of autoantibody response to SARS-CoV-2 infection. *J. Transl. Med.* **2021**, *19*, 524–537. [CrossRef]

166. Wehbe, Z.; Hammoud, S.H.; Yassine, H.M.; Fardoun, M.; El-Yazbi, A.F.; Eid, A.H. Molecular and biological mechanisms underlying gender differences in COVID-19 severity and mortality. *Front. Immunol.* **2021**, *12*, 659339. [CrossRef]

167. de Mol, J.; Kuiper, J.; Tsiantoulas, D.; Foks, A.C. The dynamics of B cell aging in health and disease. *Front. Immunol.* **2021**, *12*, 733566. [CrossRef]

168. Cusick, M.F.; Libbey, J.E.; Fujinami, R.S. Molecular mimicry as a mechanism of autoimmune disease. *Clin. Allergy Immunol.* **2012**, *42*, 102–111. [CrossRef]

169. Ma, H.; Murphy, C.; Loscher, C.E.; O'Kennedy, R. Autoantibodies—Enemies, and/or potential allies? *Front. Immunol.* **2022**, *13*, 953726. [CrossRef]

170. Liang, Z.; Dong, X.; Zhang, Z.; Zhang, Q.; Zhao, Y. Age-related thymic involution: Mechanisms and functional impact. *Aging Cell* **2022**, *21*, e13671. [CrossRef]

171. Ruan, P.; Wang, S.; Yang, M.; Wu, H. The ABC-associated immunosenescence and lifestyle interventions in autoimmune disease. *Rheumatol. Immunol. Res.* **2022**, *3*, 128–135. [CrossRef]

172. Zhou, D.; Borsa, M.; Simon, A.K. Hallmarks and detection techniques of cellular senescence and cellular ageing in immune cells. *Aging Cell* **2021**, *20*, e13316. [CrossRef]

173. Mouat, I.C.; Goldberg, E.; Horwitz, M.S. Age-associated B cells in autoimmune diseases. *Cell. Mol. Life Sci.* **2022**, *79*, 402–436. [CrossRef]

174. Hooper, B.; Whittingham, S.; Mathews, J.D.; Mackay, I.R.; Curnow, D.H. Autoimmunity in a rural community. *Clin. Exp. Immunol.* **1972**, *12*, 79–87.

175. Shu, S.; Nisengard, R.J.; Hale, W.L.; Beutner, E.H. Incidence and titers of antinuclear, antismooth muscle, and other autoantibodies in blood donors. *J. Lab. Clin. Med.* **1975**, *86*, 259–265.

176. Comarmond, C.; Lorin, V.; Marques, C.; Maciejewski-Duval, A.; Joher, N.; Planchais, C.; Touzot, M.; Biard, L.; Hieu, T.; Quiniou, V.; et al. TLR9 signalling in HCV-associated atypical memory B cells triggers Th1 and rheumatoid factor autoantibody responses. *J. Hepatol.* **2019**, *71*, 908–919. [CrossRef] [PubMed]

177. Rubtsova, K.; Rubtsov, A.V.; van Dyk, L.F.; Kappler, J.W.; Marrack, P. T-box transcription factor T-bet, a key player in a unique type of B-cell activation essential for effective viral clearance. *Proc. Natl. Acad. Sci. USA* **2013**, *110*, E3216–E3224. [CrossRef] [PubMed]

178. Johnson, J.L.; Rosenthal, R.L.; Knox, J.J.; Myles, A.; Naradikian, M.S.; Madej, J.; Kostiv, M.; Rosenfeld, A.M.; Meng, W.; Christensen, S.R.; et al. The transcription factor T-bet resolves memory B cell subsets with distinct tissue distributions and antibody specificities in mice and humans. *Immunity* **2020**, *52*, 842–855.e6. [CrossRef]

179. Austin, J.W.; Buckner, C.M.; Kardava, L.; Wang, W.; Zhang, X.; Melson, V.A.; Swanson, R.G.; Martins, A.J.; Zhou, J.Q.; Hoehn, K.B.; et al. Overexpression of T-bet in HIV infection is associated with accumulation of B cells outside germinal centers and poor affinity maturation. *Sci. Transl. Med.* **2019**, *11*, eaax0904. [CrossRef]

180. Woodruff, M.C.; Ramonell, R.P.; Nguyen, D.C.; Cashman, K.S.; Saini, A.S.; Haddad, N.S.; Ley, A.M.; Kyu, S.; Howell, J.C.; Ozturk, T.; et al. Extrafollicular B cell responses correlate with neutralizing antibodies and morbidity in COVID-19. *Nat. Immunol.* **2020**, *21*, 1506–1516. [CrossRef] [PubMed]

181. Britanova, O.V.; Putintseva, E.V.; Shugay, M.; Merzlyak, E.M.; Turchaninova, M.A.; Staroverov, D.B.; Bolotin, D.A.; Lukyanov, S.; Bogdanova, E.A.; Mamedov, I.Z.; et al. Age-related decrease in TCR repertoire diversity measured with deep and normalized sequence profiling. *J. Immunol.* **2014**, *192*, 2689–2698. [CrossRef]

182. Hale, B.G. Autoantibodies targeting type I interferons: Prevalence, mechanisms of induction, and association with viral disease susceptibility. *Eur. J. Immunol.* **2023**, e2250164. [CrossRef]

183. Knight, V.; Merkel, P.A.; O'Sullivan, M.D. Anticytokine autoantibodies: Association with infection and immune dysregulation. *Antibodies* **2016**, *5*, 3. [CrossRef]

184. Wolff, A.S.; Sarkadi, A.K.; Maródi, L.; Kärner, J.; Orlova, E.; Oftedal, B.E.; Kisand, K.; Oláh, E.; Meloni, A.; Myhre, A.G.; et al. Anti-cytokine autoantibodies preceding onset of autoimmune polyendocrine syndrome type I features in early childhood. *J. Clin. Immunol.* **2013**, *33*, 1341–1348. [CrossRef]

185. Mourão, L.C.; Cardoso-Oliveira, G.P.; Braga, É.M. Autoantibodies and malaria: Where we stand? Insights into pathogenesis and protection. *Front. Cell. Infect. Microbiol.* **2020**, *10*, 262–272. [CrossRef]

186. Ortona, E.; Malorni, W. Long COVID: To investigate immunological mechanisms and sex/gender related aspects as fundamental steps for tailored therapy. *Eur. Respir. J.* **2022**, *59*, 2102245. [CrossRef] [PubMed]

187. Shaw, E.R.; Matzinger, P. Transient autoantibodies to danger signals. *Front. Immunol.* **2023**, *14*, 1046300. [CrossRef] [PubMed]

188. Pagano, S.; Gaertner, H.; Cerini, F.; Mannic, T.; Satta, N.; Teixeira, P.C.; Cutler, P.; Mach, F.; Vuilleumier, N.; Hartley, O. The human autoantibody response to apolipoprotein A-I is focused on the C-terminal helix: A new rationale for diagnosis and treatment of cardiovascular disease? *PLoS ONE* **2015**, *10*, e0132780. [CrossRef] [PubMed]

189. de Jonge, H.; Iamele, L.; Maggi, M.; Pessino, G.; Scotti, C. Anti-cancer auto-antibodies: Roles, applications and open issues. *Cancers* **2021**, *13*, 813. [CrossRef]

190. Morgulchik, N.; Athanasopoulou, F.; Chu, E.; Lam, Y.; Kamaly, N. Potential therapeutic approaches for targeted inhibition of inflammatory cytokines following COVID-19 infection-induced cytokine storm. *Interface Focus* **2021**, *12*, 20210006. [CrossRef] [PubMed]

191. Santer, D.M.; Li, D.; Ghosheh, Y.; Zahoor, M.A.; Prajapati, D.; Hansen, B.E.; Tyrrell, D.L.J.; Feld, J.J.; Gehring, A.J. Interferon-λ treatment accelerates SARS-CoV-2 clearance despite age-related delays in the induction of T cell immunity. *Nat. Commun.* **2022**, *13*, 6992–7004. [CrossRef] [PubMed]

192. Metz-Zumaran, C.; Kee, C.; Doldan, P.; Guo, C.; Stanifer, M.L.; Boulant, S. Increased sensitivity of SARS-CoV-2 to type III interferon in human intestinal epithelial cells. *J. Virol.* **2022**, *96*, e0170521. [CrossRef] [PubMed]

Review

Interleukins, Chemokines, and Tumor Necrosis Factor Superfamily Ligands in the Pathogenesis of West Nile Virus Infection

Emna Benzarti [1,2], Kristy O. Murray [1,2,3,4] and Shannon E. Ronca [1,2,3,4,5,*]

1   Department of Pediatrics, Division of Tropical Medicine, Baylor College of Medicine and Texas Children's Hospital, Houston, TX 77030, USA
2   William T. Shearer Center for Human Immunobiology, Texas Children's Hospital, Houston, TX 77030, USA
3   National School of Tropical Medicine, Baylor College of Medicine, Houston, TX 77030, USA
4   Department of Immunology and Microbiology, Baylor College of Medicine, Houston, TX 77030, USA
5   Department of Molecular Virology and Microbiology, Baylor College of Medicine, Houston, TX 77030, USA
*   Correspondence: ronca@bcm.edu; Tel.: +1-832-824-7595

Abstract: West Nile virus (WNV) is a mosquito-borne pathogen that can lead to encephalitis and death in susceptible hosts. Cytokines play a critical role in inflammation and immunity in response to WNV infection. Murine models provide evidence that some cytokines offer protection against acute WNV infection and assist with viral clearance, while others play a multifaceted role WNV neuropathogenesis and immune-mediated tissue damage. This article aims to provide an up-to-date review of cytokine expression patterns in human and experimental animal models of WNV infections. Here, we outline the interleukins, chemokines, and tumor necrosis factor superfamily ligands associated with WNV infection and pathogenesis and describe the complex roles they play in mediating both protection and pathology of the central nervous system during or after virus clearance. By understanding of the role of these cytokines during WNV neuroinvasive infection, we can develop treatment options aimed at modulating these immune molecules in order to reduce neuroinflammation and improve patient outcomes.

Keywords: West Nile virus; cytokines; interleukins; chemokines; tumor necrosis factor superfamily ligands; infection model

Citation: Benzarti, E.; Murray, K.O.; Ronca, S.E. Interleukins, Chemokines, and Tumor Necrosis Factor Superfamily Ligands in the Pathogenesis of West Nile Virus Infection. *Viruses* **2023**, *15*, 806. https://doi.org/10.3390/v15030806

Academic Editors: Caijun Sun and Feng Ma

Received: 28 February 2023
Revised: 15 March 2023
Accepted: 17 March 2023
Published: 22 March 2023

1. Introduction

West Nile virus (WNV) is a positive-sense, single-stranded RNA virus belonging to the Japanese encephalitis serocomplex, genus *Flavivirus*, family *Flaviviridae* [1]. Its life cycle mainly involves birds and mosquitoes, whereas humans, horses, and other vertebrates are considered incidental hosts [2]. The WNV genome is translated into a single polypeptide and co- and post-translationally processed into ten proteins: three structural (capsid C, membrane precursor prM, and envelope E), which form the virion; seven nonstructural proteins (NS1, NS2A, NS2B, NS3, NS4A, 2K, NS4B, and NS5) involved in the viral replication cycle, evasion of host innate immunity, and WNV pathogenesis [3]; and one peptide 2K, which plays a role in rearranging cytoplasmic membranes and Golgi trafficking of the NS4A protein [4]. The susceptibility to WNV is highly variable among its hosts [5]. The majority of WNV infections in humans are either asymptomatic or mild, presenting with headache, weakness, and/or fever [6]. However, a small percentage of WNV-infected patients (less than 1% [7]) will develop neuroinvasive disease, including meningitis, encephalitis, or acute flaccid paralysis, for whom death occurs in 10–30% of cases [8,9]. Long-term physical and neurocognitive sequelae, including weakness, fatigue, myalgia, memory or hearing loss, depression, and motor dysfunction, may also occur in 30 to 60% of patients that develop clinical disease [9–11]. Although currently regarded as a top priority zoonotic disease for the US population [12], there are no standard treatment guidelines outside of supportive care, nor is there an FDA-approved drug or vaccine available for the treatment or prevention of WNV neuroinvasive disease, respectively [8].

WNV pathogenesis is characterized by three phases: (1) an early phase of skin infection and spread to the local draining lymph nodes following a bite from an infected mosquito, (2) viral dissemination to peripheral organs, and (3) invasion of the central nervous system (CNS) [13]. To fight the invasion of WNV, the mammalian host mobilizes three lines of defense: the skin and the innate immunity at the early phase followed by the adaptive (humoral and cellular) immunity at later stages [13,14].

Cytokines are signaling proteins that are expressed by many immune and nonimmune mammalian cells (Figure 1). Their induction and regulation are tightly linked to WNV replication during the early phase of infection [15–20]. While they are engaged in all three lines of defense against WNV, they also contribute to immune-mediated tissue damage in the brain. Among these cytokines, interleukins (ILs), chemokines, and tumor necrosis factor superfamily (TNFSF) ligands are major players in immunity against WNV, as evidenced by transcriptome profiling of WNV-infected cells and tissues using DNA microarrays or RNA sequencing [21]. Several reviews have shed light on their role in flaviviral infections in general [22–24] and in specific flaviviral diseases, including dengue [25,26] and Zika [27] viruses. Closely related flaviviruses elicit different immunomodulation profiles in their hosts [28–30] and differentially antagonize antiviral pathways [31]; however, WNV pathogenesis appears to have unique aspects compared with other neurotropic viruses [28,32], which will be discussed throughout this review. Therefore, it is important to address the role of cytokines in the specific context of WNV infection.

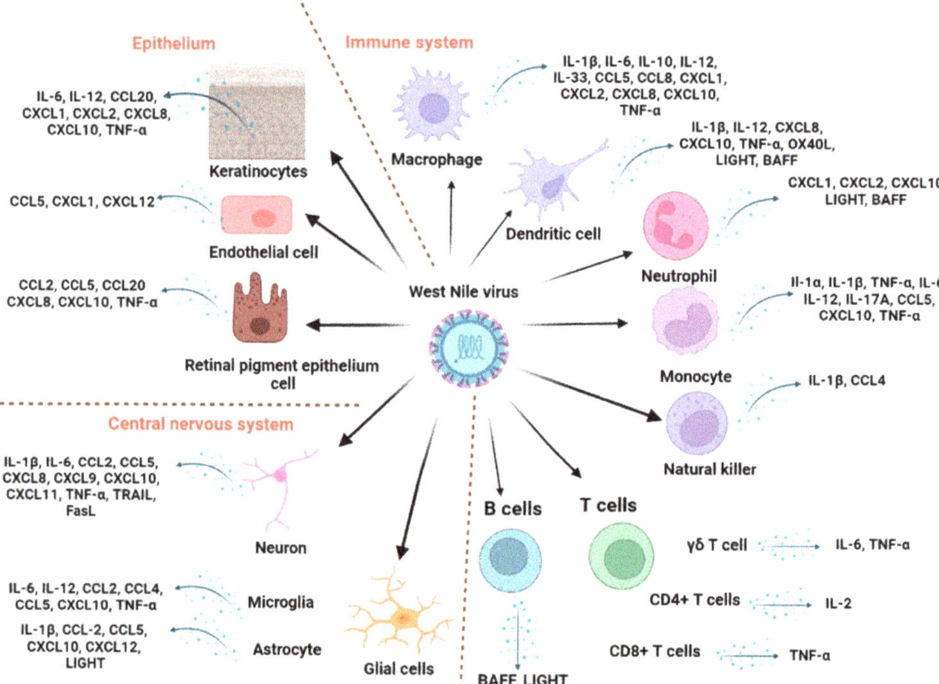

Figure 1. Cellular targets of West Nile virus and corresponding cytokine response in mammals. The illustration was created in Biorender.com. Abbreviations: BAFF: B-cell activating factor; FasL: Fas ligand; TNF-α: tumor necrosis factor -α, TRAIL: TNF-related apoptosis-inducing ligand.

There is mounting evidence that the host immune response, driven by cytokines, plays a pivotal role in the pathogenesis of WNV and the outcome of the disease. First, clinical data support that diverse cytokine profiles, depending on the sex [33], health condition [34], and human polymorphisms in these immune-coding genes [35–37] correlate with different outcomes of the infection, thus inciting the use of related biomarkers to predict the severity of WNV disease in a clinical setting [33]. Secondly, using cytokines as agonists or blocking their effects via pharmaceutical or genetic means in murine models demonstrated their ability to completely alter WNV-associated disease phenotypes [16,38–49]. Therefore, an improved understanding of the participation of cytokines in the pathogenesis of WNV may not only help optimize diagnosis and prognosis, but also guide research of immune-modulatory strategies to treat WNV-induced neurological disease.

In this review, we summarize findings from clinical studies as well as experiments conducted over the past two decades using in vitro and in vivo WNV infection models to recapitulate ILs, chemokines, and TNFSF ligands that participate in WNV infection, indicate those of a known relevance in WNV pathogenesis, and identify candidates needing further research to undercover their pertinence as therapeutic targets.

2. Interleukins (ILs) in WNV Infection

ILs are proteins that modulate cell growth, differentiation, and activation during antiviral response [50]. WNV induces the release of at least 22 ILs in mammalian hosts (Table S1). To date, IL-1β, IL-6, IL-10, IL-12, IL-17A, IL-22, and IL-23 have been directly investigated (Table 1), while little information is available about the remaining ILs involved in the immune response to WNV infection.

Table 1. Summary of interleukins, chemokines, chemokine receptors and tumor necrosis factor ligands triggered following West Nile virus (WNV) infection, whose pathogenesis has been studied in mice models in vivo.

	Survival Following Lethal WNV Challenge [1]	Role in WNV Pathogenesis [1]	References
		Interleukins	
IL-1 β	+	Langerhans cell migration to the draining LNs. Protective CNS-intrinsic immune response and leukocytes migration to the CNS. Synaptic deficits and spatial learning defects during recovery.	[16,18,32,51–65]
IL-6	N.E.	N/A	[19,20,52,65–75]
IL-10	−	Viral replication in the periphery and in the CNS and downregulation of IL-12/23 p40 and TNF-α in the CNS.	[19,32,61,64,68,73,76–79]
IL-12	N.E.	No role of IL-12 p35 in brain infiltration or homing of leukocytes.	[66,70,76,80–84]
IL-17 A	+	CD8+T cell cytotoxicity.	[43,76,85]
IL-22	−	WNV entry into the CNS via neutrophils Intrinsic control of viral replication in the brain CXCL1, CXCL5 and Cxcr2 expression in the brain	[42,61,76,86]
IL-23	+	Brain infiltration and homing of leukocytes	[76,82]
		CC chemokines	
CCL2	+/−	Monocytes migration and differentiation into DCs in the skin and LNs. Monocytosis and monocytes trafficking to the brain.	[19,20,32,44,45,47,56,64,65,69,76,77,84,85,87–92]

Table 1. *Cont.*

	Survival Following Lethal WNV Challenge [1]	Role in WNV Pathogenesis [1]	References
CCL7	+	Monocytosis, recruitment of neutrophils and CD8+ T cells into the CNS Viral clearance from the brain.	[44,45,47,76,85,91]
C chemokine receptors			
Ccr2	+	Monocytosis, monocytes migration to the brain and viral clearance from the CNS.	[44,76]
Ccr5	+	Leukocyte trafficking to the CNS/control of the BBB permeabilityViral clearance in the brain	[32,93]
Ccr7	+	DCs and T cell trafficking to the LNs Control of WNV-infected myeloid cells infiltration into the CNS.	[46]
CX chemokine			
CXCL10	+	Recruitment of CD8+ T-cells into the CNS.	[19,20,32,47–49,56,57,64,65,68,70,71,75–77,80,84,85,87,88,91,92,94–98]
CX chemokine receptors			
Cxcr2	−	N/A	[49]
Cxcr3	+	CD8+ T cells control of WNV infection within the cerebellum	[48,99]
Cxcr4	−	Downregulation of T cells trafficking to the brain	[100]
CX3C Chemokine			
CX3CL1	N.E.	Monocytes (microglial precursor) recruitment to the brain.	[32]
CX3C Chemokine receptor			
Cx3cr1	N.E.	N/A	[32,46,90]
Tumor necrosis factor superfamily ligands			
TNF-α	+/−	No effect in Langerhans cell migration to the draining lymph nodes. Regulation of leukocyte infiltration in the CNS. Down-regulation of neuronal Cxcr3 and subsequent neuronal apoptosis.	[17,19,20,32,52,55,57,61,62,64–70,72–77,84,87,88,95,99,101–108]
TRAIL	+	CD8+ T cells-mediated viral clearance in the CNS.	[38,69,109]
FasL	+/N.E.	CD8+ T cells-mediated viral clearance in the CNS.	[39,59,76,77,110]
CD40L	+	Efficient production of neutralizing antibodies, trafficking of CD8+ T cells into the brain, and control of WNV replication in the CNS.	[111]
BAFF	+	Viral clearance from sera, spleen, and brain.	[41,112]

[1] Data from experiments using mice (*Mus musculus*) in vivo as WNV-infection models. (+): Enhances survival rates; N.E: No effect (no difference between cytokine-deficient and control groups); (−): Enhances mortality rates; N/A: Not available. Abbreviations: BAFF: B-cell activating factor; BBB: Blood–brain barrier; CNS: Central nervous system; LNs: Lymph nodes; TNF-α: Tumor necrosis factor -α, TRAIL: TNF-related apoptosis-inducing ligand.

2.1. Interleukin-1 Family

Currently, 11 cytokines are considered members of the IL-1 family: IL-1α, IL-1β, the IL-1 receptor antagonist [IL-1ra], IL-18, IL-33, IL-36α, IL-36β, IL-36γ, IL-36ra, IL-37, and IL-38 [113,114]. Among them, IL-1α [19,76,87,115], IL-1β [10,16,51–53,80,94,116], IL-1ra [16,29,77], IL-18 [117], and IL-33 [118] are known to be released in response to WNV infection.

IL-1 is an extremely potent inflammatory cytokine, triggered in response to WNV infection both in vitro and in vivo, in the periphery and in the CNS (Table S1). The role of IL-1 in WNV infection has been studied mainly through murine models deficient in IL-1R1 and unable to respond to IL-1α, IL-1β, or IL-1ra [16,54,66,115]. IL-1R1 signaling conferred protection to mice against WNV disease and mortality [16]. During early WNV encephalitis, IL-1R1 controlled viral replication and subsequent apoptosis within neurons [16,115]. Further, IL-1 controlled leukocyte infiltration as well as T cell responses in the CNS [54,66,115] and restrained inflammation by downregulating pro-inflammatory cytokines, such as TNF-α and IL-6 [16] and chemokines, such as CCL2 and CCL5 [16,51,54]. Intracranial injection of WNV in wild-type C57BL/6 mice led to paradoxical results concerning the direct effect of IL-1 on viral replication in the brain. In fact, while IL-1 did not directly impact viral replication within the CNS in some studies [51,53,115], IL-1 was found to mediate CNS-intrinsic virus restriction in another study [16]. The disparities between these studies despite the use of the same model of infection could be in part explained by the differences in the viral strains used to infect the mice.

The pro-inflammatory IL-1α and anti-inflammatory IL-1ra cytokine expression patterns and roles during WNV infection are still not clear. In human sera, IL-1ra expression was variable in WNV-infected presymptomatic and asymptomatic donors [29], but upregulated during acute WNV infection [16,34]. While no human study has yet reported IL-1α regulation during WNV natural infection, IL-1α modulation during WNV infection varies across studies using experimental models (Table S1).

IL-1β is a key player in early, acute, and severe WNV pathogenesis. Indeed, IL-1β is one of the earliest cytokines detected following an infected mosquito bite in murine models [22,62,119]. Further, this cytokine mediated epidermal dendritic cell (DC) and Langerhans cell migration from the epidermis to the local draining lymph nodes [22,62,119]. In mouse brains, IL-1β was secreted during the acute phase, mainly by infiltrating/resident macrophages [54,63] and even later during recovery, mainly by astrocytes [63]. Current evidence demonstrates that IL-1β plays a dual role during WNV-induced disease, being protective during the acute phase, and driving neurological sequela in the long term. Mice deficient in both IL-1β signaling and Apoptosis-associated speck-like protein containing C-terminal caspase recruitment domain (ASC), which induces caspase-1-dependent inflammasome activation and IL-1β production, were found to increase WNV viral titers and disease severity specifically in the CNS [16,51]. Further, in humans with a history of asymptomatic or severe WNV infection, decreased IL-1β induction in their peripheral blood mononuclear cells and macrophages was a hallmark of severe disease [94]. In blood donors who tested positive for WNV RNA after routine blood screening, IL-1β was upregulated in the plasma for six months after their initial blood donation and correlated inversely with WNV RNA loads [16]. In mice, such a sustained IL-1β overexpression, induced specifically by the NOD-like receptor-pyrin-containing proteins 3 (NLRP3) inflammasome cleavage in astrocytes after the recovery period of WNV encephalitis [16,22,63,119], resulted in defective spatial learning and synaptic recovery [53,63]. Thus, improper NLRP3 inflammasome activation and IL-1β secretion in the brain is currently considered as a plausible mechanism for the development of long-term neurological sequela after WNV infection [120].

IL-18 is also a pro-inflammatory cytokine produced following inflammasome activation [121,122]. WNV infection of human primary monocyte-derived DCs or transformed human neuroblastoma cell line (SK-N-SH, ATCC HTB-11™) did not increase IL-18 production [52,117]. However, IL-18 was upregulated in spleen and lung tissues of WNV-infected mice [76]. IL-18 is suggested to further immunopathogenesis of DENV [123], but no investigations have been conducted yet to test this during WNV infection.

Another member of the IL-1 family, IL-33, was upregulated in splenic macrophages of WNV-infected mice [118]. IL-33 signaling through the ST2 receptor can trigger pro-inflammatory and anti-inflammatory responses [124,125]. Generally, in viral infections, IL-33 is considered a protective agent through enhancing CD8+ T cell responses [126] and attenuating viral encephalitis by downregulating iNOS expression in the CNS [127].

Therefore, with this understanding, promoting the activity or production of this cytokine during WNV infection could present therapeutic advantages. More work is needed to investigate the functions of this cytokine in the context of WNV infection.

2.2. Interleukin 6 Family

IL-6 is a pleiotropic cytokine involved in many biological processes, including immune responses, hematopoiesis, bone metabolism, and embryonic development [128]. It is one of the most important cytokines during a viral infection [129], and studies using different experimental models describe IL-6 changes during WNV infection (Table S1). Human cytokine studies following WNV infection highly suggest an important role for IL-6. Acute infection in humans could induce a high synthesis of IL-6 in the CSF [130] and the serum [130,131] of patients with WNV fever and WNV neuroinvasive disease. Further, in another study, IL-6 levels in the serum were lower in healthy viremic individuals compared to uninfected individuals before and after IgM seroconversion [124]. IL-6 prolonged expression has been reported in the serum of individuals who experience severe long-term fatigue following symptomatic WNV infection [132]; however, no studies have yet been performed to support the causal relationship between IL-6 levels and WNV-associated disease severity in humans. A single in vivo study investigated the involvement of IL-6 in WNV infection [67] and described that, when infected with WNV, IL-6-deficient mice exhibited similar mortality rates as wild-type mice [67]. Further work is needed to clarify whether this is due to a minor role of this cytokine in WNV infection or to the specific experimental conditions used in this study.

2.3. Interleukin 17 Family

Currently, there are 6 inflammatory cytokines that represent the IL-17 family, namely IL-17A, IL-17B, IL-17C, IL-17D, IL-17E, and IL-17F [133]. Among these, IL-17A, a pro-inflammatory cytokine, is upregulated in vitro [43] and in vivo [43,76,85] following WNV infection (Table S1). In humans, in the absence of symptoms, increased IL-17 levels were found in WNV-infected individuals when compared to levels from non-infected blood donors [29]. On the contrary, very low serum levels of IL-17A, as well as complete absence of IL-17A expression in the CSF [130], were found in both febrile and neuroinvasive disease patients. In four serologically confirmed WNV patients with persistent post-infectious symptoms, no increase in IL-17 could be detected [125]. These findings, suggesting a link between IL-17A expression and a favorable outcome of WNV human infection, are supported by one in vivo study in mice, wherein they found that IL-17A facilitated WNV clearance by inducing expression of cytotoxic mediator genes and promoting CD8+ T cell cytotoxicity [43].

2.4. Interleukin 12 Family

The IL-12 family includes four members: IL-12, IL-23, IL-27, and IL-35 [134], among which IL-12 and IL-23 are upregulated in vivo following WNV infection (Table S1). IL-12 is composed of two covalently linked subunits, p40 and p35, which form when combined with the bioactive IL-12p70 [134]. IL-23 also comprises two subunits, p19 and p40, and the latter is shared with IL-12 [134]. Currently, no human studies have highlighted IL-23 changes following WNV infection, but IL-12 was reported to be highly expressed in presymptomatic and asymptomatic WNV-infected blood donors [29] and unchanged in symptomatic WNV-infected blood donors during the early phase of infection [34]. Cytokine analyses in WNV-infected individuals confirmed that IL-12p70 could be overexpressed in the serum for months [132] and even years [135] after infection. Mice deficient in individual subunits of IL-12 (p35) or IL-23 (p19) or the shared p40 subunit were used to determine the specific role of each cytokine. Animals deficient in IL-12p40 or IL-23p19, but not IL-12p35, had decreased leukocyte homing to the brains and increased mortality, supporting the importance of IL-23 in protective immune cell infiltration and homing during the acute phase of infection [82].

More research is warranted to cast light on the participation of these cytokines during the recovery phase from WNV infection.

2.5. Interleukin 10 Family

The IL-10 family of cytokines contains IL-10, IL-19, IL-20, IL-22, IL-24, IL-26, IL-28, and IL-29 [114], among which IL-10 and IL-22 expression are upregulated in models of WNV infection (Table S1). Increased levels of IL-10 were detected in the plasma of acute, viremic [124], asymptomatic blood donors diagnosed with WNV [29]. However, no significant difference in the IL-10 in serum [130,131] and CSF [130] samples from patients with WNV fever and WNV neuroinvasive disease was found [130]. Genetic or pharmacologic blockade of IL-10 signaling helped to increase survival after WNV lethal challenge in mice [79], and additional studies corroborate a pathogenic role of IL-10 in acute WNV infection. First, previous sensitization to salivary proteins delivered by multiple *A. aegypti* bites resulted in an increased IL-10 expression associated with an aggravated disease [136]. Second, in mice infected with a hamster-derived WNV strain, a decreased IL-10 production correlated with lower frequency of virus persistence in the spleen compared to that of WNV NY99-infected mice [78]. The only study to investigate IL-22 described a minimal effect in the periphery, but mice deficient in IL-22 were more resistant to lethal WNV infection, and IL-22 promoted early entry of virus-carrying neutrophils into the CNS by regulating chemotaxis (mainly via Cxcr2 signaling) at the blood–brain barrier (BBB) [42].

3. Chemokines in WNV Infection

Chemokines are chemotactic cytokines that bind to G protein-coupled receptors to direct cell movement during homeostasis and inflammation [114]. These proteins are divided into four subfamilies: C chemokine, CC chemokine, CXC chemokine, and CX3C chemokine, based on the number and positioning of conserved N-terminal cysteine residues [114]. Changes in the expression of chemokines and their receptors have been observed in response to WNV infection in mammalian models (Table S2). Studies focusing on chemokines receptors, including Ccr2, Ccr5, Ccr7, Cxcr2, Cxcr3, Cxcr4 and Cx3cr1 in WNV-infected models have helped define the importance of chemokines in a time- and organ-specific manner (Table 1). However, each of these receptors can be bound to several chemokines, and few reports concerning the participation of chemokines in WNV infection exist to date, including CCL2, CCL7 and CXCL10 (Table 1). Thus, the precise importance of individual chemokines during WNV infection demands additional research.

3.1. CC Chemokines

3.1.1. CCL2, CCL7 and CCL12 (Ccr2 Agonists)

Ccr2 and its ligands play important roles in monocytes mobilization under inflammatory conditions [137], and they can be induced following experimental WNV infections (Table S2). Ccr2 agonist CCL2 is highly expressed during human WNV infections. The CCL2 gene expression is upregulated in the brain tissues of patients that succumb to WNV encephalomyelitis [56]. Accordingly, CCL2 production was significantly elevated in the serum of WNV-infected patients [124], with male blood donors having higher levels of CCL2 than the female blood donors in the post-IgM phase [33]. Further, elevation of CCL2 in the post-IgM seroconversion phase was associated with improved symptom outcomes following WNV infection [33,34]. During WNV infection, Ccr2 activation induced monocytosis dependent on CCL2 and CCL7, but not CCL12, and protected mice from lethal challenge mainly by regulating blood monocyte levels [44]. CCL2 mediated monocyte migration to the infected dermis and the draining lymph node, as well as their return from the blood to the bone marrow and their differentiation into DCs during the early phase in WNV-infected mice [89]. CCL2 also mediated accumulation of inflammatory monocytes into the brain, and their differentiation to microglia decreased survival, thus playing a pathogenic role in WNV encephalitis [90]. However, in another study using murine models, CCL2 was only partly involved in monocyte recruitment and did not play a pivotal role in survival following lethal challenge [45]. In contrast, CCL7 deficiency resulted in increased viral burden in the brain, enhanced mortality, and delayed migration of neutrophils and CD8+ T cells into the CNS [45]. CCL7 was significantly decreased in humans with a worse outcome as compared to those with a better outcome in the post-IgM phase [34]. Whereas the role of CCL2 in WNV pathogenesis remains unresolved, CCL7 appears to have favorable effects, improving the outcome of WNV infection.

3.1.2. CCL3, CCL4, and CCL5 (Ccr5 Agonists)

CCR5 and its interaction with chemokine ligands mediate chemotactic activity in leukocytes and are involved in hematopoiesis and immune response [138]. CCL3, 4 and 5 chemokines binding the chemokine receptor Ccr5 could not be detected in human sera during the early and late phases of infection [34,124] but were strongly induced in the CNS of mice after experimental WNV infection (Table S2). In humans, Ccr5 deficiency did not predispose to WNV infection, but once infected, patients could be particularly susceptible to present early and late clinical manifestations if their Ccr5 function was missing or blocked [35,36,139]. In line with these findings, studies in mice describe that Ccr5 deficiency led to increased symptomatic disease and mortality after subcutaneous infection with WNV, although Ccr5 was not required for cell-mediated immunity in the periphery [32,93]. WNV-infected Ccr5-/- mice had a significantly decreased ability to recruit antiviral mononuclear cells specifically into their WNV-infected brain, increased BBB permeability, and elevated levels of Ccr5 ligands [32,93]. The individual roles of the Ccr5 ligands remain unclear, as no in vivo models have been applied to address their contribution in WNV pathogenesis. One in vitro study described that induction of CCL5 in response to WNV infection was not sufficient to promote leukocyte transmigration across the endothelial layer in a model of the BBB containing both endothelial cells and astrocytes [140].

3.1.3. CCL19 and CCL21 (Ccr7 Agonists)

Interactions between Ccr7 and its cognate ligands are involved in the induction of inflammatory and T cell responses [141]. WNV infection in murine models supports that Ccr7 and ligands CCL19 and CCL21 could be upregulated at the gene levels [64,76,110] and contribute to the host resistance against WNV. Chemokine receptor Ccr7 was essential for survival following a WNV lethal challenge in mice [46]. Further, Ccr7 was required for myeloid cells' infiltration into the lymph nodes and restricted their entry into the brain, assisting in viral clearance and decreasing pathological effects of an excessive cytokine production [46].

3.2. CXC Chemokines

3.2.1. CXCL1-3, CXCL6-8 (Cxcr2 Agonists)

Cxcr2 plays a nonredundant role in mediating trafficking of neutrophils, which are suggested as a carrier of WNV in the blood [49]. Mice deficient in Cxcr2 had a similar mortality rate as wild-type mice, although their time to death was delayed [49]. Cxcr2 binds to CXCL1, CXCL2, CXCL3, CXCL5, CXCL6, CXCL7, and CXCL8 [114], which are all modulated by WNV infection (Table S2). CXCL8 is upregulated in WNV-infected primary human cultures [88] and cell lines [18,52,57,75,88,98] as well as brain and spinal cord samples from experimentally infected Rhesus monkeys (*Macaca mulatta*) [77]. Genes associated with CXCL8 production and upregulation were induced in New Zealand white rabbits (*Oryctolagus cuniculus*) [68]. CXCL8 is also detected at high levels in WNV-infected individuals [131,135]. Further, patients with more severe symptoms during the early phase of infection had significantly higher CXCL8 expression in their serum when compared to WNV-negative controls [34]. These findings suggest an important role of this cytokine in the pathogenesis of natural infections in humans. Yet, to date, no in vivo studies investigate this observation. This can be explained by the lack of true CXCL8 homologs in mice [142], which are currently the most widely used animal models to study WNV pathogenesis.

Using alternative WNV models of infection that have an ortholog CXCL8 gene, such as rabbits [70,143–146] and nonhuman primates, [77,147] will be necessary to circumvent this issue.

3.2.2. CXCL9 and CXCL10 (Cxcr3 Ligands)

Cxcr3 and its ligands are responsible for T-cell trafficking, activation, differentiation, and functions [48]. WNV natural infection in humans can induce high levels of CXCL9 [124] and CXCL10 [34,124,131,135] in the serum. Likewise, these chemokines were elevated following WNV infection in various experimental models (Table S2). Evidence from these models suggests that Cxcr3 signaling can have multifaced roles during WNV infection. In vitro, downregulation of neuronal CXCR3 signaling through TNF receptor 1 (TNFR1) decreased CXCL10 and resultant apoptosis following WNV infection [99]. In vivo, Cxcr3 had no effect in WNV replication or clearance in peripheral lymphoid tissues [47]. However, CXCL10, but not CXCL9, and its cognate receptor Cxcr3 were required for survival after lethal WNV challenge and regulated the CD8+ T cells migration and clearing of WNV infection in the brain compared to control mice [47,48]. This can explain the evidence of both protective and deleterious effects of CXCL10 in humans. In fact, higher susceptibility to WNV in blood donors was marked by lower levels of CXCL10/IP-10 during the post-IgM phase [33,34]. Importantly, analysis of autopsied neural tissues from humans with WNV encephalomyelitis revealed upregulation of CXCL10-coding gene [56] and symptom development was positively correlated with CXCL10/IP-10 production during the earliest phase of disease [34]. In later stages, significantly higher serum levels of CXCL10 were detected in patients with prolonged post-infection fatigue (>6 months) after symptomatic WNV infection [132]. Thus, CXCL10 transition from driving protective immune responses to deleterious ones needs further research as a possible therapeutic target.

3.2.3. CXCL12 (Cxcr4 Ligand)

Cxcr4 is the most widely expressed chemokine receptor and is involved in cell migration, hematopoiesis, and cell homing [148]. Changes in the expression of Cxcr4 and its canonical ligands CXCL12 can be induced following experimental WNV infection (Table S2), whereas their expression patterns in WNV-infected patients is still unclear. Current evidence from experimental infections suggests that CXCL12 favors WNV neuropathogenesis. In fact, CXCL12 expression, which was mediated by IL-1 at the CNS microvasculature [54], restricted entry of T cells at the BBB and prevented virus-specific CD8+ T cells from clearing WNV within the CNS parenchyma, resulting in enhanced mortality in a murine model of infection [100].

3.3. CX3C Chemokines

Chemokine CX3CL1 and its receptor CX3CR1 can exert pro-inflammatory or anti-inflammatory responses [149]. Their coding genes were upregulated in vivo following WNV infection in B6129PF2 and C57BL6/J mice [32] and Rhesus monkeys [77] (Table S2). Investigation in a murine model did not support their having a role that aids in survival against WNV infection [32].

4. Tumor Necrosis Factor Superfamily Ligands

Interactions between TNFSF ligands and their cognate receptors control the survival, proliferation, differentiation, and functions of immune cells [111]. Among the TNFSF ligands, TNF-α [72,99,103–105], Fas ligand (FasL) [39,76,110], TNF-related apoptosis-inducing ligand (TRAIL) [69,109,110], CD40L [85,111], B-cell activating factor (BAFF) [112], TNF-related weak inducer of apoptosis (TWEAK) [85], OX40L [117], and tumor necrosis factor superfamily member 14 (LIGHT) [76,150] are implicated in WNV pathogenesis (Table S3).

4.1. Tumor Necrosis Factor α

TNF-α, a cytokine having pro- and anti-inflammatory properties [151], has inconsistent expression patterns following WNV infection in humans. Human sera analysis during WNV fever and WNV neuroinvasive disease showed no detectable change in TNF-α expression [131], but others describe a significant TNF-α upregulation in WNV-infected patients during the acute phase [124,130] and even long after the virus had presumably been cleared by the immune system [125]. TNF-α was significantly higher in individuals with a history of WNV infection and subsequent development of chronic kidney disease compared to healthy individuals [135]. In accordance with the latter human reports, almost all studies using experimental models describe increased TNF-α during WNV infection (Table S3). Studies investigating the importance of TNF-α in the pathogenesis of WNV infection indicate that this cytokine has a limited role in controlling WNV infection in peripheral organs [38,104], with no consensus regarding the signaling cascade and contribution to control WNV in the CNS (Table S3). For instance, TNF-α receptor 1 (TNF-R1) signaling was suggested downstream of Toll-like receptor (TLR)-3, as TLR3 deficiency led to impaired TNF-α production during WNV infection in microglia [67], but the same observation did not occur in bone marrow-derived DCs [152]. Whereas in one study, mice deficient in TNF-R1 had a mortality rate significantly greater than that in wild-type mice after WNV challenge [104], the opposite phenomenon was observed in another study using the same model [67]. The former study suggested that TNF-α interaction with TNF-R1 protected mice against WNV infection by regulating migration of inflammatory cells into the brain during acute infection [104], while the latter suggested that TNFα could be responsible for early WNV neuroinvasion due to increased permeability of the BBB [67]. Immunization of mice with salivary gland components led to early production of TNF-α following WNV infection, which aligned with a delay in CNS infection and significantly lower WNV brain titers compared to mock-immunized mice [153], suggesting a protective role during WNV encephalitis. However, another study described higher TNF-α levels that corroborated an increased pathogenicity of neuroinvasive WNV variants compared to non-neuroinvasive variants in mice, [73] and TNF-α was involved in WNV-induced neurotoxicity [52]. Additional research is necessary.

4.2. TRAIL and FasL

TRAIL and FasL activate apoptosis through cell surface death receptors [111]. These cytokines are upregulated on the gene levels using in vivo models, including mice (Table S3). In mouse models, TRAIL contributes to disease resolution [38], while the role of FasL remains elusive [28,39]. In mice, TRAIL genetic deficiency increased susceptibility to lethal WNV challenge, and CD8+ T cells encountered difficulty in clearing WNV from neurons [38]. WNV induced expression of Fas in neurons, functional FasL was required to protect IFNγ-deficient C57BL/6 mice from lethal WNV infection, and CD8+ T cells utilized FasL to restrict WNV infection in neurons [39]. However, another study using the same mice deficient in either Fas or FasL did not find differences in mortality or viral burden in the brain [28]. Inconsistent results from these studies could be attributed to the differences in the viral strains (WNV 3000.0259 strain [39] versus WNV Sarafend strain [28]) as well as the route of animal infection (footpad [39] versus intravenous route [28]).

4.3. CD40L

CD40L is a modulator of a wide range of humoral and cellular immune responses [111] and is regulated by WNV infection in the mouse brain [64]. In mice, CD40-CD40L interactions were required for protection from lethal WNV challenge, efficient antibody production by B cells, and T-cell migration across the BBB [40]. While there is evidence to suggest a role for CD40L in WNV infection, more research is needed.

4.4. BAFF

BAFF is required for peripheral B-cell survival and homeostasis and is upregulated in mice neutrophils and DCs after WNV challenge [112]. BAFF signaling was essential for survival against lethal WNV infection in mice [41]. BAFF from DCs, not neutrophils, helped to sustain or promote B-cell humoral responses to WNV, since WNV-specific antibody responses were decreased in mice lacking BAFF expression on DCs [112]. Further, BAFF receptor-deficient mice were susceptible to WNV infection but could develop sustained protective immunity when treated with immune sera from a wild-type mouse with antibodies to WNV [41].

5. Conclusions

Cytokine characterization represents a major advance in our understanding of the overall regulation of WNV-driven immune responses. Cytokine signaling of IL-1β, IL-23, IL-17A, CCL7, CXCL10, TRAIL, CD40L, and BAFF provides protection against acute WNV infection in mice; IL-10 and IL-22 aid in WNV pathogenicity; IL-6 and IL-12 had no apparent effect during infection; and CCL2, TNF-α, and FasL roles remain elusive. Determining the exact function of a particular cytokine can be challenging and underscores the most important messages from this review: First, the biological context, such as the cellular source, the target, the phase of the immune response, and the presence or absence of other cytokines influences their expression pattern and function. Experimental conditions, varying across the studies, such as viral strains or passages, laboratory investigation techniques, and time-points of sample harvesting, might also explain inconsistent, sometimes paradoxical results regarding the roles of cytokines during WNV infection. Second, the outcomes of WNV infection depend not only on viral clearance but also on the extent of the inflammatory response driven by cytokines. In fact, WNV infections in humans and laboratory animals provide evidence that pro-inflammatory cytokines, such as IL-1β, TNF-α, IL-12p70, CXCL10, and IL-6, can be chronically elevated after WNV is cleared. This indicates that an effective treatment against WNV neuroinvasive disease should include anti-inflammatory drugs to treat the exacerbated inflammatory response during the acute phase and to prevent long-term neurological sequalae, as these cytokines are linked to neuronal injury in several neurodegenerative diseases [154]. Future studies are critical to understanding how regulation of these cytokines can improve the course of illness. This can be accomplished by studying existing drugs or small molecules against the aforementioned cytokines, as well as development of new therapeutics that interfere with these cytokine pathways. Finally, this review highlights the need for additional research into these cytokines, considering the biological importance they maintain, which will help identify immunomodulatory therapeutic targets against WNV neuroinvasive disease. For example, alternative infection models should be developed for the study of CXCL8, hampered to date by the lack of true homologs in rats and mice. More work aimed at dissecting the roles of important cytokines depicted from clinical human studies, such as IL-15, CCL8, CCL11, CCL13, and CCL20 is warranted to understand their contribution to the immunopathogenesis of WNV infection.

Supplementary Materials: The following supporting information can be downloaded at: https://www.mdpi.com/article/10.3390/v15030806/s1, Table S1: Interleukins involved in West Nile virus pathogenesis; Table S2: Chemokines involved in West Nile virus pathogenesis; Table S3: Tumor necrosis factor superfamily ligands involved in West Nile virus pathogenesis.

Author Contributions: Conceptualization, E.B., K.O.M. and S.E.R.; methodology, E.B. and S.E.R.; writing—original draft preparation, E.B., K.O.M. and S.E.R.; writing—review and editing, E.B., K.O.M. and S.E.R. All authors have read and agreed to the published version of the manuscript.

Funding: Authors' effort was funded by The Brockman Foundation.

Institutional Review Board Statement: Not applicable.

Informed Consent Statement: Not applicable.

Data Availability Statement: Not applicable.

Conflicts of Interest: The authors declare no conflict of interest.

References

1. Chambers, T.J.; Hahn, C.S.; Galler, R.; Rice, C.M. Flavivirus Genome Organization, Expression, and Replication. *Annu. Rev. Microbiol.* **1990**, *44*, 649–688. [CrossRef]
2. McLean, R.G.; Ubico, S.R.; Docherty, D.E.; Hansen, W.R.; Sileo, L.; McNamara, T.S. West Nile Virus Transmission and Ecology in Birds. *Ann. N. Y. Acad. Sci.* **2001**, *951*, 54–57. [CrossRef] [PubMed]
3. Chen, S.; Wu, Z.; Wang, M.; Cheng, A. Innate Immune Evasion Mediated by Flaviviridae Non-Structural Proteins. *Viruses* **2017**, *9*, 291. [CrossRef] [PubMed]
4. Roosendaal, J.; Westaway, E.G.; Khromykh, A.; Mackenzie, J.M. Regulated Cleavages at the West Nile Virus NS4A-2K-NS4B Junctions Play a Major Role in Rearranging Cytoplasmic Membranes and Golgi Trafficking of the NS4A Protein. *J. Virol.* **2006**, *80*, 4623–4632. [CrossRef]
5. Samuel, C.E. Host Genetic Variability and West Nile Virus Susceptibility. *Proc. Natl. Acad. Sci. USA* **2002**, *99*, 11555–11557. [CrossRef] [PubMed]
6. Bai, F.; Ashley Thompson, E.; Vig, P.J.S.; Arturo Leis, A. Current Understanding of West Nile Virus Clinical Manifestations, Immune Responses, Neuroinvasion, and Immunotherapeutic Implications. *Pathogens* **2019**, *8*, 193. [CrossRef] [PubMed]
7. Petersen, L.R.; Brault, A.C.; Nasci, R.S.; Infectious, Z.; Services, H.; Collins, F. West Nile Virus: Review of the Literature. *JAMA* **2013**, *310*, 308–315. [CrossRef]
8. Alli, A.; Ortiz, J.F.; Atoot, A.; Atoot, A.; Millhouse, P.W. Management of West Nile Encephalitis: An Uncommon Complication of West Nile Virus. *Cureus* **2021**, *13*, e13183. [CrossRef]
9. Ouhoumanne, N.; Lowe, A.M.; Fortin, A.; Kairy, D.; Vibien, A.; K-Lensch, J.; Tannenbaum, T.N.; Milord, F. Morbidity, Mortality and Long-Term Sequelae of West Nile Virus Disease in Québec. *Epidemiol. Infect.* **2018**, *146*, 867–874. [CrossRef]
10. Fulton, C.D.M.; Beasley, D.W.C.; Bente, D.A.; Dineley, K.T. Long-Term, West Nile Virus-Induced Neurological Changes: A Comparison of Patients and Rodent Models. *Brain Behav. Immun. Health* **2020**, *7*, 100105. [CrossRef]
11. Weatherhead, J.E.; Miller, V.E.; Garcia, M.N.; Hasbun, R.; Salazar, L.; Dimachkie, M.M.; Murray, K.O. Long-Term Neurological Outcomes in West Nile Virus-Infected Patients: An Observational Study. *Am. J. Trop. Med. Hyg.* **2015**, *92*, 1006–1012. [CrossRef] [PubMed]
12. Tebas, P.; Spitsin, S.; Barrett, J.S.; Tuluc, F.; Elci, O.; Korelitz, J.J.; Wagner, W.; Winters, A.; Kim, D.; Catalano, R.; et al. Reduction of Soluble CD163, Substance P, Programmed Death 1 and Inflammatory Markers: Phase 1B Trial of Aprepitant in HIV-1-Infected Adults. *AIDS* **2015**, *29*, 931–939. [CrossRef] [PubMed]
13. Suthar, M.S.; Diamond, M.S.; Gale, M., Jr. West Nile Virus Infection and Immunity. *Nat. Rev. Microbiol.* **2013**, *11*, 115–128. [CrossRef]
14. Shrestha, B.; Diamond, M.S. Role of CD8 + T Cells in Control of West Nile Virus Infection. *J. Virol.* **2004**, *78*, 8312–8321. [CrossRef] [PubMed]
15. Zimmerman, M.G.; Bowen, J.R.; McDonald, C.E.; Pulendran, B.; Suthar, M.S. West Nile Virus Infection Blocks Inflammatory Response and T Cell Costimulatory Capacity of Human Monocyte-Derived Dendritic Cells. *J. Virol.* **2019**, *93*, e00664-19. [CrossRef]
16. Ramos, H.J.; Lanteri, M.C.; Blahnik, G.; Negash, A.; Suthar, M.S.; Brassil, M.M.; Sodhi, K.; Treuting, P.M.; Busch, M.P.; Norris, P.J.; et al. IL-1β Signaling Promotes CNS-Intrinsic Immune Control of West Nile Virus Infection. *PLoS Pathog.* **2012**, *8*, e1003039. [CrossRef]
17. Martina, B.E.E.; Koraka, P.; van den Doel, P.; Rimmelzwaan, G.F.; Haagmans, B.L.; Osterhaus, A.D.M.E. DC-SIGN Enhances Infection of Cells with Glycosylated West Nile Virus in Vitro and Virus Replication in Human Dendritic Cells Induces Production of IFN-α and TNF-α. *Virus Res.* **2008**, *135*, 64–71. [CrossRef]
18. Kong, K.; Wang, X.; Anderson, J.; Fikrig, E.; Montgomery, R.R. West Nile Virus Attenuates Activation of Primary Human Macrophages. *Viral Immunol.* **2008**, *21*, 78–82. [CrossRef]
19. Kumar, M.; Roe, K.; O'Connell, M.; Nerurkar, V.R. Induction of Virus-Specific Effector Immune Cell Response Limits Virus Replication and Severe Disease in Mice Infected with Non-Lethal West Nile Virus Eg101 Strain. *J. Neuroinflamm.* **2015**, *12*, 178. [CrossRef]
20. Cheeran, M.C.J.; Hu, S.; Sheng, W.S.; Rashid, A.; Peterson, P.K.; Lokensgard, J.R. Differential Responses of Human Brain Cells to West Nile Virus Infection. *J. NeuroVirol.* **2005**, *11*, 512–524. [CrossRef]
21. Kosch, R.; Delarocque, J.; Claus, P.; Becker, S.C.; Jung, K. Gene Expression Profiles in Neurological Tissues during West Nile Virus Infection: A Critical Meta-Analysis. *BMC Genom.* **2018**, *19*, 530. [CrossRef] [PubMed]
22. Hassert, M.; Brien, J.D.; Pinto, A.K. The Temporal Role of Cytokines in Flavivirus Protection and Pathogenesis. *Curr. Clin. Microbiol. Rep.* **2019**, *6*, 25–33. [CrossRef]
23. Pan, Y.; Cai, W.; Cheng, A.; Wang, M.; Yin, Z.; Jia, R. Flaviviruses: Innate Immunity, Inflammasome Activation, Inflammatory Cell Death, and Cytokines. *Front. Immunol.* **2022**, *13*, 829433. [CrossRef]
24. Bardina, S.V.; Lim, J.K. The Role of Chemokines in the Pathogenesis of Neurotropic Flaviviruses. *Immunol. Res.* **2012**, *54*, 121–132. [CrossRef]

25. Lee, Y.H.; Leong, W.Y.; Wilder-Smith, A. Markers of Dengue Severity: A Systematic Review of Cytokines and Chemokines. *J. Gen. Virol.* **2016**, *97*, 3103–3119. [CrossRef]
26. Kuczera, D.; Assolini, J.P.; Tomiotto-Pellissier, F.; Pavanelli, W.R.; Silveira, G.F. Highlights for Dengue Immunopathogenesis: Antibody-Dependent Enhancement, Cytokine Storm, and Beyond. *J. Interferon Cytokine Res.* **2018**, *38*, 69–80. [CrossRef]
27. Maucourant, C.; Queiroz, G.A.N.; Samri, A.; Grassi, M.F.R.; Yssel, H.; Vieillard, V. Zika Virus in the Eye of the Cytokine Storm. *Eur. Cytokine Netw.* **2019**, *30*, 74–81. [CrossRef] [PubMed]
28. Wang, Y.; Lobigs, M.; Lee, E.V.A.; Müllbacher, A. Exocytosis and Fas Mediated Cytolytic Mechanisms Exert Protection from West Nile Virus Induced Encephalitis in Mice. *Immunol. Cell Biol.* **2004**, *82*, 170–173. [CrossRef]
29. Fares-Gusmao, R.; Rocha, B.C.; Sippert, E.; Lanteri, M.C.; Áñez, G.; Rios, M. Differential Pattern of Soluble Immune Markers in Asymptomatic Dengue, West Nile and Zika Virus Infections. *Sci. Rep.* **2019**, *9*, 17172. [CrossRef]
30. Clark, D.C.; Brault, A.C.; Hunsperger, E. The Contribution of Rodent Models to the Pathological Assessment of Flaviviral Infections of the Central Nervous System. *Arch. Virol.* **2015**, *157*, 1423–1440. [CrossRef]
31. Zimmerman, M.G.; Bowen, J.R.; McDonald, C.E.; Young, E.; Baric, R.S.; Pulendran, B.; Suthar, M.S. STAT5: A Target of Antagonism by Neurotropic Flaviviruses. *J. Virol.* **2019**, *93*, e00665-19. [CrossRef] [PubMed]
32. Glass, W.G.; Lim, J.K.; Cholera, R.; Pletnev, A.G.; Gao, J.L.; Murphy, P.M. Chemokine Receptor CCR5 Promotes Leukocyte Trafficking to the Brain and Survival in West Nile Virus Infection. *J. Exp. Med.* **2005**, *202*, 1087–1098. [CrossRef] [PubMed]
33. Hoffman, K.W.; Lee, J.J.; Foster, G.A.; Krysztof, D.; Stramer, S.L.; Lim, J.K. Sex Differences in Cytokine Production Following West Nile Virus Infection: Implications for Symptom Manifestation. *Pathog. Dis.* **2019**, *77*, ftz016. [CrossRef] [PubMed]
34. Hoffman, K.W.; Sachs, D.; Bardina, S.V.; Michlmayr, D.; Rodriguez, C.A.; Sum, J.; Foster, G.A.; Krysztof, D.; Stramer, S.L.; Lim, J.K. Differences in Early Cytokine Production Are Associated with Development of a Greater Number of Symptoms Following West Nile Virus Infection. *J. Infect. Dis.* **2016**, *214*, 634–643. [CrossRef]
35. Lim, J.K.; Louie, C.Y.; Glaser, C.; Jean, C.; Johnson, B.; Johnson, H.; McDermott, D.H.; Murphy, P.M. Genetic Deficiency of Chemokine Receptor CCR5 Is a Strong Risk Factor for Symptomatic West Nile Virus Infection: A Meta-Analysis of 4 Cohorts in the US Epidemic. *J. Infect. Dis.* **2008**, *197*, 262–265. [CrossRef]
36. Glass, W.G.; Mcdermott, D.H.; Lim, J.K.; Lekhong, S.; Yu, S.F.; Frank, W.A.; Pape, J.; Cheshier, R.C.; Murphy, P.M. CCR5 Deficiency Increases Risk of Symptomatic West Nile Virus Infection. *J. Exp. Med.* **2006**, *203*, 35–40. [CrossRef]
37. Rituparna, D.; Kerry, G.; Charles, M.; Feng, Q.; Lin, L.; Yan, S.; Ruth, R.M.; Mark, L.; Richard, B. Association between High Expression Macrophage Migration Inhibitory Factor (MIF) Alleles and West Nile Virus Encephalitis. *Cytokine* **2016**, *78*, 51–54. [CrossRef]
38. Shrestha, B.; Pinto, A.K.; Green, S.; Bosch, I.; Diamond, M.S. CD8+ T Cells Use TRAIL To Restrict West Nile Virus Pathogenesis by Controlling Infection in Neurons. *J. Virol.* **2012**, *86*, 8937–8948. [CrossRef] [PubMed]
39. Shrestha, B.; Diamond, M.S. Fas Ligand Interactions Contribute to CD8 + T-Cell-Mediated Control of West Nile Virus Infection in the Central Nervous System. *J. Virol.* **2007**, *81*, 11749–11757. [CrossRef]
40. Sitati, E.; Mccandless, E.E.; Klein, R.S.; Diamond, M.S. CD40-CD40 Ligand Interactions Promote Trafficking of CD8 + T Cells into the Brain and Protection against West Nile Virus Encephalitis. *J. Virol.* **2007**, *81*, 9801–9811. [CrossRef]
41. Giordano, D.; Draves, K.E.; Young, L.B.; Roe, K.; Bryan, M.A.; Dresch, C.; Richner, J.M.; Diamond, M.S.; Gale, M.; Clark, E.A. Protection of Mice Deficient in Mature B Cells from West Nile Virus Infection by Passive and Active Immunization. *PLoS Pathog.* **2017**, *13*, e1006743. [CrossRef] [PubMed]
42. Wang, P.; Bai, F.; Zenewicz, L.A.; Dai, J.; Gate, D.; Cheng, G.; Yang, L.; Qian, F.; Yuan, X.; Montgomery, R.R.; et al. IL-22 Signaling Contributes to West Nile Encephalitis Pathogenesis. *PLoS ONE* **2012**, *7*, e44153. [CrossRef] [PubMed]
43. Acharya, D.; Wang, P.; Paul, A.; Dai, J.; Gate, D.; Lowery, J.; Stokic, D.; Leis, A.; Flavell, R.; Town, T.; et al. Interleukin-17A Promotes CD8+ T Cell Cytotoxicity to Facilitate West Nile Virus Clearance. *J. Virol.* **2017**, *91*, e01529-16. [CrossRef] [PubMed]
44. Lim, J.K.; Obara, C.J.; Rivollier, A.; Pletnev, A.G.; Kelsall, B.L.; Murphy, P.M. Chemokine Receptor CCR2 Is Critical for Monocyte Accumulation and Survival in West Nile Virus Encephalitis. *J. Immunol.* **2011**, *186*, 471–478. [CrossRef]
45. Bardina, S.V.; Michlmay, D.; Hoffman, K.W.; Obara, C.J.; Sum, J.; Charo, I.F.; Lu, W.; Pletnev, A.G.; Lim, J.K. Differential Roles of Chemokines CCL2 and CCL7 in Monocytosis and Leukocyte Migration during West Nile Virus Infection. *Physiol. Behav.* **2015**, *195*, 4306–4318. [CrossRef]
46. Bardina, S.V.; Brown, J.A.; Michlmayr, D.; Hoffman, K.W. Chemokine Receptor Ccr7 Restricts Fatal West Nile Virus Encephalitis. *J. Virol.* **2017**, *91*, e02409-16. [CrossRef] [PubMed]
47. Klein, R.S.; Lin, E.; Zhang, B.; Luster, A.D.; Tollett, J.; Samuel, M.A.; Engle, M.; Diamond, M.S. Neuronal CXCL10 Directs CD8 + T-Cell Recruitment and Control of West Nile Virus Encephalitis. *J. Virol.* **2005**, *79*, 11457–11466. [CrossRef] [PubMed]
48. Zhang, B.; Chan, Y.K.; Lu, B.; Diamond, M.S.; Klein, R.S. CXCR3 Mediates Region-Specific Antiviral T Cell Trafficking within the Central Nervous System during West Nile Virus Encephalitis. *J. Immunol.* **2008**, *180*, 2641–2649. [CrossRef]
49. Bai, F.; Kong, K.; Dai, J.; Qian, F.; Zhang, L.; Brown, C.R.; Fikrig, E.; Montgomery, R.R. A Paradoxical Role for Neutrophils in the Pathogenesis of West Nile Virus. *J. Infect. Dis.* **2010**, *8031*, 1804–1812. [CrossRef]
50. Brocker, C.; Carpenter, C.; Nebert, D.W.; Vasiliou, V. Evolutionary Divergence and Functions of the Human Acyl-CoA Thioesterase Gene (ACOT) Family. *Hum. Genom.* **2010**, *4*, 411–420. [CrossRef]

51. Kumar, M.; Roe, K.; Orillo, B.; Muruve, D.A.; Nerurkar, V.R.; Gale, M.; Verma, S. Inflammasome Adaptor Protein Apoptosis-Associated Speck-Like Protein Containing CARD (ASC) Is Critical for the Immune Response and Survival in West Nile Virus Encephalitis. *J. Virol.* **2013**, *87*, 3655–3667. [CrossRef] [PubMed]
52. Kumar, M.; Verma, S.; Nerurkar, V.R. Pro-Inflammatory Cytokines Derived from West Nile Virus (WNV)-Infected SK-N-SH Cells Mediate Neuroinflammatory Markers and Neuronal Death. *J. Neuroinflamm.* **2010**, *7*, 73. [CrossRef]
53. Soung, A.L.; Davé, V.A.; Garber, C.; Tycksen, E.D.; Vollmer, L.L.; Klein, R.S. IL-1 Reprogramming of Adult Neural Stem Cells Limits Neurocognitive Recovery after Viral Encephalitis by Maintaining a Proinflammatory State. *Brain Behav. Immun.* **2022**, *99*, 383–396. [CrossRef]
54. Durrant, D.M.; Daniels, B.P.; Klein, R.S. IL-1R1 Signaling Regulates CXCL12-Mediated T Cell Localization and Fate within the CNS during West Nile Virus Encephalitis. *J. Immunol.* **2014**, *193*, 4095–4106. [CrossRef] [PubMed]
55. Riccetti, S.; Sinigaglia, A.; Desole, G.; Nowotny, N.; Trevisan, M.; Barzon, L. Modelling West Nile Virus and Usutu Virus Pathogenicity in Human Neural Stem Cells. *Viruses* **2020**, *12*, 882. [CrossRef] [PubMed]
56. van Marle, G.; Antony, J.; Ostermann, H.; Dunham, C.; Hunt, T.; Halliday, W.; Maingat, F.; Urbanowski, M.D.; Hobman, T.; Peeling, J.; et al. West Nile Virus-Induced Neuroinflammation: Glial Infection and Capsid Protein-Mediated Neurovirulence. *J. Virol.* **2007**, *81*, 10933–10949. [CrossRef]
57. Qian, F.; Chung, L.; Zheng, W.; Bruno, V.; Alexander, R.P.; Wang, Z.; Wang, X.; Kurscheid, S.; Zhao, H.; Fikrig, E.; et al. Identification of Genes Critical for Resistance to Infection by West Nile Virus Using RNA-Seq Analysis. *Viruses* **2013**, *5*, 1664–1681. [CrossRef]
58. Yao, Y.; Strauss-Albee, D.M.; Zhou, J.Q.; Malawista, A.; Garcia, M.N.; Murray, K.O.; Blish, C.A.; Montgomery, R.R. The Natural Killer Cell Response to West Nile Virus in Young and Old Individuals with or without a Prior History of Infection. *PLoS ONE* **2017**, *12*, e0172625. [CrossRef]
59. Aarreberg, L.D.; Wilkins, C.; Ramos, H.J.; Green, R.; Davis, M.A.; Chow, K.; Gale, M. Interleukin-1 β Signaling in Dendritic Cells Induces Antiviral Interferon Responses. *mBio* **2018**, *9*, e00342-18. [CrossRef]
60. Uddin, M.J.; Suen, W.W.; Bosco-Lauth, A.; Hartwig, A.E.; Hall, R.A.; Bowen, R.A.; Bielefeldt-Ohmann, H. Kinetics of the West Nile Virus Induced Transcripts of Selected Cytokines and Toll-like Receptors in Equine Peripheral Blood Mononuclear Cells. *Vet. Res.* **2016**, *47*, 61. [CrossRef]
61. Bielefeldt-Ohmann, H.; Bosco-lauth, A.; Hartwig, A.; Uddin, M.J.; Barcelon, J.; Suen, W.W.; Wang, W.; Hall, R.A.; Bowen, R.A. Microbial Pathogenesis Characterization of Non-Lethal West Nile Virus (WNV) Infection in Horses: Subclinical Pathology and Innate Immune Response. *Microb. Pathog.* **2017**, *103*, 71–79. [CrossRef] [PubMed]
62. Byrne, S.N.; Halliday, G.M.; Johnston, L.J.; King, N.J.C. Interleukin-1β but Not Tumor Necrosis Factor Is Involved in West Nile Virus-Induced Langerhans Cell Migration from the Skin in C57BL/6 Mice. *J. Investig. Dermatol.* **2001**, *117*, 702–709. [CrossRef] [PubMed]
63. Garber, C.; Vasek, M.J.; Vollmer, L.L.; Sun, T.; Jiang, X.; Klein, R.S. Astrocytes Decrease Adult Neurogenesis during Virus-Induced Memory Dysfunction via Interleukin-1. *Nat. Immunol.* **2018**, *19*, 151–161. [CrossRef] [PubMed]
64. Kumar, M.; Nerurkar, V.R. Integrated Analysis of MicroRNAs and Their Disease Related Targets in the Brain of Mice Infected with West Nile Virus. *Virology* **2014**, *452–453*, 143–151. [CrossRef]
65. Lim, S.M.; van den Ham, H.J.; Oduber, M.; Martina, E.; Zaaraoui-Boutahar, F.; Roose, J.M.; van IJcken, W.F.J.; Osterhaus, A.D.M.E.; Andeweg, A.C.; Koraka, P.; et al. Transcriptomic Analyses Reveal Differential Gene Expression of Immune and Cell Death Pathways in the Brains of Mice Infected with West Nile Virus and Chikungunya Virus. *Front. Microbiol.* **2017**, *8*, 1556. [CrossRef]
66. Xiea, G.; Weltea, T.; Wanga, J.; Whitemanb, M.C.; Wickerb, J.A.; Saxenaa, V.; Conga, Y.; Barretta, A.D.T.; Wang, T. A West Nile Virus NS4B-P38G Mutant Strain Induces Adaptive Immunity via TLR7-MyD88-Dependent and Independent Signaling Pathways. *Vaccine* **2013**, *31*, 4143–4151. [CrossRef]
67. Wang, T.; Town, T.; Alexopoulou, L.; Anderson, J.F.; Fikrig, E.; Flavell, R.A. Toll-like Receptor 3 Mediates West Nile Virus Entry into the Brain Causing Lethal Encephalitis. *Nat. Med.* **2004**, *10*, 1366–1373. [CrossRef]
68. Suen, W.W.; Uddin, M.J.; Prow, N.A.; Bowen, R.A.; Hall, R.A.; Bielefeldt-Ohmann, H. Tissue-Specific Transcription Profile of Cytokine and Chemokine Genes Associated with Flavivirus Control and Non-Lethal Neuropathogenesis in Rabbits. *Virology* **2016**, *494*, 1–14. [CrossRef]
69. Quick, E.D.; Leser, J.S.; Clarke, P.; Tyler, K.L. Activation of Intrinsic Immune Responses and Microglial Phagocytosis in an Ex Vivo Spinal Cord Slice Culture Model of West Nile Virus Infection. *J. Virol.* **2014**, *88*, 13005–13014. [CrossRef]
70. Uddin, M.J.; Suen, W.W.; Prow, N.A.; Hall, R.A.; Bielefeldt-Ohmann, H. West Nile Virus Challenge Alters the Transcription Profiles of Innate Immune Genes in Rabbit Peripheral Blood Mononuclear Cells. *Front. Vet. Sci.* **2015**, *2*, 76. [CrossRef]
71. Shirato, K.; Miyoshi, H.; Kariwa, H.; Takashima, I. The Kinetics of Proinflammatory Cytokines in Murine Peritoneal Macrophages Infected with Envelope Protein-Glycosylated or Non-Glycosylated West Nile Virus. *Virus Res.* **2006**, *121*, 11–16. [CrossRef] [PubMed]
72. Saxenaa, V.; Weltea, T.; Baob, X.; Xiea, G.; Wanga, J.; Higgsc, S.; Teshd, R.B.; Wang, T. A Hamster-Derived West Nile Virus Strain Is Highly Attenuated and Induces a Differential Proinflammatory Cytokine Response in Two Murine Cell Lines. *Virus Res.* **2013**, *167*, 179–187. [CrossRef] [PubMed]

73. Sapkal, G.N.; Harini, S.; Ayachit, V.M.; Fulmali, P.V.; Mahamuni, S.A.; Bondre, V.P.; Gore, M.M. Neutralization Escape Variant of West Nile Virus Associated with Altered Peripheral Pathogenicity and Differential Cytokine Profile. *Virus Res.* **2011**, *158*, 130–139. [CrossRef] [PubMed]

74. Fang, H.; Welte, T.; Zheng, X.; Chang, G.-J.J.; Holbrook, M.R.; Soong, L.; Wang, T. γδ T Cells Promote the Maturation of Dendritic Cells During West Nile Virus Infection. *FEMS Immunol. Med. Microbiol.* **2010**, *59*, 71–80. [CrossRef] [PubMed]

75. Garcia, M.; Alout, H.; Diop, F.; Damour, A.; Bengue, M.; Weill, M.; Missé, D.; Lévêque, N.; Bodet, C. Innate Immune Response of Primary Human Keratinocytes to West Nile Virus Infection and Its Modulation by Mosquito Saliva. *Front. Cell. Infect. Microbiol.* **2018**, *8*, 387. [CrossRef]

76. Peña, J.; Plante, J.A.; Carillo, A.C.; Roberts, K.K.; Smith, J.K.; Juelich, T.L.; Beasley, D.W.C.; Freiberg, A.N.; Labute, M.X.; Naraghi-Arani, P. Multiplexed Digital MRNA Profiling of the Inflammatory Response in the West Nile Swiss Webster Mouse Model. *PLoS Negl. Trop. Dis.* **2014**, *8*, e3216. [CrossRef]

77. Maximova, O.A.; Sturdevant, D.E.; Kash, J.C.; Kanakabandi, K.; Xiao, Y.; Minai, M.; Moore, I.N.; Taubenberger, J.; Martens, C.; Cohen, J.I.; et al. Virus Infection of the Cns Disrupts the Immune-Neural-Synaptic Axis via Induction of Pleiotropic Gene Regulation of Host Responses. *Elife* **2021**, *10*, e62273. [CrossRef]

78. Saxena, V.; Xie, G.; Li, B.; Farris, T.; Welte, T.; Gong, B.; Boor, P.; Wu, P.; Tang, S.J.; Tesh, R.; et al. A Hamster-Derived West Nile Virus Isolate Induces Persistent Renal Infection in Mice. *PLoS Negl. Trop. Dis.* **2013**, *7*, e2275. [CrossRef]

79. Bai, F.; Town, T.; Qian, F.; Wang, P.; Kamanaka, M.; Connolly, T.M.; Gate, D.; Montgomery, R.R.; Flavell, R.A.; Fikrig, E. IL-10 Signaling Blockade Controls Murine West Nile Virus Infection. *PLoS Pathog.* **2009**, *5*, e1000610. [CrossRef]

80. Xie, G.; Luo, H.; Tian, B.; Mann, B.; Bao, X.; Mcbride, J.; Tesh, R.; Barrett, A.D.; Wang, T. A West Nile Virus NS4B-P38G Mutant Strain Induces Cell Intrinsic Innate Cytokine Responses in Human Monocytic and Macrophage Cells. *Vaccine* **2015**, *33*, 869–878. [CrossRef]

81. Graham, J.B.; Swarts, J.L.; Wilkins, C.; Thomas, S.; Green, R.; Sekine, A.; Voss, K.M.; Ireton, R.C.; Mooney, M.; Choonoo, G.; et al. A Mouse Model of Chronic West Nile Virus Disease. *PLoS Pathog.* **2016**, *12*, e1005996. [CrossRef] [PubMed]

82. Town, T.; Bai, F.; Wang, T.; Kaplan, A.; Qian, F.; Montgomery, R.; Anderson, J.; Flavell, R.A.; Fikrig, E. Toll-like Receptor 7 Mitigates Lethal West Nile Encephalitis via Interleukin 23-Dependent Immune Cell Infiltration and Homing. *Immunity* **2009**, *30*, 242–253. [CrossRef] [PubMed]

83. Welte, T.; Reagan, K.; Fang, H.; Machain-Williams, C.; Zheng, X.; Mendell, N.; Chang, G.J.J.; Wu, P.; Blair, C.D.; Wang, T. Toll-like Receptor 7-Induced Immune Response to Cutaneous West Nile Virus Infection. *J. Gen. Virol.* **2009**, *90*, 2660–2668. [CrossRef] [PubMed]

84. Luo, H.; Winkelmann, E.R.; Zhu, S.; Ru, W.; Mays, E.; Silvas, J.A.; Vollmer, L.L.; Gao, J.; Peng, B.H.; Bopp, N.E.; et al. Peli1 Facilitates Virus Replication and Promotes Neuroinflammation during West Nile Virus Infection. *J. Clin. Investig.* **2018**, *128*, 4980–4991. [CrossRef]

85. Clarke, P.; Leser, J.S.; Bowen, R.A.; Tylera, K.L. Virus-Induced Transcriptional Changes in the Brain Include the Differential Expression of Genes Associated with Interferon, Apoptosis, Interleukin 17 Receptor A, and Glutamate Signaling as Well as Flavivirus-Specific Upregulation of TRNA Synthetases. *mBio* **2014**, *5*, e00902-14. [CrossRef]

86. Bourgeois, M.A.; Denslow, N.D.; Seino, K.S.; Barber, D.S.; Long, M.T. Gene Expression Analysis in the Thalamus and Cerebrum of Horses Experimentally Infected with West Nile Virus. *PLoS ONE* **2011**, *6*, e24371. [CrossRef]

87. Garcia-tapia, D.; Hassett, D.E.; Mitchell, W.J., Jr.; Johnson, G.C.; Kleiboeker, S.B. West Nile Virus Encephalitis: Sequential Histopathological and Immunological Events in a Murine Model of Infection. *J. NeuroVirol.* **2007**, *13*, 130–138. [CrossRef]

88. Munoz-Erazo, L.; Natoli, R.; Provis, J.M.; Madigan, M.C.; Jonathan, N.; King, C. Microarray Analysis of Gene Expression in West Nile Virus–Infected Human Retinal Pigment Epithelium. *Mol. Vis.* **2012**, *18*, 730–743.

89. Davison, A.M.; King, N.J.C. Accelerated Dendritic Cell Differentiation from Migrating Ly6C Lo Bone Marrow Monocytes in Early Dermal West Nile Virus Infection. *J. Immunol.* **2011**, *186*, 2382–2396. [CrossRef]

90. Getts, D.R.; Terry, R.L.; Getts, M.T.; Marcus, M.; Rana, S.; Shrestha, B.; Radford, J.; van Rooijen, N.; Campbell, I.L.; King, N.J.C. Ly6c+ "Inflammatory Monocytes" Are Microglial Precursors Recruited in a Pathogenic Manner in West Nile Virus Encephalitis. *J. Exp. Med.* **2008**, *205*, 2319–2337. [CrossRef]

91. Michlmayr, D.; McKimmie, C.S.; Pingen, M.; Haxton, B.; Mansfield, K.; Johnson, N.; Fooks, A.R.; Graham, G.J. Defining the Chemokine Basis for Leukocyte Recruitment during Viral Encephalitis. *J. Virol.* **2014**, *88*, 9553–9567. [CrossRef]

92. Vidaña, B.; Johnson, N.; Fooks, A.R.; Sánchez-Cordón, P.J.; Hicks, D.J.; Nuñez, A. West Nile Virus Spread and Differential Chemokine Response in the Central Nervous System of Mice: Role in Pathogenic Mechanisms of Encephalitis. *Transbound. Emerg. Dis.* **2020**, *67*, 799–810. [CrossRef] [PubMed]

93. Durrant, D.M.; Daniels, B.P.; Pasieka, T.; Dorsey, D.; Klein, R.S. CCR5 Limits Cortical Viral Loads during West Nile Virus Infection of the Central Nervous System. *J. Neuroinflamm.* **2015**, *12*, 233. [CrossRef] [PubMed]

94. Qian, F.; Goel, G.; Meng, H.; Wang, X.; You, F.; Devine, L.; Raddassi, K.; Garcia, M.N.; Murray, K.O.; Bolen, C.R.; et al. Systems Immunology Reveals Markers of Susceptibility to West Nile Virus Infection. *Clin. Vaccine Immunol.* **2015**, *22*, 6–16. [CrossRef] [PubMed]

95. Suthar, M.S.; Brassil, M.M.; Blahnik, G.; McMillan, A.; Ramos, H.J.; Proll, S.C.; Belisle, S.E.; Katze, M.G.; Gale, M. A Systems Biology Approach Reveals That Tissue Tropism to West Nile Virus Is Regulated by Antiviral Genes and Innate Immune Cellular Processes. *PLoS Pathog.* **2013**, *9*, e1003168. [CrossRef]

96. Shirato, K.; Kimura, T.; Mizutani, T.; Kariwa, H.; Takashima, I. Different Chemokine Expression in Lethal and Non-Lethal Murine West Nile Virus Infection. *J. Med. Virol.* **2004**, *74*, 507–513. [CrossRef]

97. Huang, B.; West, N.; Vider, J.; Zhang, P.; Griffiths, R.E.; Wolvetang, E.; Burtonclay, P.; Warrilow, D. Inflammatory Responses to a Pathogenic West Nile Virus Strain. *BMC Infect. Dis.* **2019**, *19*, 912. [CrossRef]

98. Silva, M.C.; Guerrero-plata, A.; Gilfoy, F.D.; Garofalo, R.P.; Mason, P.W. Differential Activation of Human Monocyte-Derived and Plasmacytoid Dendritic Cells by West Nile Virus Generated in Different Host Cells. *J. Virol.* **2007**, *81*, 13640–13648. [CrossRef]

99. Zhang, B.; Patel, J.; Croyle, M.; Diamond, M.S.; Klein, R.S. TNF-α-Dependent Regulation of CXCR3 Expression Modulates Neuronal Survival during West Nile Virus Encephalitis. *J. Neuroimmunol.* **2010**, *224*, 28–38. [CrossRef]

100. McCandless, E.E.; Zhang, B.; Diamond, M.S.; Klein, R.S. CXCR4 Antagonism Increases T Cell Trafficking in the Central Nervous System and Improves Survival from West Nile Virus Encephalitis. *Proc. Natl. Acad. Sci. USA* **2008**, *105*, 11270–11275. [CrossRef]

101. García-Nicolás, O.; Lewandowska, M.; Ricklin, M.E.; Summerfield, A. Monocyte-Derived Dendritic Cells as Model to Evaluate Species Tropism of Mosquito-Borne Flaviviruses. *Front. Cell. Infect. Microbiol.* **2019**, *9*, 117. [CrossRef]

102. Yeung, A.W.S.; Wu, W.; Freewan, M.; Stocker, R.; King, N.J.C.; Thomas, S.R. Flavivirus Infection Induces Indoleamine Macrophages via Tumor Necrosis Factor and NF-Kappa B. *J. Leukoc. Biol.* **2012**, *91*, 657–666. [CrossRef] [PubMed]

103. Aguilar-Valenzuela, R.; Netland, J.; Seo, Y.; Bevan, M.J.; Grakoui, A. Dynamics of Tissue-Specific CD8+ T Cell Responses during West Nile Virus Infection. *J. Virol.* **2018**, *92*, e00014-18. [CrossRef]

104. Shrestha, B.; Zhang, B.; Purtha, W.E.; Klein, R.S.; Diamond, M.S. Tumor Necrosis Factor Alpha Protects against Lethal West Nile Virus Infection by Promoting Trafficking of Mononuclear Leukocytes into the Central Nervous System. *J. Virol.* **2008**, *82*, 8956–8964. [CrossRef]

105. Cheng, Y.; King, N.J.C.; Kesson, A.M. The Role of Tumor Necrosis Factor in Modulating Responses of Murine Embryo Fibroblasts by Flavivirus, West Nile B. *Virology* **2004**, *329*, 361–370. [CrossRef] [PubMed]

106. Szretter, K.J.; Samuel, M.A.; Gilfillan, S.; Fuchs, A.; Colonna, M.; Diamond, M.S. The Immune Adaptor Molecule SARM Modulates Tumor Necrosis Factor Alpha Production and Microglia Activation in the Brainstem and Restricts West Nile Virus Pathogenesis. *J. Virol.* **2009**, *83*, 9329–9338. [CrossRef] [PubMed]

107. Getts, D.R.; Matsumoto, I.; Mu, M.; Getts, T.; Radford, J.; Shrestha, B.; Campbell, I.L.; King, N.J.C. Role of IFN-γ in an Experimental Murine Model of West Nile Virus-Induced Seizures. *J. Neurochem.* **2007**, *103*, 1019–1030. [CrossRef]

108. Shrestha, B.; Wang, T.; Samuel, M.A.; Whitby, K.; Craft, J.; Fikrig, E.; Diamond, M.S. Gamma Interferon Plays a Crucial Early Antiviral Role in Protection against West Nile Virus Infection. *J. Virol.* **2006**, *80*, 5338–5348. [CrossRef]

109. O'Neal, J.T.; Upadhyay, A.A.; Wolabaugh, A.; Patel, N.B.; Bosinger, S.E.; Suthar, M.S. West Nile Virus-Inclusive Single-Cell RNA Sequencing Reveals Heterogeneity in the Type I Interferon Response within Single Cells. *J. Virol.* **2019**, *93*, e01778-18. [CrossRef]

110. Kumar, M.; Belcaid, M.; Nerurkar, V.R. Identification of Host Genes Leading to West Nile Virus Encephalitis in Mice Brain Using RNA-Seq Analysis. *Sci. Rep.* **2016**, *6*, 26350. [CrossRef]

111. Sonar, S.; Girdhari, L. Role of Tumor Necrosis Factor Superfamily in Neuroinflammation and Autoimmunity. *Front. Immunol.* **2015**, *6*, 364. [CrossRef] [PubMed]

112. Giordano, D.; Kuley, R.; Draves, K.E.; Roe, K.; Holder, U.; Giltiay, V.; Clark, E.A. B Cell Activating Factor (BAFF) Produced by Neutrophils and Dendritic Cells Is Regulated Differently and Has Distinct Roles in Ab Responses and Protective Immunity against West Nile Virus. *J. Immunol.* **2020**, *204*, 1508–1520. [CrossRef] [PubMed]

113. Fields, J.K.; Günther, S.; Sundberg, E.J. Structural Basis of IL-1 Family Cytokine Signaling. *Front. Immunol.* **2019**, *10*, 1412. [CrossRef] [PubMed]

114. Turner, M.D.; Nedjai, B.; Hurst, T.; Pennington, D.J. Cytokines and Chemokines: At the Crossroads of Cell Signalling and Inflammatory Disease. *Biochim. Biophys. Acta Mol. Cell Res.* **2014**, *1843*, 2563–2582. [CrossRef] [PubMed]

115. Durrant, D.M.; Robinette, M.L.; Klein, R.S. IL-1R1 Is Required for Dendritic Cell-Mediated T Cell Reactivation within the CNS during West Nile Virus Encephalitis. *J. Exp. Med.* **2013**, *210*, 503–516. [CrossRef]

116. Weltea, T.; Xiea, G.; Wickerb, J.A.; Whitemanb, M.C.; Lib, L.; Rachamallua, A.; Barretta, A.; Wang, T. Immune Responses to An Attenuated West Nile Virus NS4B- P38G Mutant Strain. *Vaccine* **2011**, *29*, 4853–4861. [CrossRef]

117. Kovats, S.; Turner, S.; Simmons, A.; Powe, T.; Chakravarty, E.; Alberola-Ila, J. West Nile Virus-Infected Human Dendritic Cells Fail to Fully Activate Invariant Natural Killer T Cells. *Clin. Exp. Immunol.* **2016**, *186*, 214–226. [CrossRef] [PubMed]

118. Bryan, M.A.; Giordano, D.; Draves, K.E.; Green, R.; Gale, M.; Clark, E.A. Splenic Macrophages Are Required for Protective Innate Immunity against West Nile Virus. *PLoS ONE* **2018**, *13*, e0191690. [CrossRef]

119. Garcia, M.; Wehbe, M.; Lévêque, N.; Bodet, C. Skin Innate Immune Response to Flaviviral Infection. *Eur. Cytokine Netw.* **2017**, *28*, 41–51. [CrossRef]

120. Zheng, D.; Kern, L.; Elinav, E. The NLRP6 Inflammasome. *Immunology* **2021**, *162*, 281–289. [CrossRef]

121. Zheng, D.; Liwinski, T.; Elinav, E. Inflammasome Activation and Regulation: Toward a Better Understanding of Complex Mechanisms. *Cell Discov.* **2020**, *6*, 36. [CrossRef] [PubMed]

122. Dinarello, C.A. Overview of the IL-1 Family in Innate Inflammation and Acquired Immunity. *Immunol. Rev.* **2018**, *28*, 8–27. [CrossRef] [PubMed]

123. Nanda, J.D.; Ho, T.S.; Satria, R.D.; Jhan, M.K.; Wang, Y.T.; Lin, C.F. IL-18: The Forgotten Cytokine in Dengue Immunopathogenesis. *J. Immunol. Res.* **2021**, *2021*, 8214656. [CrossRef]

124. Tobler, L.H.; Cameron, M.J.; Lanteri, M.C.; Prince, H.E.; Danesh, A.; Persad, D.; Lanciotti, R.S.; Norris, P.J.; Kelvin, D.J.; Busch, M.P. Interferon and Interferon-Induced Chemokine Expression Is Associated with Control of Acute Viremia in West Nile Virus-Infected Blood Donors. *J. Infect. Dis.* **2008**, *198*, 979–983. [CrossRef] [PubMed]
125. Leis, A.A.; Grill, M.F.; Goodman, B.P.; Sadiq, S.B.; Sinclair, D.J.; Vig, P.J.S.; Bai, F. Tumor Necrosis Factor-Alpha Signaling May Contribute to Chronic West Nile Virus Post-Infectious Proinflammatory State. *Front. Med.* **2020**, *7*, 164. [CrossRef]
126. Bonilla, W.V.; Fröhlich, A.; Senn, K.; Kallert, S.; Fernandez, M.; Fallon, P.G.; Klemenz, R.; Nakae, S.; Adler, H.; Merkler, D. The Alarmin Interleukin-33 Drives. *Science* **2012**, *335*, 984–989. [CrossRef] [PubMed]
127. Franca, R.F.O.; Costa, R.S.; Silva, J.R.; Peres, R.S.; Mendonça, L.R.; Colón, D.F.; Alves-Filho, J.C.; Cunha, F.Q. IL-33 Signaling Is Essential to Attenuate Viral-Induced Encephalitis Development by Downregulating INOS Expression in the Central Nervous System. *J. Neuroinflamm.* **2016**, *13*, 159. [CrossRef] [PubMed]
128. Hirano, T. IL-6 in Inflammation, Autoimmunity and Cancer. *Int. Immunol.* **2021**, *33*, 127–148. [CrossRef]
129. Velazquez-Salinas, L.; Verdugo-Rodriguez, A.; Rodriguez, L.L.; Borca, M.V. The Role of Interleukin 6 during Viral Infections. *Front. Microbiol.* **2019**, *10*, 1057. [CrossRef]
130. Zidovec-Lepej, S.; Vilibic-Cavlek, T.; Barbic, L.; Ilic, M.; Savic, V.; Tabain, I.; Ferenc, T.; Grgic, I.; Gorenec, L.; Bogdanic, M.; et al. Antiviral Cytokine Response in Neuroinvasive and Non-Neuroinvasive West Nile Virus Infection. *Viruses* **2021**, *13*, 342. [CrossRef]
131. Venter, M.; Burt, F.J.; Blumberg, L.; Fickl, H.; Paweska, J.; Swanepoel, R. Cytokine Induction after Laboratory-Acquired West Nile Virus Infection. *New Engl. J. Med.* **2009**, *360*, 1260–1262. [CrossRef] [PubMed]
132. Garcia, M.N.; Hause, A.M.; Walker, C.M.; Orange, J.S.; Hasbun, R.; Murray, K.O. Evaluation of Prolonged Fatigue Post-West Nile Virus Infection and Association of Fatigue with Elevated Antiviral and Proinflammatory Cytokines. *Viral Immunol.* **2014**, *27*, 327–333. [CrossRef] [PubMed]
133. Milovanovic, J.; Arsenijevic, A.; Stojanovic, B.; Kanjevac, T.; Arsenijevic, D.; Radosavljevic, G.; Milovanovic, M.; Arsenijevic, N. Interleukin-17 in Chronic Inflammatory Neurological Diseases. *Front. Immunol.* **2020**, *11*, 947. [CrossRef] [PubMed]
134. Yung Peng, R.C.; Rose Khavari, N.D. Interleukin 12 (IL-12) Family Cytokines: Role in Immune Pathogenesis and Treatment of CNS Autoimmune Disease. *Physiol. Behav.* **2017**, *176*, 139–148. [CrossRef]
135. Hansen, M.; Nolan, M.S.; Gorchakov, R.; Hasbun, R.; Murray, K.O.; Ronca, S.E. Unique Cytokine Response in West Nile Virus Patients Who Developed Chronic Kidney Disease: A Prospective Cohort Study. *Viruses* **2021**, *13*, 311. [CrossRef]
136. Schneider, B.S.; McGee, C.E.; Jordan, J.M.; Stevenson, H.L.; Soong, L.; Higgs, S. Prior Exposure to Uninfected Mosquitoes Enhances Mortality in Naturally-Transmitted West Nile Virus Infection. *PLoS ONE* **2007**, *2*, e1171. [CrossRef]
137. Chu, H.X.; Arumugam, T.V.; Gelderblom, M.; Magnus, T.; Drummond, G.R.; Sobey, C.G. Role of CCR2 in Inflammatory Conditions of the Central Nervous System. *J. Cereb. Blood Flow Metab.* **2014**, *34*, 1425–1429. [CrossRef]
138. Venuti, A.; Pastori, C.; Lopalco, L. The Role of Natural Antibodies to CC Chemokine Receptor 5 in HIV Infection. *Front. Immunol.* **2017**, *8*, 1358. [CrossRef]
139. Lim, J.K.; McDermott, D.H.; Lisco, A.; Foster, G.A.; Krysztof, D.; Follmann, D.; Stramer, S.L.; Murphy, P.M. CCR5 Deficiency Is a Risk Factor for Early Clinical Manifestations of West Nile Virus Infection but Not for Viral Transmission. *J. Infect. Dis.* **2010**, *201*, 178–185. [CrossRef]
140. Hussmann, K.L.; Fredericksen, B.L. Differential Induction of CCL5 by Pathogenic and Non-Pathogenic Strains of West Nile Virus in Brain Endothelial Cells and Astrocytes. *J. Gen. Virol.* **2014**, *95*, 862–867. [CrossRef]
141. Yan, Y.; Chen, R.; Wang, X.; Hu, K.; Huang, L.; Lu, M.; Hu, Q. CCL19 and CCR7 Expression, Signaling Pathways, and Adjuvant Functions in Viral Infection and Prevention. *Front. Cell Dev. Biol.* **2019**, *7*, 212. [CrossRef] [PubMed]
142. de Oliveira, S.; Reyes-Aldasoro, C.C.; Candel, S.; Renshaw, S.A.; Mulero, V.; Calado, Â. Cxcl8 (IL-8) Mediates Neutrophil Recruitment and Behavior in the Zebrafish Inflammatory Response. *J. Immunol.* **2013**, *190*, 4349–4359. [CrossRef] [PubMed]
143. Goczalik, I.; Ulbricht, E.; Hollborn, M.; Raap, M.; Uhlmann, S.; Weick, M.; Pannicke, T.; Wiedemann, P.; Bringmann, A.; Reichenbach, A.; et al. Expression of CXCL8, CXCR1, and CXCR2 in Neurons and Glial Cells of the Human and Rabbit Retina. *Investig. Ophthalmol. Vis. Sci.* **2008**, *49*, 4578–4589. [CrossRef]
144. Mikolka, P.; Kopincova, J.; Kosutova, P.; Kolomaznik, M.; Calkovska, A.; Mokra, D. Anti-IL-8 Antibody Potentiates the Effect of Exogenous Surfactant in Respiratory Failure Caused by Meconium Aspiration. *Exp. Lung Res.* **2018**, *44*, 40–50. [CrossRef] [PubMed]
145. Suen, W.W.; Uddin, M.J.; Wang, W.; Brown, V.; Adney, D.R.; Broad, N.; Prow, N.A.; Bowen, R.A.; Hall, R.A.; Bielefeldt-Ohmann, H. Experimental West Nile Virus Infection in Rabbits: An Alternative Model for Studying Induction of Disease and Virus Control. *Pathogens* **2015**, *4*, 529–558. [CrossRef] [PubMed]
146. Suen, W.W.; Imoda, M.; Thomas, A.W.; Nasir, N.N.B.M.; Tearnsing, N.; Wang, W.; Bielefeldt-Ohmann, H. An Acute Stress Model in New Zealand White Rabbits Exhibits Altered Immune Response to Infection with West Nile Virus. *Pathogens* **2019**, *8*, 195. [CrossRef]
147. Koubourli, D.V.; Yaparla, A.; Popovic, M.; Grayfer, L. Amphibian (*Xenopus laevis*) Interleukin-8 (CXCL8): A Perspective on the Evolutionary Divergence of Granulocyte Chemotaxis. *Front. Immunol.* **2018**, *9*, 58. [CrossRef]
148. Bianchi, M.E.; Mezzapelle, R. The Chemokine Receptor CXCR4 in Cell Proliferation and Tissue Regeneration. *Front. Immunol.* **2020**, *11*, 2109. [CrossRef]

149. Lee, M.; Lee, Y.; Song, J.; Lee, J.; Chang, S.Y. Tissue-Specific Role of CX3 CR1 Expressing Immune Cells and Their Relationships with Human Disease. *Immune Netw.* **2018**, *18*, e5. [CrossRef]

150. Koh, W.L.; Ng, M.L. Molecular Mechanisms of West Nile Virus Pathogenesis in Brain Cells. *Emerg. Infect. Dis.* **2005**, *11*, 629–632. [CrossRef]

151. Zakharova, M.; Ziegler, H.K. Paradoxical Anti-Inflammatory Actions of TNF-α: Inhibition of IL-12 and IL-23 via TNF Receptor 1 in Macrophages and Dendritic Cells. *J. Immunol.* **2005**, *175*, 5024–5033. [CrossRef] [PubMed]

152. Xie, G.; Luo, H.; Pang, L.; Peng, B.; Winkelmann, E.; McGruder, B.; Hesse, J.; Whiteman, M.; Campbell, G.; Milligan, G.N.; et al. Dysregulation of Toll-Like Receptor 7 Compromises Innate and Adaptive T Cell Responses and Host Resistance to an Attenuated West Nile Virus Infection in Old Mice. *J. Virol.* **2016**, *90*, 1333–1344. [CrossRef]

153. Machain-Williams, C.; Reagan, K.; Wang, T.; Zeidner, N.S.; Blair, C.D. Immunization with Culex Tarsalis Mosquito Salivary Gland Extract Modulates West Nile Virus Infection. *Viral. Immunol.* **2013**, *26*, 84–92. [CrossRef] [PubMed]

154. Guzman-Martinez, L.; Maccioni, R.B.; Andrade, V.; Navarrete, L.P.; Pastor, M.G.; Ramos-Escobar, N. Neuroinflammation as a Common Feature of Neurodegenerative Disorders. *Front. Pharmacol.* **2019**, *10*, 1008. [CrossRef] [PubMed]

Review

Emerging Role of Interferon-Induced Noncoding RNA in Innate Antiviral Immunity

Jie Min [1,2], Wenjun Liu [1,2,3,*] and Jing Li [1,2,*]

1   CAS Key Laboratory of Pathogenic Microbiology and Immunology, Institute of Microbiology, Chinese Academy of Sciences, Beijing 100101, China

2   Savaid Medical School, University of Chinese Academy of Sciences, Beijing 100049, China

3   Institute of Microbiology, Center for Biosafety Mega-Science, Chinese Academy of Sciences, Beijing 100101, China

*   Correspondence: liuwj@im.ac.cn (W.L.); lj418@163.com (J.L.); Tel./Fax: +86-10-6480-7503 (J.L.)

Abstract: Thousands of unique noncoding RNAs (ncRNAs) exist within the genomes of higher eukaryotes. Upon virus infection, the host generates interferons (IFNs), which initiate the expression of hundreds of interferon-stimulated genes (ISGs) through IFN receptors on the cell surface, establishing a barrier as the host's antiviral innate immunity. With the development of novel RNA-sequencing technology, many IFN-induced ncRNAs have been identified, and increasing attention has been given to their functions as regulators involved in the antiviral innate immune response. IFN-induced ncRNAs regulate the expression of viral proteins, IFNs, and ISGs, as well as host genes that are critical for viral replication, cytokine and chemokine production, and signaling pathway activation. This review summarizes the complex regulatory role of IFN-induced ncRNAs in antiviral innate immunity from the above aspects, aiming to improve understanding of ncRNAs and provide reference for the basic research of antiviral innate immunity.

Keywords: noncoding RNA; interferon; interferon-stimulated gene; antiviral innate immunity

Citation: Min, J.; Liu, W.; Li, J. Emerging Role of Interferon-Induced Noncoding RNA in Innate Antiviral Immunity. *Viruses* **2022**, *14*, 2607. https://doi.org/10.3390/v14122607

Academic Editors: Caijun Sun and Feng Ma

Received: 20 October 2022
Accepted: 21 November 2022
Published: 23 November 2022

Publisher's Note: MDPI stays neutral with regard to jurisdictional claims in published maps and institutional affiliations.

Following Crick's central dogma [1,2], genetic information is passed from DNA to RNA, which is then translated into proteins. Over many years, proteins have been widely recognized as the final products that perform the main functions encoded by genes, though protein-coding genes occupy <2% of the human genome. As sequencing technology advances, other products of the human genome are coming into view, revealing the "dark energy" of DNA. According to the NIH ENCODE project results, approximately 80% of the human genome is endowed with biochemical functions and, interestingly, about 62% of genes are transcribed into noncoding RNAs (ncRNAs) [3]. Upon viral infection or interferon (IFN) stimulation, many ncRNAs are generated to regulate various vital cellular processes, including pre-transcriptional, transcriptional, and post-transcriptional regulatory processes. Numerous ncRNAs are reported to be involved in antiviral activities, most of which are microRNAs (miRNAs), long ncRNAs (lncRNAs), or circular RNAs (circRNAs).

IFN was originally discovered as an infection-inhibiting factor. Subsequently, IFN-specific inducible mRNAs and proteins were identified. The canonical interpretation of IFN-mediated antiviral innate immune responses suggests that IFNs induce the transcription of IFN-stimulated genes (ISGs), which further alter the proteome in cells and establish an antiviral cellular state [4]. However, according to recent studies, IFN-mediated antiviral innate immune responses may be a more complex process than transcriptional induction of ISGs. Initially, researchers mainly detected IFN-induced gene expression by surveying IFN-stimulated transcriptome changes through microarray technology, which enabled the detection of the enrichment of select mRNAs through poly(A) affinity or arrays used to probe for transcripts of encoded proteins [5–7]. Thus, they were unable to capture ncRNAs that were not polyadenylated, such as miRNAs, circRNAs, piwi-interacting RNAs (piRNAs), and enhancer RNAs (eRNAs, which are lncRNAs) [7–10]. Polyadenylated lncRNAs have been frequently dismissed as "garbage" because they do not encode proteins. However, an increasing number of studies indicate that IFNs can induce some ncRNAs to participate in antiviral innate immunity [11–15]. These results demonstrate the biogenesis of ncRNAs and the role of ncRNAs, enhancing our understanding of these molecules and serving as fundamental concepts to guide antiviral innate immunity.

1. The Antiviral Innate Immune Response of the Host

Innate immunity is the first and most rapidly formed line of defense against viral invasion. Host cells recognize pathogen-associated molecular patterns (PAMPs) through pattern recognition receptors (PRRs). After viral nucleic acid is recognized by host cells, a series of signaling pathways is activated, stimulating IFN expression and secretion. IFNs induce hundreds of ISGs, including antiviral proteins and ncRNAs, which activate antiviral innate immune defenses.

1.1. The Pattern of Host Recognition of the Virus

According to their location in a cell, PRRs are divided into two classes. One is composed of a family of toll-like receptors (TLRs) located on the endosomal membrane. A major member of the TLR family, TLR3, recognizes double-stranded RNA, while TLR7 and TLR8 recognize single-stranded RNA, and TLR9 recognizes nonmethylated CpG DNA [16]. The other class consists of three nucleic acid receptors that are widely expressed in the cytoplasm. They include AIM2-like receptors (ALRs) and cyclic GMP-AMP synthase (cGAS), which recognize DNA viruses, and retinoic acid-like receptors (RLRs), which recognize double-stranded RNA or 5′ppp-modified single-stranded RNA [17–19].

1.2. IFN Signaling Pathways

IFNs are members of the class II cytokine family and are classified into three categories based on their molecular structure, antigenicity, and cell source. Type I IFNs are mainly composed of IFN-α produced by leukocytes and fibroblast-generated IFN-β. IFN-ε, IFN-ω, and IFN-κ also belong to this family. Type III IFNs include IFN-λ1, IFN-λ2, IFN-λ3, and IFN-λ4, which have antiviral mechanisms similar to those of type I IFNs [20–22]. The type II IFN family, also known as the immune IFN family, has only one member, IFN-γ, which is mainly produced by natural killer (NK) cells, T lymphocytes, and antigen-presenting cells (APCs) [23].

As shown in Figure 1, type I IFNs elicit signals through IFN-α/β receptor subunit 1/2 (IFNAR1/2), and different subtypes exhibit inconsistent tissue expression and binding affinity for IFNAR1/2 receptors, leading to various biological reactions [24]. Type III IFNs bind IL-10R2 of the interleukin-10 receptor and type III IFN-λ receptor 1 (IFNLR1) to activate the JAK-STAT signaling [25,26]. Type I IFN receptors are extensively distributed on the surface of nucleated cells. The type III IFN receptor complex IL-10R2 is ubiquitous, but the IFNLR1 subunit is only expressed in specific cells and tissues. Research has revealed that IFN-III receptors are also distributed in greater quantities in epithelial cells than in other cells and are involved in mucosal immunity [27]. In addition, with either autocrine or paracrine function, three different types of IFNs bind to high-affinity receptors (IFNAR2, IFNGR1, and IFNLR1) on the surface of the cell, recruiting low-affinity receptor subunits (IFNAR1, IFNGR2, and IL-10R2) to produce complexes that can transmit signals, further inducing the expression of hundreds of antiviral proteins encoded by ISGs and the transcription of antiviral effector ncRNAs, thereby activating antiviral innate immune defenses [28].

Several groups simultaneously found that a variety of ncRNAs are induced by type I, II, and III IFNs [29–32]. For example, STAT3 homodimers indirectly suppress proinflammatory gene expression by facilitating suppressors of cytokine signaling 3 (SOCS3) expression [33,34] and miR221/222 expression [35–37], or induce hitherto unknown transcriptional repressors. These studies strongly suggest that there is a close relationship between IFN-mediated antiviral innate immunity and ncRNAs.

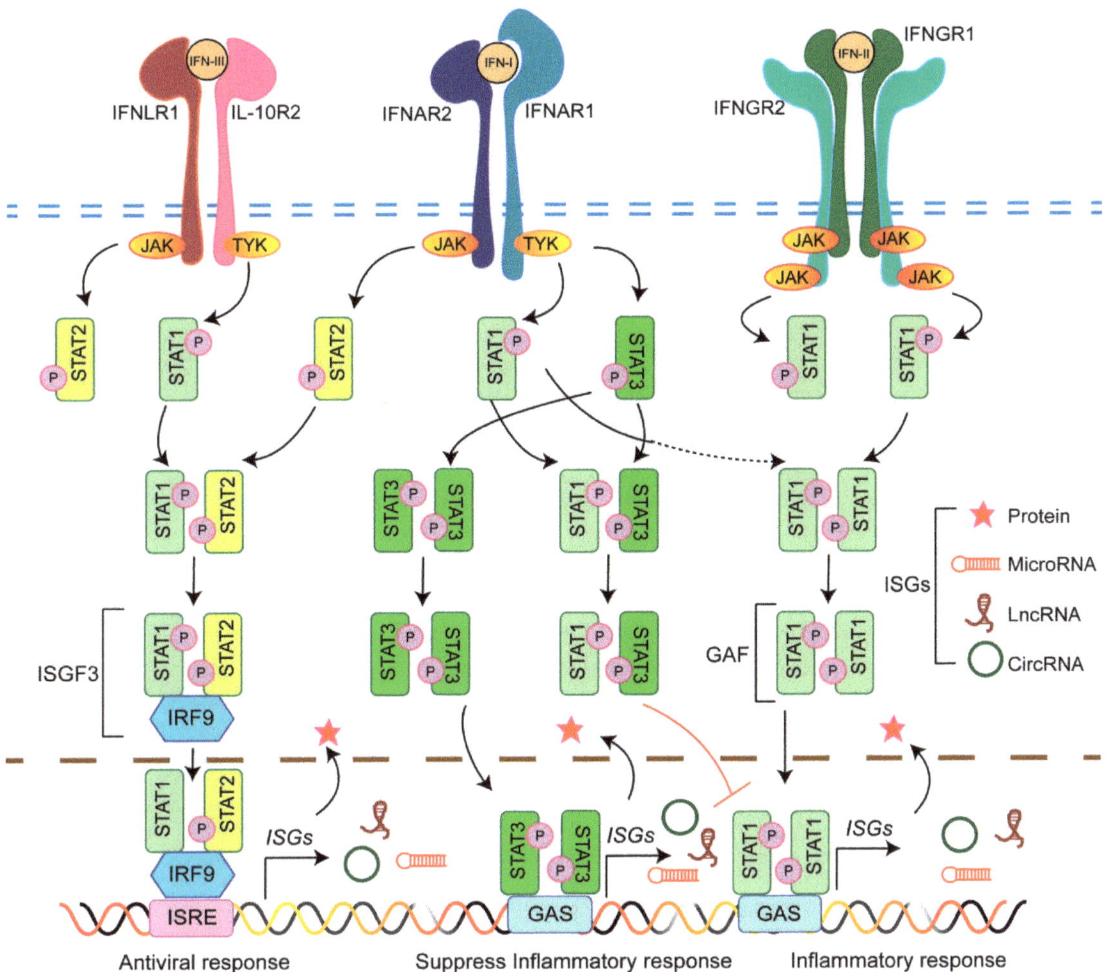

Figure 1. Interferon (IFN)-stimulated gene (ISG) expression. The three types of IFNs trigger similar downstream signal transduction and transcriptional reactions, albeit these reactions are mediated through different receptors. Type I IFN and type III IFN receptors actively phosphorylate JAK1 and TYK2, which phosphorylate the specific intracellular tyrosine residues of the receptors. STAT1/2/3 are recruited and phosphorylated. Type II IFN receptors trigger the phosphorylation of bound JAK1 and JAK2, and then phosphorylate receptor chains to induce the recruitment and phosphorylation of STAT1. STAT1 and 2 form a heterodimer that recruits IRF9 to form IFN-stimulated gene factor 3 (ISGF3), which is translocated into the nucleus and binds to IFN-stimulated response element (ISRE) sequences to trigger antiviral gene expression. STAT3 homodimers suppress proinflammatory gene expression. STAT1 homodimers bind to the gamma-activated sequence (GAS) to activate the inflammatory response. Phosphorylated STAT3 sequesters activated STAT1 to form heterodimers and prevent STAT1 from forming functional homodimers that trigger downstream gene expression. Gamma-activated factor (GAF).

1.3. Features and Functions of ISGs

Approximately 2000 ISGs in humans and mice have been identified to date. However, most of these have not been functionally characterized [38]. Analyses of >300 known ISGs indicate that they mainly play an immunomodulatory role. They can directly induce antiviral activity, especially by enhancing host cell recognition of and response to pathogens, and can negatively regulate IFN signaling pathway activity. To successfully infect a host, a virus needs to enter a target cell to replicate, assemble, and proliferate. ISGs produced by host cells are directly involved in the antiviral response. These ISGs target the key steps of virus infection and inhibit virus replication.

In addition, most PRRs that recognize viruses and the molecules involved in IFN signaling are ISGs, such as protein kinase R (PKR), cGAS, and the IFN regulator 1/3/7/9 (IRF1/3/7/9). ISGs are expressed at a basal level when not stimulated by exogenous factors. However, once a pathogen is recognized, downstream signaling pathways are activated and the expression of many ISGs is induced, including molecules involved in other PRRs and IFN signaling pathways. If not quickly eliminated or overactivated, IFNs produce signals that may lead to chronic inflammation or autoimmune diseases. Thus, host cells activate and produce a series of ISGs that negatively regulate IFN signaling to maintain host immune balance. Furthermore, post-translational modification of the virus and host proteins exerts an impact on viral replication. For example, ISG15 is one of the most highly expressed ISGs, which can inhibit herpes virus, influenza virus, and coxsackievirus B3 infection [39]. ISG15 is a 15-kDa ubiquitin-like protein that plays a role in the post-translational modification of hundreds of viral and host proteins [40,41].

2. The Role of NcRNAs in Antiviral Innate Immunity Regulation

Each stage of the viral life cycle is affected by the antiviral proteins encoded by ISGs. According to the life cycle of the virus, antiviral ISGs can be broadly divided into three categories: ISGs that encode proteins that prevent virus entry into cells, inhibit virus replication and translation, or prevent virus release from cells (Figure 2). Moreover, some ISGs can participate in antiviral innate immunity through their transcribed ncRNAs [42,43]. These ncRNAs usually regulate transcription, splicing, and nucleic acid degradation by acting as a decoy, and they modulate translation via RNA–DNA, RNA–RNA, or RNA–protein interactions. IFN-induced ncRNAs can also be used as novel antiviral effectors to participate in antiviral innate immunity.

NcRNAs are roughly sorted into short ncRNAs (fewer than 200 nt) and lncRNAs (more than 200 nt). A number of miRNAs and lncRNAs have been reported to be involved in antiviral activity. The miRNAs constitute a class of short ncRNAs of ~22 nt in length that participate in different biological processes through the post-transcriptional regulation of genes. The lncRNAs are grouped into intronic lncRNAs, long intergenic lncRNAs (lincRNAs), antisense lncRNAs, pseudogene lncRNAs, and eRNAs based on their position in the genome or relationship to mRNA. In addition, emerging circRNAs are classified as lncRNAs that are generated by back splicing and 3' to 5' end self-ligation. The lncRNAs can act as miRNA sponges, scaffold for macromolecules, decoy proteins, and transcription regulators. Notably, most ncRNAs do not encode proteins, but encode small functional polypeptides. Interestingly, we note that a class of ncRNAs regulated by IFN may be involved in antiviral innate immunity. Therefore, it is meaningful to classify IFN-induced ncRNAs as ISG complement products into interferon-stimulated non-coding RNA (ISR).

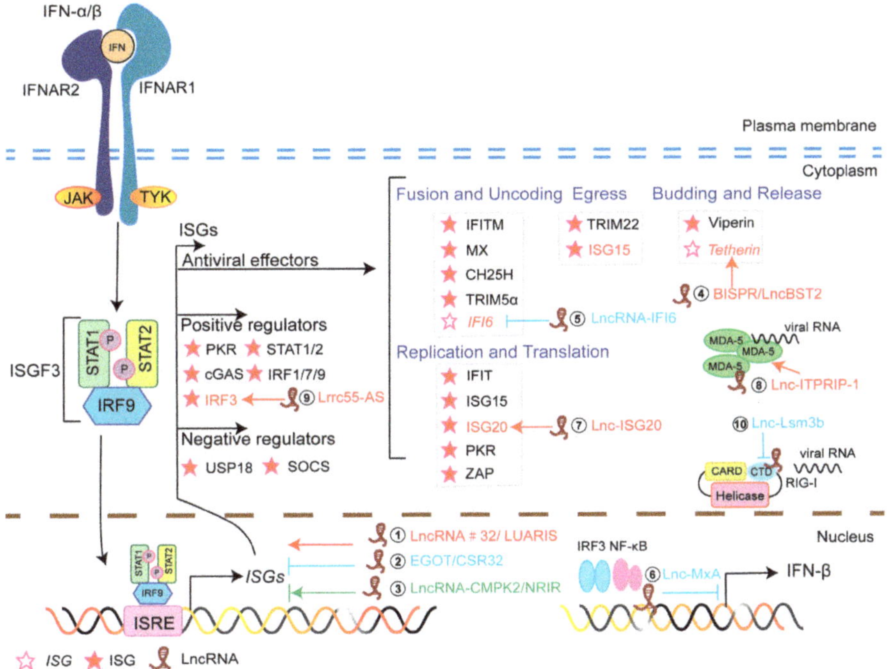

Figure 2. Antiviral innate immunity induced through ISGs. ISG products interfere with different stages of viral life cycles. Long noncoding RNAs (lncRNAs) act as ISGs in the regulation of antiviral innate immunity. ① lncRNA #32/LUARIS positively regulates the expression of ISGs by interacting with ATF2 and hnRNPU. ② lncRNA EGOT/CSR32 inhibits the expression of ISGs. ③ lncRNA CMPK2/NRIR suppresses ISGs transcription in hepatocytes or epithelial cells, and up-regulates the expression of ISGs in human monocytes. ④ BISPR/lncBST2 promotes the expression of the antiviral ISG BST2/tetherin. ⑤ Through its spatial domain, lncRNA IFI6 regulates histone modification at the IFI6 promoter, inhibiting the expression of IFI6. ⑥ lncRNA MxA inhibits the transcription of IFN-β. ⑦ As an endogenous competitive RNA, lnc-ISG20 binds to the microRNA (miRNA) miR-326, releasing ISG20 mRNA from the miRNA. ⑧ lncRNA ITPRIP-1 promotes the inhibitory effect of MDA5 by facilitating the binding of MDA5 to viral RNA. ⑨ lncRNA Lrrc55-AS supports PME-1-mediated demethylation and inactivation of PP2A, enhancing IRF3 phosphorylation and signaling. ⑩ lncRNA Lsm3b binds RIG-I, restricting the conformational change in RIG-I protein and thus preventing downstream signaling. Red arrow stands for promoting; blue arrow stands for inhibiting; green arrow stands for promoting or inhibiting depending on the cell type. The lncRNAs ①–⑥ are involved in antiviral factor transcription, and ⑦–⑩ affect the antiviral factors post-transcription.

2.1. Effect of IFN-Induced miRNAs

The effect of miRNA on mRNA translation and stability is well established. Previous studies have attempted to explore the role of miRNAs in viral infection, and found that the interaction between miRNA and the RNA viral genome directly regulates viral pathogenesis [44]. To date, two outcomes of these interactions have been identified, both of which directly affect virus replication. One is to prevent viral replication by inhibiting viral translation; the other is to maintain the stability of viral RNA to enhance replication. Moreover, miRNA-mediated changes in protein expression alter the host's response to infection.

2.1.1. IFN-Induced MiRNAs Directly Affect Virus Replication

Furthermore, IFN-induced miRNAs directly target viral transcripts. IFN-β-regulated miRNAs substantially reduce hepatitis C virus (HCV) replication and are sufficient to impose an antiviral cellular state. A customized microarray containing 245 human and mouse miRNAs was used to characterize IFN-regulated miRNAs in Huh7 cells, and the expression of 30 miRNAs was found to be either induced or repressed. The seed sequences of eight up-regulated miRNAs (miR1, miR30, miR128, miR196, miR296, miR351, miR431, and miR448) were complementary to those in HCV RNA, which was a surprising result. In addition, IFN inhibits miR122 expression and thus positively controls HCV replication [13]. Notably, miR-122 is critical for maximum HCV replication, and strategies based on blocking this miRNA to prevent HCV replication have shown promise in both in vivo models and preliminary clinical trials [45]. Furthermore, miR122 can target hepatitis B transcripts [46], whereas miR29 targets human immunodeficiency virus (HIV) transcripts [47]. Both miR122 and miR29 were previously identified as IFN-regulated miRNAs.

2.1.2. IFN-Induced MiRNAs Mediate the Expression of Host Proteins

Altering miRNA transcription or activity may lead to important consequences for IFN responses, and recent reports suggest that IFN stimulation can affect cellular miRNA expression [15]. For example, IFN-γ-activated STAT1 induces the expression of miR-155 via directly binding to the miR-155 promoter [48]. STAT1 binds to the miR-146a promoter, leading to the inhibition of miR-146a expression [49]. In addition, miR-155 expression induced by viral infection targets SOCS1 (a STAT inhibitor) to inhibit SOCS1 activity and generates feedback to promote IFN-I-mediated antiviral activity [48,50].

MiRNAs may indirectly down-regulate the expression of ISGs, thereby antagonizing the IFN-I response. For example, miR-221 expression, caused by up-regulation of miR-221 expression in peritoneal macrophages induced by viral infection, negatively regulates the innate antiviral response to vesicular stomatitis virus (VSV) [35]. In addition, STAT3 stimulates miR-221/222 expression, which targets PDLIM2 to stabilize and increase STAT3 levels [36]. Further, inhibition of miR-221/222 up-regulates the expression of components of the IFN-α signaling pathway (including STAT1, STAT2, IRF9, and several ISGs) in human glioma cell lines [37]. Hence, miR-221/222 can indirectly regulate ISG expression by affecting the balance of STAT1/STAT2 verses STAT3 homodimer signaling.

2.2. Effect of IFN-Induced LncRNAs

The regulatory effect of IFN-induced lncRNAs on antiviral proteins has been confirmed. In response to viral infection or IFN stimulation, many lncRNAs are produced, which regulate various activities as shown in Figure 2 and Table 1. The roles of these IFN-induced lncRNAs in regulating ISG *cis*-transcription, ISG *trans*-expression, ISG translation, and IFN expression are reviewed below.

Table 1. Interferon (IFN)-induced lncRNAs act as ISGs and are involved in the host antiviral innate immune response.

lncRNAs	Description	Stimuli	Regulation	Effect on Viral Replication	Characteristics/Functions	References
LncRNA CMPK2/NRIR	Negative regulatory factor of ISG response in hepatocytes or epithelial cells; positive regulator of the LPS-induced IFN response in human monocytes.	IFN-α, IFN-γ, HCV, LPS	Up	+/−	LncRNA CMPK2/NRIR inhibits ISG (CMPK2, CXCL10, IFIT3, IFITM1, ISG15, Viperin, and IFITM3) transcription by forming RNA–protein complexes, interacting with chromatin during remodeling, or transcription factors in hepatocytes or epithelial cells; NRIR upregulates the expression of IFN-I stimulated genes (CXCL10, MX1, IFITM3, and ISG15) in monocytes.	[29,51,52]

Table 1. *Cont.*

lncRNAs	Description	Stimuli	Regulation	Effect on Viral Replication	Characteristics/Functions	References
BISPR/lncRNA BST2	Positive regulator of ISG response.	IFN-α2, IFN-λ, IAV (PR8ΔNS1), VSV(M51R), HCV	Up	−	Promotes the expression of the antiviral ISG BST2/tetherin.	[31,53]
EGOT/CSR32	IFN signaling pathway negative regulator; induced by NF-κB after PKR or RIG-I activation.	IFN-α2, poly(I:C), IAV, HCV, SFV	Up	+	Inhibits the expression of ISGs (GBP1, ISG15, Mx1, BST2, ISG56, IFI6, and IFITM1).	[54]
LncRNA ISG20/ NONHSAG017802	Positive regulator of the ISG response; has the same chromosomal location as ISG20; most of the sequences are the same.	IFN-β, IAV, SeV, poly(I:C)	Up	−	Lnc-ISG20, as an endogenous competitive RNA that binds to miR-326, releasing ISG20 mRNA, and inhibiting IAV replication.	[42]
LncRNA MxA/ NONHSAG032905	Negative IFN signaling pathway regulator; in the MxA locus.	IFN-β, IAV, SeV, poly(I:C)	Up	+	Lnc-MxA negatively regulates the RIG-I-mediated antiviral immune response, inhibits the transcription of IFN-β by combining with the IFN-β promoter to form a lncRNA–DNA triplex.	[43]
LncRNA IFI6/lncRNA RP11-288L9.4	Negative regulator of the ISG response; overlaps with the antisense strand of IFI6 within intron 1 and is located in the IFI6 gene in the human genome.	IFN-α, HCV	Up	+	Through its spatial domain (large right arm), it regulates histone modification at the IFI6 promoter, inhibiting the expression of IFI6 and promoting HCV infection.	[11]
LncRNA ISR	Within the BAHCC1 locus.	IFN-β, IAV	Up	−	It relies on RIG-I signaling and inhibits IAV replication; however, the specific mechanism is still unclear.	[12]
LncRNA #32/LUARIS	Positive IFN signaling pathway regulator.	IFN-β, poly(I: C)	Down	−	It positively regulates the expression of IRF7, CCL5, CXCL11, OASL, RSAD2, and IP-10 by interacting with ATF2 and hnRNPU.	[55]
LncRNA ITPRIP-1	Cofactors of MDA5.	IFN-α, HCV, HSV, SeV, VSV	Up	−	Promotes the inhibitory effect of MDA5 on HCV replication by facilitating the binding of MDA5 to viral RNA.	[56]
LncRNA Lrrc55-AS	Positive regulator of IFN-I production.	IFN-β, poly(I:C), HSV, LPS,	Up	−	It supports PME-1-mediated demethylation and inactivation of PP2A, enhancing IRF3 phosphorylation and signaling.	[57]
LncRNA Lsm3b	Negative IFN signaling pathway regulator; multivalent structural motifs and long-stem structure.	IFN-α, IFN-β, VSV, SeV, HSV	Up	+	Its binding restricts the conformational change of the RIG-I protein and prevents downstream signaling, terminating the production of type I IFNs.	[58]
PSMB8-AS1	Near PSMB8 and TAP1.	IAV and IFN-β	Up	+	Repressed PSMB8-AS1 reduces IAV replication and proliferation.	[59]
AC015849.2	Near Chemokine (C-C Motif) Ligand 5 (CCL5) and TATA Box Binding Protein (TBP)-Associated Factor (TAF15).	IAV and IFN-β	Up	NA	NA	[59]
RP-1-7H24.1	Near OAS2, OAS3 and TRIM25.	IAV and IFN-β	Up	NA	NA	[59]

Table 1. *Cont.*

lncRNAs	Description	Stimuli	Regulation	Effect on Viral Replication	Characteristics/Functions	References
CTD-2639E6.9	An intergenic lncRNA (lincRNA).	IAV and IFN-β	Up	NA	NA	[59]
PSOR1C3	A sense intronic lncRNA that lies within introns and does not overlap with exons; near POU5F1 and HLA-C.	IAV and IFN-β	Up	NA	NA	[59]
AC007283.5	3 prime overlapping lncRNA that overlaps the 3′-UTR of a protein-coding locus on the same strand; near CASP10 and CFLAR.	IAV and IFN-β	Up	NA	NA	[59]
RP11-670E13.5	Near OAS2, OAS3 and TRIM25.	IAV and IFN-β	Up	NA	NA	[59]
LncISG15	Near ISG15.	IFN-α2, IFN-λ, IAV (PR8ΔNS1), VSV (M51R), HCV	Up	NA	NA	[31]
ISR2	Located at the end of the GBP gene cluster, adjacent to GBP6, and is a pseudogene of GBP1.	IFN-α2, IFN-β, HCV, IAV (PR8ΔNS1), HIV	Up	NA	NA	[32]
ISR8	Near IRF1.	IFN-α2, IFN-β, HCV, IAV (PR8ΔNS1)	Up	NA	NA	[32]
ISR12	Near IL-16.	IFN-α2, IFN-β, TNFα, LPS, poly(I:C)	Up	NA	NA	[32]

2.2.1. IFN-Induced LncRNAs Regulate ISGs *cis*-Transcription

IFN-induced lncRNAs can be co-expressed with adjacent ISGs to regulate the latter in *cis*. Some lncRNAs originate from bidirectional promoters shared with ISGs or are located nearby (<2 kb) immune-related protein-coding genes [60]. For example, Tetherin is a transmembrane protein encoded by the ISG bone marrow stromal antigen 2 (*BST2*), and through the unique topology of this protein, HIV-1 viral particles can be trapped for subsequent degradation in endosomes or lysosomes, preventing viral budding [61]. ISG-encoded BST2 can also inhibit the release of progeny influenza viruses [62]. The expression of lnc*BST2* depends on the JAK-STAT signaling pathway. The lncRNAs *BST2* and BST2 share the same promoter and are co-expressed upon IFN stimulation or influenza infection to promote the expression of the antiviral protein ISG BST2/tetherin [31,53].

In contrast, lncRNA *IFI6* negatively regulates IFI6 promoter function by modifying histones to inhibit IFI6 expression, thereby increasing HCV replication [11]. The lncRNA *IFI6* is an *IFI6*-transcribed lncRNA with expression induced by IFN-α or HCV. However, the lncRNA *IFI6* does not exert its regulatory effect through the JAK-STAT signaling pathway.

The loci of IFN- and influenza A virus (IAV)-inducible (PR8ΔNS1), ISR2, ISR8, and lncISG15 are close to those of ISGs GBP1, IRF1, and ISG15, respectively [31,32]. After IAV infection, the expression levels of ISR2, ISR8, and lncISG15 are significantly increased, and the expression levels of downstream genes display the same up-regulated trend [31,32]. In addition, the expression of ISR12 is strongly induced by IFN at late times, ISR12 is located upstream of IL-6, and the ISR12 promoter has a binding site for NF-κB [32]. These studies reveal an important role for ISR2, ISR8, ISR12 and ISG15 in viral infection and IFN regulation. However, the underlying mechanisms remain uncertain.

2.2.2. IFN-Induced LncRNAs Regulate ISGs *trans*-Expression

IFN-induced lncRNAs also can regulate ISGs in *trans*. For example, lncRNA #32 leverages the interaction between hnRNPU and ATF2 to regulate ISG expression and positively regulate the host antiviral response. Interestingly, the lncRNA is negatively regulated by IFN-β [55]. IFN-α2 or the IAV can induce lncRNA EGOT activity. The EGOT lncRNA negatively regulates antiviral response by inhibiting the expression of ISGs (GBP1, ISG15, and Mx1) [54].

In addition, IFN-induced lncRNAs differentially affect the expression of ISGs in different types of cells. The lncRNA NRIR is also known as the lncRNA *CMPK2*. LncRNA *CMPK2* activity is induced by IFN or HCV in hepatocytes, and it is a negative regulatory factor of ISG-induced inhibition of HCV replication [29]. Hantavirus (HTNV) infects epithelial cells in the same manner that HCV infects hepatocytes, and NRIR is a negative regulator of infection promoters. For example, IFITM3 inhibits HTNV infection, but IFITM3 action is negatively regulated by NRIR [51]. Thus, lncRNA CMPK2 has high cell- and induction stimulus-specificity. The expression of repressed ISGs in hepatocytes or epithelial cells is re-established by up-regulation lncRNA *CMPK2* induced in monocytes to promote lipopolysaccharide (LPS)- induced activation of IFN signaling [52]. Together, these results confirm the role of lncRNA *CMPK2*/NRIR as a negative regulator (hepatocytes and epithelial cells) and a positive regulator (monocytes) of the IFN response.

2.2.3. IFN-Induced LncRNAs Regulate ISGs Translation

As competitive endogenous RNAs (ceRNAs), lncRNAs can trap miRNAs. Some lncR-NAs contain miRNA-targeted mRNA sequences, and lncRNAs contain miRNA-targeted mRNA sequences that are similar to miRNA-targeted mRNA sequences, which can promote the release of mRNA from miRNA and thus antagonize miRNA function. Our study shows that IFN-β- or IAV-induced lncRNA ISG20 is a ceRNA. As a common miRNA of the ISG20 gene and lncRNA ISG20, miR-326 acts on the 3′ untranslated region (UTR) of ISG20 mRNA to suppress ISG20 translation. We found that lnc-ISG20 binds to miR-326 and acts as a ceRNA, which ablates the inhibitory effect of miR-326 on ISG20 mRNA, increases ISG20 protein levels, and inhibits influenza virus replication [42].

2.2.4. IFN-Induced lncRNAs Regulate IFN Expression

IFN-induced lncRNAs exert regulatory effects by modulating the transcription of target genomic loci in *trans* to inhibit IFN signaling pathway activation. Our previous study reports that ISG *MxA* transcribes a lncRNA after IAV infection or IFN-β stimulation [43]. MxA acts as a broad-spectrum antiviral by trapping viral components early during infection and preventing the virus from reaching host cells [63–66]. We found that a lncRNA–DNA triplex was formed by lnc-MXA binding to the IFN-β promoter, interfering with IRF3 and p65 binding to the IFN-β promoter, inhibiting the transcription of IFN-β in *trans*, negatively regulating RIG-I-mediated antiviral immune response, and promoting IAV proliferation [43]. IFN-induced lncRNA participates in antiviral innate immunity by inhibiting RIG-I activation. lncRNA Lsm3b binds RIG-I monomers to compete with viral RNA, and the feedback generated at the late stage of the innate reaction leads to RIG-I-induced inactivation of innate immune function. Mechanistically, lncRNA Lsm3b binding limits conformational changes in the RIG-I protein and prevents downstream signaling, terminating type I IFN production [58].

In addition, lncRNAs can be cofactors to mediate interactions between macromolecules, thereby inhibiting viral replication. Binding of lncITPRIP-1 to MDA5 promotes MDA5 oligomerization and enhances the association between MDA5 and HCV RNA, resulting in downstream MAVS that promotes signaling and produces IFN that inhibits HCV replication [56]. LncRNA Lrrc55-AS binds to phosphatase methylesterase-1 (PME-1) and promotes the interaction between PME-1 and phosphatase PP2A (an IRF3 signaling inhibitor). LncLrrc55-AS supports PME-1-mediated demethylation and PP2A inactivation, which enhances IRF3 phosphorylation and signaling pathways, promoting IFN-I production and inhibiting viral replication [57].

Moreover, there are many IFN-induced lncRNAs whose functions remain unknown. For instance, the effect of lncRNA ISR is dependent on the effect of IAV or IFN-α-induced

RIG-I, and lncRNA ISR inhibits IAV replication, but the specific mechanism remains unclear [12]. Further, seven candidate IAV- and IFN-up-regulated lncRNAs have been identified: AC015849.2, RP-1-7H24.1, PSMB8-AS1, CTD-2639E6.9, PSOR1C3, AC007283.5, and RP11-670E13.5. Among them, inhibition of PSMB8-AS1 reduces IAV replication and proliferation [59].

2.3. Effect of CircRNAs

CircRNA is a newly discovered class of single-stranded ncRNAs that can inhibit the activity of other RNAs or RNA-binding proteins. With the continuous development of sequencing technology, many circRNAs have been identified in animals, plants, and humans. Notably, circRNAs display specific expression patterns dependent on cell type, tissue type, and developmental stage, revealing their significant regulatory role in gene expression. More importantly, circRNAs sponge miRNA, an important regulator of gene expression. For example, it is reported that circ-Vav3 expression is up-regulated in chicken hepatomas infected by avian leukemia virus J-subunit, causing sponging of gga-miR-375 to eliminate the effect of the miRNA on its target gene, YAP1, increasing YAP1 expression, and inducing the epithelial-mesenchymal transition [67]. Additionally, the alternative splicing of mRNA before it forms a circRNA affects protein production. When ectopically expressed in Kaposi's sarcoma herpes virus (KSHV) infected cells, the circRNA hsa_circ_0001400, which is induced by KSHV infection, regulates viral gene expression without affecting viral genome replication [68]. CircRNAs can also be translated into small peptides. Carcinogenic human papillomavirus (HPV)-induced circRNAs contain part of the E7 oncogene, which is modified by N6-methyladenosine (m6A), preferentially localized in the cytoplasm, related to polysomes, and ultimately translated into the E7 oncoprotein to facilitate cancer cell growth [69]. CircRNAs impair autophagy and participate in viral infection. After infection with H1N1-subtyped IAV, the expression of circGATAD2A is up-regulated, but overexpression of circGATAD2A impairs autophagy and promotes IAV replication [70]. CircRNA produced in vitro activates RIG-I-mediated innate immune responses and confers protection against viral infection [71].

CircRNAs have antiviral effects similar to ISGs. Because circRNAs are derived from pre-mRNA, they are transcribed ISGs, and many ISG exon- or intron-specific circRNAs have been identified [72]. CircRNAs also regulate IFN signaling to participate in antiviral innate immunity. A novel intron splicing circRNA, AIVR, is up-regulated by IAV infection, which acts as a miR-330-3p sponge to release CREBBP mRNA [73]. Thus, increased cellular expression of CREBBP can accelerate IFN-β production. However, whether IFN can induce AIVR in IAV-infected A549 cells needs further study. In addition, IAV-PR8 induces circRNA_0082633 by JAK-STAT signaling activation, enhancing ISRE promoter activity to up-regulate IFNB1 mRNA levels, promoting type I IFN signaling [74]. Importantly, hundreds of circRNAs may be involved in M1 macrophage activation after IFN-γ stimulation [75].

3. Discussion

3.1. Validation of IFN-Induced NcRNAs Complements the Traditional Definition of ISGs

IFNs trigger the transcription of ISGs, change the protein composition of host cells, and mediate the antiviral state. Although IFN-mediated antiviral innate immunity has been studied for many years, more research is needed to understand IFN-regulated antiviral mechanisms. Many recent discoveries in infectomics based on contemporary sequencing and proteomic techniques have changed our understanding of IFN-related biology [4,30,76]. In a narrow sense, ISG expression is up-regulated at the transcriptional level in response to stimulation with IFN, a direct target of the IFN signaling pathway. ISGs have traditionally been thought to function through the proteins they encode, but recent studies have shown that they can also exert antiviral functions through the ncRNAs they transcribe [13,29,30,42,43]. Peng et al. [77] detected that widespread differentially expressed (DE) lncRNAs respond to virus infection and regulate the host's innate immune response by sequencing analysis. Interestingly, most up-regulated DE lncRNAs after viral infection are also significantly up-regulated by IFN-α treatment [30,77]. Therefore, promoter analysis and expression correlation studies suggest that these lncRNAs might be ISGs [30]. For example, a recent study used ChIPBase v2.0 to reveal that the expression of lncRNAs in SARS-CoV-2-infected cells might regulate STAT1, STAT3, and IFN regulatory factors [78]. These studies have updated our understanding of the biological landscape of IFNs and lncRNAs.

Andrew et al. investigated the presence of IFN-induced ncRNAs in human cells. They found hundreds of DE ncRNAs, but by analyzing ISG-encoded transcript types, a total of 90% and 92% of up-and down-regulated genes were classified as protein-coding genes, respectively [79]. Along these lines, we propose that IFN-induced ncRNAs complement ISG products. As genome annotation becomes more complete and more ncRNAs are discovered, a deeper understanding of the role of ncRNAs in antiviral innate immunity will become possible. Of note, the expression of many ISGs is induced by IFNs, viral infection, double-stranded RNA, and signaling pathways [80]. Therefore, studying IFN-induced ISGs and ISGs triggered by pathogenic microorganisms will help to discover ISGs and reveal their antiviral mechanisms.

3.2. Diversity of IFN-Induced NcRNAs

The dozens of IFNs are classified into three categories. The classification of IFNs may lead to the diversity of IFN-induced ncRNAs. As seen in Figure 1, IFNs activate multiple signaling pathways through different receptors. Different signaling pathways stimulated by IFNs may induce different ncRNAs. Although the type I and type III IFN genes are distinct and bind to different receptors, they are induced by the same pathologic route of infection and activate the expression of relevant antiviral, antiproliferative, and immunoregulatory genes. Although IFN-γ is derived only from immune cells, its receptor is expressed at moderate or low levels in almost all cell types [23]. The structure of type II IFNs is independent of that of type I and III IFNs, but it induces the same transcription of some ISGs as that induced by type I IFNs [81]. Moreover, type I IFN signaling initiates IFN-γ signaling [82]. In addition, the distribution of receptors in different cells may be the reason ncRNAs exhibit specific expression patterns that depend on cell or tissue type. These results suggest that IFN-induced ncRNAs are diverse, but some of them can be induced by all types of IFNs, which may play a more important role in antiviral immunity responses.

The expression of ncRNAs largely depends on the cell type and is tightly controlled by various cellular signaling pathways. Because different IFNs induce the expression of different proteins, different IFNs specifically induce the expression of various ncRNAs. For example, numerous ISG-encoded proteins inhibit HCV replication. However, certain ISG-encoded proteins, such as ISG15 and USP18, promote HCV replication. Thus, ncRNAs act as positive or negative regulators to orchestrate a complex network and balance the IFN response, similar to the effect of ISG-encoded proteins. Furthermore, high-throughput sequencing of the transcriptome reveals that sequences in ncRNAs are similar to those reported to be in ISGs, suggesting that they originate from the transcription of the same genes. The ncRNA network of ISGs may help extend our understanding of the innate immune system against viruses. Whether IFN-stimulated ncRNAs interact with IFN-stimulated proteins to play a synergistic or antagonistic role in innate immunity remains to be further investigated.

3.3. Specificity of NcRNAs Induced by a Viral Infection

The characterization of ncRNAs not only provides a novel model for host defense in mammalian cells, but has also enabled the identification of novel components that can be added to the repertoire of antiviral effectors. Similar to the lncRNA *VIN* [83], these ncRNAs are induced only by specific viruses and are not affected by a general viral or IFN response, which is associated with differences in virulence. Additionally, some ncRNAs are not expressed at the same level when stimulated by different viruses, and some play opposite roles in the antiviral response. For example, lncRNA *NEAT1* is up-regulated by HIV-1 [84,85], Herpes simplex virus (HSV) [86,87], HTNV [88], hepatitis D virus [89], and IAV [86], but down-regulated by severe dengue virus [90]. In general, up-regulation of lncRNA *NEAT1* expression appears to be a general response to viral infection. However, by promoting HSV infectivity, lncRNA *NEAT1* inhibits the replication of other viruses. Whether the consequences promote or inhibit viral infection depends on downstream mechanisms.

3.4. Others Potential Effects of NcRNAs on Host Antiviral Innate Immunity

Many recent advances have been made in identifying ncRNA modifications and have been used to study their effects on lncRNAs and circRNAs, in addition to ribosomal RNAs (rRNAs) and tRNAs [91]. The methylation of adenosine at the N6 position in mammalian RNA is a highly abundant RNA base modification that promotes efficient initiation of translation of circular RNA proteins, participates in cellular activities, or performs other functions [92]. Therefore, we wondered whether different RNA modifications of IFN-stimulated ncRNAs could affect ncRNA function in antiviral immunity. IFN-stimulated ncRNAs are translated whether or not they participate in antiviral immunity through peptides. However, whether IFN-stimulated ncRNAs directly exert antiviral effects by targeting viral proteins remain to be determined. Both circRNAs and lncRNAs interact with miRNAs to perform various biological functions. Numerous circRNA–miRNA–mRNA interactions have been identified in viral infection, but the functions of these interaction networks remain to be further explored. Furthermore, whether IFNs induce certain other ncRNAs, such as snRNAs, rRNAs, piRNAs, and/or tRNAs, to participate in antiviral innate immunity is a worthy direction for future research. Indeed, determining whether IFNs induce these RNAs to participate in antiviral innate immunity and whether they form interaction networks warrants further investigation.

4. Conclusions

Given the role of IFN-stimulated ncRNAs in antiviral immunity, which enriches our understanding of IFN-mediated antiviral innate immunity, we were interested in determining the function of IFN-induced antiviral ncRNAs as ISGs. The IFN-stimulated miRNAs can directly target viral transcript or indirectly affect host protein expression to take part in antiviral innate immunity. The IFN-stimulated lncRNAs regulate gene expression on translation in *cis* or *trans*. Notably, most ncRNAs do not encode proteins, but some encode small polypeptides. We were also interested in ncRNA, which encodes peptides that add to antiviral innate immunity. Compared to other types of ncRNAs, lncRNAs exhibit surprisingly more functions. However, there are still many unanswered questions, such as whether the evolution of lncRNAs and miRNAs is conserved. In order to solve these outstanding problems, future studies need to focus on lncRNA biology and immune-related lncRNA functions in vivo. In addition, emerging circRNAs are classified as lncRNAs generated by back splicing and 3′ to 5′ end self-ligation. However, few reports address the basic mechanisms of IAV infection and replication, or the regulatory mechanisms by which circRNAs affect the interaction between IAV and host cells. Therefore, research on circRNA with respect to antiviral innate immune responses induced by influenza virus, the molecular basis of virus RNA binding proteins, the circRNA transcription regulatory mechanisms, analysis of circRNA effects on cell function, and determining the network(s) involved in signaling pathways and regulation are worthy of future research. We expect to find that circRNAs regulate cell functions and discover the mechanisms by which they influence viral replication and pathogenicity.

Author Contributions: J.L. conceived and designed the manuscript. J.M. and J.L. prepared the manuscript and completed its revision. W.L. suggested many of the ideas presented in this study. All authors have read and agreed to the published version of the manuscript.

Funding: This work was supported by grants from the National Natural Science Foundation of China (32170143, 32061123001, 31970153). J.L. is supported by the Youth Innovation Promotion Association of CAS (2019091).

Institutional Review Board Statement: Not applicable.

Informed Consent Statement: Not applicable.

Data Availability Statement: Not applicable.

Conflicts of Interest: The authors declare no conflict of interest.

References

1. Crick, F. Central dogma of molecular biology. *Nature* **1970**, *227*, 561–563. [CrossRef]
2. Crick, F.H. On protein synthesis. *Symp. Soc. Exp. Biol.* **1958**, *12*, 138–163. [PubMed]
3. Consortium, E.P. An integrated encyclopedia of DNA elements in the human genome. *Nature* **2012**, *489*, 57–74. [CrossRef] [PubMed]
4. Schneider, W.M.; Chevillotte, M.D.; Rice, C.M. Interferon-stimulated genes: A complex web of host defenses. *Annu. Rev. Immunol.* **2014**, *32*, 513–545. [CrossRef]
5. Zhang, X.; Yang, W.; Wang, X.; Zhang, X.; Tian, H.; Deng, H.; Zhang, L.; Gao, G. Identification of new type I interferon-stimulated genes and investigation of their involvement in IFN-beta activation. *Protein Cell* **2018**, *9*, 799–807. [CrossRef]
6. Der, S.D.; Zhou, A.; Williams, B.R.; Silverman, R.H. Identification of genes differentially regulated by interferon alpha, beta, or gamma using oligonucleotide arrays. *Proc. Natl. Acad. Sci. USA* **1998**, *95*, 15623–15628. [CrossRef] [PubMed]
7. Rani, M.R.; Shrock, J.; Appachi, S.; Rudick, R.A.; Williams, B.R.; Ransohoff, R.M. Novel interferon-beta-induced gene expression in peripheral blood cells. *J. Leukoc. Biol.* **2007**, *82*, 1353–1360. [CrossRef]
8. Sarasin-Filipowicz, M.; Oakeley, E.J.; Duong, F.H.; Christen, V.; Terracciano, L.; Filipowicz, W.; Heim, M.H. Interferon signaling and treatment outcome in chronic hepatitis C. *Proc. Natl. Acad. Sci. USA* **2008**, *105*, 7034–7039. [CrossRef]
9. Indraccolo, S.; Pfeffer, U.; Minuzzo, S.; Esposito, G.; Roni, V.; Mandruzzato, S.; Ferrari, N.; Anfosso, L.; Dell'Eva, R.; Noonan, D.M.; et al. Identification of genes selectively regulated by IFNs in endothelial cells. *J. Immunol.* **2007**, *178*, 1122–1135. [CrossRef]
10. de Veer, M.J.; Holko, M.; Frevel, M.; Walker, E.; Der, S.; Paranjape, J.M.; Silverman, R.H.; Williams, B.R. Functional classification of interferon-stimulated genes identified using microarrays. *J. Leukoc. Biol.* **2001**, *69*, 912–920. [CrossRef]
11. Liu, X.; Duan, X.; Holmes, J.A.; Li, W.; Lee, S.H.; Tu, Z.; Zhu, C.; Salloum, S.; Lidofsky, A.; Schaefer, E.A.; et al. A Long Noncoding RNA Regulates Hepatitis C virus Infection Through Interferon Alpha-Inducible Protein 6. *Hepatology* **2019**, *69*, 1004–1019. [CrossRef]
12. Pan, Q.; Zhao, Z.; Liao, Y.; Chiu, S.H.; Wang, S.; Chen, B.; Chen, N.; Chen, Y.; Chen, J.L. Identification of an Interferon-Stimulated Long Noncoding RNA (LncRNA ISR) Involved in Regulation of Influenza A virus Replication. *Int. J. Mol. Sci.* **2019**, *20*, 5118. [CrossRef]
13. Pedersen, I.M.; Cheng, G.; Wieland, S.; Volinia, S.; Croce, C.M.; Chisari, F.V.; David, M. Interferon modulation of cellular microRNAs as an antiviral mechanism. *Nature* **2007**, *449*, 919–922. [CrossRef] [PubMed]
14. Schmitt, M.J.; Philippidou, D.; Reinsbach, S.E.; Margue, C.; Wienecke-Baldacchino, A.; Nashan, D.; Behrmann, I.; Kreis, S. Interferon-gamma-induced activation of Signal Transducer and Activator of Transcription 1 (STAT1) up-regulates the tumor suppressing microRNA-29 family in melanoma cells. *Cell Commun. Signal.* **2012**, *10*, 41. [CrossRef]
15. Yang, C.H.; Yue, J.; Fan, M.; Pfeffer, L.M. IFN induces miR-21 through a signal transducer and activator of transcription 3-dependent pathway as a suppressive negative feedback on IFN-induced apoptosis. *Cancer Res.* **2010**, *70*, 8108–8116. [CrossRef] [PubMed]
16. Kawai, T.; Akira, S. The role of pattern-recognition receptors in innate immunity: Update on Toll-like receptors. *Nat. Immunol.* **2010**, *11*, 373–384. [CrossRef] [PubMed]
17. Chan, Y.K.; Gack, M.U. RIG-I-like receptor regulation in virus infection and immunity. *Curr. Opin. Virol.* **2015**, *12*, 7–14. [PubMed]
18. Chen, Q.; Sun, L.; Chen, Z.J. Regulation and function of the cGAS-STING pathway of cytosolic DNA sensing. *Nat. Immunol.* **2016**, *17*, 1142–1149. [CrossRef] [PubMed]
19. Chan, Y.K.; Gack, M.U. Viral evasion of intracellular DNA and RNA sensing. *Nat. Rev. Microbiol.* **2016**, *14*, 360–373. [CrossRef] [PubMed]
20. Kotenko, S.V.; Gallagher, G.; Baurin, V.V.; Lewis-Antes, A.; Shen, M.; Shah, N.K.; Langer, J.A.; Sheikh, F.; Dickensheets, H.; Donnelly, R.P. IFN-lambdas mediate antiviral protection through a distinct class II cytokine receptor complex. *Nat. Immunol.* **2003**, *4*, 69–77. [CrossRef] [PubMed]
21. Sheppard, P.; Kindsvogel, W.; Xu, W.; Henderson, K.; Schlutsmeyer, S.; Whitmore, T.E.; Kuestner, R.; Garrigues, U.; Birks, C.; Roraback, J.; et al. IL-28, IL-29 and their class II cytokine receptor IL-28R. *Nat. Immunol.* **2003**, *4*, 63–68. [CrossRef]
22. Prokunina-Olsson, L.; Muchmore, B.; Tang, W.; Pfeiffer, R.M.; Park, H.; Dickensheets, H.; Hergott, D.; Porter-Gill, P.; Mumy, A.; Kohaar, I.; et al. A variant upstream of IFNL3 (IL28B) creating a new interferon gene IFNL4 is associated with impaired clearance of hepatitis C virus. *Nat. Genet.* **2013**, *45*, 164–171. [CrossRef] [PubMed]

23. Alspach, E.; Lussier, D.M.; Schreiber, R.D. Interferon gamma and Its Important Roles in Promoting and Inhibiting Spontaneous and Therapeutic Cancer Immunity. *Cold Spring Harb. Perspect. Biol.* **2019**, *11*, a028480. [CrossRef] [PubMed]

24. van Boxel-Dezaire, A.H.; Rani, M.R.; Stark, G.R. Complex modulation of cell type-specific signaling in response to type I interferons. *Immunity* **2006**, *25*, 361–372. [CrossRef] [PubMed]

25. Marcello, T.; Grakoui, A.; Barba-Spaeth, G.; Machlin, E.S.; Kotenko, S.V.; MacDonald, M.R.; Rice, C.M. Interferons alpha and lambda inhibit hepatitis C virus replication with distinct signal transduction and gene regulation kinetics. *Gastroenterology* **2006**, *131*, 1887–1898. [CrossRef] [PubMed]

26. Bolen, C.R.; Ding, S.; Robek, M.D.; Kleinstein, S.H. Dynamic expression profiling of type I and type III interferon-stimulated hepatocytes reveals a stable hierarchy of gene expression. *Hepatology* **2014**, *59*, 1262–1272. [CrossRef] [PubMed]

27. Galani, I.E.; Koltsida, O.; Andreakos, E. Type III interferons (IFNs): Emerging Master Regulators of Immunity. *Adv. Exp. Med. Biol.* **2015**, *850*, 1–15. [CrossRef] [PubMed]

28. Borden, E.C.; Sen, G.C.; Uze, G.; Silverman, R.H.; Ransohoff, R.M.; Foster, G.R.; Stark, G.R. Interferons at age 50: Past, current and future impact on biomedicine. *Nat. Rev. Drug Discov.* **2007**, *6*, 975–990. [CrossRef]

29. Kambara, H.; Niazi, F.; Kostadinova, L.; Moonka, D.K.; Siegel, C.T.; Post, A.B.; Carnero, E.; Barriocanal, M.; Fortes, P.; Anthony, D.D.; et al. Negative regulation of the interferon response by an interferon-induced long non-coding RNA. *Nucleic Acids Res.* **2014**, *42*, 10668–10680. [CrossRef]

30. Josset, L.; Tchitchek, N.; Gralinski, L.E.; Ferris, M.T.; Eisfeld, A.J.; Green, R.R.; Thomas, M.J.; Tisoncik-Go, J.; Schroth, G.P.; Kawaoka, Y.; et al. Annotation of long non-coding RNAs expressed in collaborative cross founder mice in response to respiratory virus infection reveals a new class of interferon-stimulated transcripts. *RNA Biol.* **2014**, *11*, 875–890. [CrossRef]

31. Barriocanal, M.; Carnero, E.; Segura, V.; Fortes, P. Long Non-Coding RNA BST2/BISPR is Induced by IFN and Regulates the Expression of the Antiviral Factor Tetherin. *Front. Immunol.* **2014**, *5*, 655. [CrossRef] [PubMed]

32. Carnero, E.; Barriocanal, M.; Segura, V.; Guruceaga, E.; Prior, C.; Borner, K.; Grimm, D.; Fortes, P. Type I Interferon Regulates the Expression of Long Non-Coding RNAs. *Front. Immunol.* **2014**, *5*, 548. [CrossRef] [PubMed]

33. Sakai, I.; Takeuchi, K.; Yamauchi, H.; Narumi, H.; Fujita, S. Constitutive expression of SOCS3 confers resistance to IFN-alpha in chronic myelogenous leukemia cells. *Blood* **2002**, *100*, 2926–2931. [CrossRef] [PubMed]

34. Qin, H.; Niyongere, S.A.; Lee, S.J.; Baker, B.J.; Benveniste, E.N. Expression and functional significance of SOCS-1 and SOCS-3 in astrocytes. *J. Immunol.* **2008**, *181*, 3167–3176. [CrossRef] [PubMed]

35. Du, H.; Cui, S.; Li, Y.; Yang, G.; Wang, P.; Fikrig, E.; You, F. MiR-221 negatively regulates innate anti-viral response. *PLoS ONE* **2018**, *13*, e0200385. [CrossRef] [PubMed]

36. Liu, S.; Sun, X.; Wang, M.; Hou, Y.; Zhan, Y.; Jiang, Y.; Liu, Z.; Cao, X.; Chen, P.; Liu, Z.; et al. A microRNA 221- and 222-mediated feedback loop maintains constitutive activation of NFkappaB and STAT3 in colorectal cancer cells. *Gastroenterology* **2014**, *147*, 847–859.e11. [CrossRef] [PubMed]

37. Zhang, C.; Han, L.; Zhang, A.; Yang, W.; Zhou, X.; Pu, P.; Du, Y.; Zeng, H.; Kang, C. Global changes of mRNA expression reveals an increased activity of the interferon-induced signal transducer and activator of transcription (STAT) pathway by repression of miR-221/222 in glioblastoma U251 cells. *Int. J. Oncol.* **2010**, *36*, 1503–1512. [CrossRef]

38. Hertzog, P.; Forster, S.; Samarajiwa, S. Systems biology of interferon responses. *J. Interferon Cytokine Res.* **2011**, *31*, 5–11. [CrossRef] [PubMed]

39. Perng, Y.C.; Lenschow, D.J. ISG15 in antiviral immunity and beyond. *Nat. Rev. Microbiol.* **2018**, *16*, 423–439. [CrossRef]

40. Farrell, P.J.; Broeze, R.J.; Lengyel, P. Accumulation of an mRNA and protein in interferon-treated Ehrlich ascites tumour cells. *Nature* **1979**, *279*, 523–525. [CrossRef]

41. Zhang, D.; Zhang, D.E. Interferon-stimulated gene 15 and the protein ISGylation system. *J. Interferon Cytokine Res.* **2011**, *31*, 119–130. [CrossRef] [PubMed]

42. Chai, W.; Li, J.; Shangguan, Q.; Liu, Q.; Li, X.; Qi, D.; Tong, X.; Liu, W.; Ye, X. Lnc-ISG20 Inhibits Influenza A Virus Replication by Enhancing ISG20 Expression. *J. Virol.* **2018**, *92*, e00539-18. [CrossRef] [PubMed]

43. Li, X.; Guo, G.; Lu, M.; Chai, W.; Li, Y.; Tong, X.; Li, J.; Jia, X.; Liu, W.; Qi, D.; et al. Long Noncoding RNA Lnc-MxA Inhibits Beta Interferon Transcription by Forming RNA-DNA Triplexes at Its Promoter. *J. Virol.* **2019**, *93*, e00786-19. [CrossRef] [PubMed]

44. Kincaid, R.P.; Sullivan, C.S. Virus-encoded microRNAs: An overview and a look to the future. *PLoS Pathog.* **2012**, *8*, e1003018. [CrossRef]

45. Haussecker, D.; Kay, M.A. miR-122 continues to blaze the trail for microRNA therapeutics. *Mol. Ther.* **2010**, *18*, 240–242. [CrossRef]

46. Chen, Y.; Shen, A.; Rider, P.J.; Yu, Y.; Wu, K.; Mu, Y.; Hao, Q.; Liu, Y.; Gong, H.; Zhu, Y.; et al. A liver-specific microRNA binds to a highly conserved RNA sequence of hepatitis B virus and negatively regulates viral gene expression and replication. *FASEB J.* **2011**, *25*, 4511–4521. [CrossRef]

47. Nathans, R.; Chu, C.Y.; Serquina, A.K.; Lu, C.C.; Cao, H.; Rana, T.M. Cellular microRNA and P bodies modulate host-HIV-1 interactions. *Mol. Cell* **2009**, *34*, 696–709. [CrossRef]

48. Kutty, R.K.; Nagineni, C.N.; Samuel, W.; Vijayasarathy, C.; Hooks, J.J.; Redmond, T.M. Inflammatory cytokines regulate microRNA-155 expression in human retinal pigment epithelial cells by activating JAK/STAT pathway. *Biochem. Biophys. Res. Commun.* **2010**, *402*, 390–395. [CrossRef]

49. Liu, Z.; Qin, Q.; Wu, C.; Li, H.; Shou, J.; Yang, Y.; Gu, M.; Ma, C.; Lin, W.; Zou, Y.; et al. Downregulated NDR1 protein kinase inhibits innate immune response by initiating an miR146a-STAT1 feedback loop. *Nat. Commun.* **2018**, *9*, 2789. [CrossRef]

50. Wang, P.; Hou, J.; Lin, L.; Wang, C.; Liu, X.; Li, D.; Ma, F.; Wang, Z.; Cao, X. Inducible microRNA-155 feedback promotes type I IFN signaling in antiviral innate immunity by targeting suppressor of cytokine signaling 1. *J. Immunol.* **2010**, *185*, 6226–6233. [CrossRef]

51. Zheng, X.-Y.; Bian, P.-Y.; Ye, C.-T.; Ye, W.; Ma, H.-W.; Tang, K.; Zhang, C.-M.; Lei, Y.-F.; Wei, X.; Wang, P.-Z. Interferon-Induced Transmembrane Protein 3 Inhibits Hantaan Virus Infection, and Its Single Nucleotide Polymorphism rs12252 Influences the Severity of Hemorrhagic Fever with Re-nal Syndrome. *Front. Immunol.* **2016**, *7*, 535.

52. Mariotti, B.; Servaas, N.H.; Rossato, M.; Tamassia, N.; Cassatella, M.A.; Cossu, M.; Beretta, L.; van der Kroef, M.; Radstake, T.; Bazzoni, F. The Long Non-coding RNA NRIR Drives IFN-Response in Monocytes: Implication for Systemic Sclerosis. *Front. Immunol.* **2019**, *10*, 100. [CrossRef] [PubMed]

53. Paliwal, D.; Joshi, P.; Panda, S.K. Hepatitis E virus (HEV) egress: Role of BST2 (Tetherin) and interferon induced long non- coding RNA (lncRNA) BISPR. *PLoS ONE* **2017**, *12*, e0187334. [CrossRef] [PubMed]

54. Carnero, E.; Barriocanal, M.; Prior, C.; Pablo Unfried, J.; Segura, V.; Guruceaga, E.; Enguita, M.; Smerdou, C.; Gastaminza, P.; Fortes, P. Long noncoding RNA EGOT negatively affects the antiviral response and favors HCV replication. *EMBO Rep.* **2016**, *17*, 1013–1028. [CrossRef]

55. Nishitsuji, H.; Ujino, S.; Yoshio, S.; Sugiyama, M.; Mizokami, M.; Kanto, T.; Shimotohno, K. Long noncoding RNA #32 con-tributes to antiviral responses by controlling interferon-stimulated gene expression. *Proc. Natl. Acad. Sci. USA* **2016**, *113*, 10388–10393.

56. Xie, Q.; Chen, S.; Tian, R.; Huang, X.; Deng, R.; Xue, B.; Qin, Y.; Xu, Y.; Wang, J.; Guo, M.; et al. Long Noncoding RNA ITPRIP-1 Positively Regulates the Innate Immune Response through Promotion of Oligomerization and Activation of MDA5. *J. Virol.* **2018**, *92*, e00507-18. [CrossRef]

57. Zhou, Y.; Li, M.; Xue, Y.; Li, Z.; Wen, W.; Liu, X.; Ma, Y.; Zhang, L.; Shen, Z.; Cao, X. Interferon-inducible cytoplasmic lncLrrc55-AS promotes antiviral innate responses by strengthening IRF3 phosphorylation. *Cell Res.* **2019**, *29*, 641–654. [CrossRef]

58. Jiang, M.; Zhang, S.; Yang, Z.; Lin, H.; Zhu, J.; Liu, L.; Wang, W.; Liu, S.; Liu, W.; Ma, Y.; et al. Self-Recognition of an Inducible Host lncRNA by RIG-I Feedback Restricts Innate Immune Response. *Cell* **2018**, *173*, 906–919.e13. [CrossRef]

59. More, S.; Zhu, Z.; Lin, K.; Huang, C.; Pushparaj, S.; Liang, Y.; Sathiaseelan, R.; Yang, X.; Liu, L. Long non-coding RNA PSMB8-AS1 regulates influenza virus replication. *RNA Biol.* **2019**, *16*, 340–353. [CrossRef]

60. Kambara, H.; Gunawardane, L.; Zebrowski, E.; Kostadinova, L.; Jobava, R.; Krokowski, D.; Hatzoglou, M.; Anthony, D.D.; Valad-khan, S. Regulation of Interferon-Stimulated Gene BST2 by a lncRNA Transcribed from a Shared Bidirectional Pro-moter. *Front. Immunol.* **2014**, *5*, 676.

61. Perez-Caballero, D.; Zang, T.; Ebrahimi, A.; McNatt, M.W.; Gregory, D.A.; Johnson, M.C.; Bieniasz, P.D. Tetherin inhibits HIV-1 release by directly tethering virions to cells. *Cell* **2009**, *139*, 499–511. [CrossRef] [PubMed]

62. Mangeat, B.; Cavagliotti, L.; Lehmann, M.; Gers-Huber, G.; Kaur, I.; Thomas, Y.; Kaiser, L.; Piguet, V. Influenza virus partially counteracts restriction imposed by tetherin/BST-2. *J. Biol. Chem.* **2012**, *287*, 22015–22029. [CrossRef] [PubMed]

63. Gao, S.; von der Malsburg, A.; Paeschke, S.; Behlke, J.; Haller, O.; Kochs, G.; Daumke, O. Structural basis of oligomerization in the stalk region of dynamin-like MxA. *Nature* **2010**, *465*, 502–506. [CrossRef] [PubMed]

64. Kane, M.; Yadav, S.S.; Bitzegeio, J.; Kutluay, S.B.; Zang, T.; Wilson, S.J.; Schoggins, J.W.; Rice, C.M.; Yamashita, M.; Hatziioannou, T.; et al. MX2 is an interferon-induced inhibitor of HIV-1 infection. *Nature* **2013**, *502*, 563–566. [CrossRef] [PubMed]

65. Goujon, C.; Moncorge, O.; Bauby, H.; Doyle, T.; Ward, C.C.; Schaller, T.; Hue, S.; Barclay, W.S.; Schulz, R.; Malim, M.H. Human MX2 is an interferon-induced post-entry inhibitor of HIV-1 infection. *Nature* **2013**, *502*, 559–562. [CrossRef] [PubMed]

66. Zimmermann, P.; Manz, B.; Haller, O.; Schwemmle, M.; Kochs, G. The viral nucleoprotein determines Mx sensitivity of influenza A viruses. *J. Virol.* **2011**, *85*, 8133–8140. [CrossRef] [PubMed]

67. Zhang, X.; Yan, Y.; Lin, W.; Li, A.; Zhang, H.; Lei, X.; Dai, Z.; Li, X.; Li, H.; Chen, W.; et al. Circular RNA Vav3 sponges gga-miR-375 to promote epithelial-mesenchymal transition. *RNA Biol.* **2019**, *16*, 118–132. [CrossRef] [PubMed]

68. Tagawa, T.; Gao, S.; Koparde, V.N.; Gonzalez, M.; Spouge, J.L.; Serquina, A.P.; Lurain, K.; Ramaswami, R.; Uldrick, T.S.; Yarchoan, R.; et al. Discovery of Kaposi's sarcoma herpesvirus-encoded circular RNAs and a human antiviral circular RNA. *Proc. Natl. Acad. Sci. USA* **2018**, *115*, 12805–12810. [CrossRef] [PubMed]

69. Zhao, J.; Lee, E.E.; Kim, J.; Yang, R.; Chamseddin, B.; Ni, C.; Gusho, E.; Xie, Y.; Chiang, C.M.; Buszczak, M.; et al. Transforming activity of an oncoprotein-encoding circular RNA from human papillomavirus. *Nat. Commun.* **2019**, *10*, 2300. [CrossRef] [PubMed]

70. Yu, T.; Ding, Y.; Zhang, Y.; Liu, Y.; Li, Y.; Lei, J.; Zhou, J.; Song, S.; Hu, B. Circular RNA GATAD2A promotes H1N1 replication through inhibiting autophagy. *Vet. Microbiol.* **2019**, *231*, 238–245. [CrossRef] [PubMed]

71. Chen, Y.G.; Kim, M.V.; Chen, X.; Batista, P.J.; Aoyama, S.; Wilusz, J.E.; Iwasaki, A.; Chang, H.Y. Sensing Self and Foreign Circular RNAs by Intron Identity. *Mol. Cell* **2017**, *67*, 228–238.e5. [CrossRef] [PubMed]

72. Glazar, P.; Papavasileiou, P.; Rajewsky, N. circBase: A database for circular RNAs. *RNA* **2014**, *20*, 1666–1670. [CrossRef] [PubMed]

73. Qu, Z.; Meng, F.; Shi, J.; Deng, G.; Zeng, X.; Ge, J.; Li, Y.; Liu, L.; Chen, P.; Jiang, Y.; et al. A Novel Intronic Circular RNA Antagonizes Influenza virus by Absorbing a microRNA That Degrades CREBBP and Accelerating IFN-beta Production. *mBio* **2021**, *12*, e0101721. [CrossRef] [PubMed]

74. Guo, Y.; Yu, X.; Su, N.; Shi, N.; Zhang, S.; Zhang, L.; Yang, L.; Zhao, L.; Guan, Z.; Zhang, M.; et al. Identification and characterization of circular RNAs in the A549 cells following Influenza A virus infection. *Vet. Microbiol.* **2022**, *267*, 109390. [CrossRef]

75. Zhang, Y.; Zhang, Y.; Li, X.; Zhang, M.; Lv, K. Microarray analysis of circular RNA expression patterns in polarized macrophages. *Int. J. Mol. Med.* **2017**, *39*, 373–379. [CrossRef]
76. Iyer, M.K.; Niknafs, Y.S.; Malik, R.; Singhal, U.; Sahu, A.; Hosono, Y.; Barrette, T.R.; Prensner, J.R.; Evans, J.R.; Zhao, S.; et al. The landscape of long noncoding RNAs in the human transcriptome. *Nat. Genet.* **2015**, *47*, 199–208. [CrossRef]
77. Peng, X.; Gralinski, L.; Armour, C.D.; Ferris, M.T.; Thomas, M.J.; Proll, S.; Bradel-Tretheway, B.G.; Korth, M.J.; Castle, J.C.; Biery, M.C.; et al. Unique signatures of long noncoding RNA expression in response to virus infection and altered innate immune signaling. *mBio* **2010**, *1*, e00206-10. [CrossRef]
78. Laha, S.; Saha, C.; Dutta, S.; Basu, M.; Chatterjee, R.; Ghosh, S.; Bhattacharyya, N.P. In silico analysis of altered expression of long non-coding RNA in SARS-CoV-2 infected cells and their possible regulation by STAT1, STAT3 and interferon regulatory factors. *Heliyon* **2021**, *7*, e06395. [CrossRef]
79. Shaw, A.E.; Hughes, J.; Gu, Q.; Behdenna, A.; Singer, J.B.; Dennis, T.; Orton, R.J.; Varela, M.; Gifford, R.J.; Wilson, S.J.; et al. Fundamental properties of the mammalian innate immune system revealed by multispecies comparison of type I interferon responses. *PLoS Biol.* **2017**, *15*, e2004086. [CrossRef]
80. Sen, G.C.; Sarkar, S.N. The interferon-stimulated genes: Targets of direct signaling by interferons, double-stranded RNA, and viruses. *Curr. Top. Microbiol. Immunol.* **2007**, *316*, 233–250. [CrossRef]
81. Lew, D.J.; Decker, T.; Darnell, J.E., Jr. Alpha interferon and gamma interferon stimulate transcription of a single gene through different signal transduction pathways. *Mol. Cell. Biol.* **1989**, *9*, 5404–5411. [CrossRef]
82. Fenner, J.E.; Starr, R.; Cornish, A.L.; Zhang, J.G.; Metcalf, D.; Schreiber, R.D.; Sheehan, K.; Hilton, D.J.; Alexander, W.S.; Hertzog, P.J. Suppressor of cytokine signaling 1 regulates the immune response to infection by a unique inhibition of type I interferon activity. *Nat. Immunol.* **2006**, *7*, 33–39. [CrossRef] [PubMed]
83. Winterling, C.; Koch, M.; Koeppel, M.; Garcia-Alcalde, F.; Karlas, A.; Meyer, T.F. Evidence for a crucial role of a host non-coding RNA in influenza A virus replication. *RNA Biol.* **2014**, *11*, 66–75. [CrossRef] [PubMed]
84. Budhiraja, S.; Liu, H.; Couturier, J.; Malovannaya, A.; Qin, J.; Lewis, D.E.; Rice, A.P. Mining the human complexome database identifies RBM14 as an XPO1-associated protein involved in HIV-1 Rev function. *J. Virol.* **2015**, *89*, 3557–3567. [CrossRef] [PubMed]
85. Zhang, Q.; Chen, C.Y.; Yedavalli, V.S.; Jeang, K.T. NEAT1 long noncoding RNA and paraspeckle bodies modulate HIV-1 posttranscriptional expression. *mBio* **2013**, *4*, e00596-12. [CrossRef] [PubMed]
86. Imamura, K.; Imamachi, N.; Akizuki, G.; Kumakura, M.; Kawaguchi, A.; Nagata, K.; Kato, A.; Kawaguchi, Y.; Sato, H.; Yoneda, M.; et al. Long noncoding RNA NEAT1-dependent SFPQ relocation from promoter region to paraspeckle mediates IL8 expression upon immune stimuli. *Mol. Cell* **2014**, *53*, 393–406. [CrossRef] [PubMed]
87. Wang, Z.; Fan, P.; Zhao, Y.; Zhang, S.; Lu, J.; Xie, W.; Jiang, Y.; Lei, F.; Xu, N.; Zhang, Y. NEAT1 modulates herpes simplex virus-1 replication by regulating viral gene transcription. *Cell. Mol. Life Sci.* **2017**, *74*, 1117–1131. [CrossRef]
88. Ma, H.; Han, P.; Ye, W.; Chen, H.; Zheng, X.; Cheng, L.; Zhang, L.; Yu, L.; Wu, X.; Xu, Z.; et al. The Long Noncoding RNA NEAT1 Exerts Antihantaviral Effects by Acting as Positive Feedback for RIG-I Signaling. *J. Virol.* **2017**, *91*, e02250-16. [CrossRef]
89. Beeharry, Y.; Goodrum, G.; Imperiale, C.J.; Pelchat, M. The Hepatitis Delta virus accumulation requires paraspeckle components and affects NEAT1 level and PSP1 localization. *Sci. Rep.* **2018**, *8*, 6031. [CrossRef]
90. Pandey, A.D.; Goswami, S.; Shukla, S.; Das, S.; Ghosal, S.; Pal, M.; Bandyopadhyay, B.; Ramachandran, V.; Basu, N.; Sood, V.; et al. Correlation of altered expression of a long non-coding RNA, NEAT1, in peripheral blood mononuclear cells with dengue disease progression. *J. Infect.* **2017**, *75*, 541–554. [CrossRef]
91. Nombela, P.; Miguel-Lopez, B.; Blanco, S. The role of m(6)A, m(5)C and Psi RNA modifications in cancer: Novel therapeutic opportunities. *Mol. Cancer* **2021**, *20*, 18. [CrossRef] [PubMed]
92. Kong, S.; Tao, M.; Shen, X.; Ju, S. Translatable circRNAs and lncRNAs: Driving mechanisms and functions of their translation products. *Cancer Lett.* **2020**, *483*, 59–65. [CrossRef] [PubMed]

Article

IFITM3 Inhibits SARS-CoV-2 Infection and Is Associated with COVID-19 Susceptibility

Fengwen Xu [1,†], Geng Wang [2,3,†], Fei Zhao [1,†], Yu Huang [1], Zhangling Fan [1], Shan Mei [1], Yu Xie [1], Liang Wei [1], Yamei Hu [1], Conghui Wang [2], Shan Cen [4], Chen Liang [5], Lili Ren [2,6,*], Fei Guo [1,*] and Jianwei Wang [2,6,*]

1   NHC Key Laboratory of Systems Biology of Pathogens, Institute of Pathogen Biology, and Center for AIDS Research, Chinese Academy of Medical Sciences & Peking Union Medical College, Beijing 100730, China
2   NHC Key Laboratory of Systems Biology of Pathogens and Christophe Mérieux Laboratory, Institute of Pathogen Biology, Chinese Academy of Medical Sciences & Peking Union Medical College, Beijing 100730, China
3   Department of Respiratory and Critical Care Medicine, Clinical Research Center for Respiratory Disease, West China Hospital, Sichuan University, Chengdu 610041, China
4   Institute of Medicinal Biotechnology, Chinese Academy of Medical Sciences & Peking Union Medical College, Beijing 100050, China
5   Lady Davis Institute, Jewish General Hospital, McGill University, Montreal, QC H3T 1E2, Canada
6   Key Laboratory of Respiratory Disease Pathogenomics, Chinese Academy of Medical Sciences and Peking Union Medical College, Beijing 100730, China
*   Correspondence: renliliipb@163.com (L.R.); guofei@ipb.pumc.edu.cn (F.G.); wangjw28@163.com (J.W.)
†   These authors contributed equally to this work.

Citation: Xu, F.; Wang, G.; Zhao, F.; Huang, Y.; Fan, Z.; Mei, S.; Xie, Y.; Wei, L.; Hu, Y.; Wang, C.; et al. IFITM3 Inhibits SARS-CoV-2 Infection and Is Associated with COVID-19 Susceptibility. *Viruses* **2022**, *14*, 2553. https://doi.org/10.3390/v14112553

Academic Editors: Caijun Sun and Feng Ma

Received: 28 September 2022
Accepted: 16 November 2022
Published: 18 November 2022

Publisher's Note: MDPI stays neutral with regard to jurisdictional claims in published maps and institutional affiliations.

Abstract: SARS-CoV-2 has become a global threat to public health. Infected individuals can be asymptomatic or develop mild to severe symptoms, including pneumonia, respiratory distress, and death. This wide spectrum of clinical presentations of SARS-CoV-2 infection is believed in part due to the polymorphisms of key genetic factors in the population. In this study, we report that the interferon-induced antiviral factor IFITM3 inhibits SARS-CoV-2 infection by preventing SARS-CoV-2 spike-protein-mediated virus entry and cell-to-cell fusion. Analysis of a Chinese COVID-19 patient cohort demonstrates that the rs12252 CC genotype of IFITM3 is associated with SARS-CoV-2 infection risk in the studied cohort. These data suggest that individuals carrying the rs12252 C allele in the IFITM3 gene may be vulnerable to SARS-CoV-2 infection and thus may benefit from early medical intervention.

Keywords: SARS-CoV-2; COVID-19; IFITM3; rs12252

1. Introduction

The coronavirus disease 2019 (COVID-19) epidemic emerged in Wuhan, China, in December 2019, and has since spread worldwide. COVID-19 is caused by a new coronavirus, severe acute respiratory syndrome coronavirus 2 (SARS-CoV-2) [1,2]. Effective drugs and vaccines against SARS-CoV-2 are the hope of halting the rapid spread of COVID-19.

The severity of COVID-19 varies widely between infected individuals. Many are asymptomatic, some show mild symptoms, and a significant number of severe cases develop pneumonia, respiratory distress, and even death [3]. It is believed the polymorphisms of key genes in part underpin this broad range of COVID-19 symptoms, and some of these genes can be antiviral effectors that are induced by interferon (IFN), given the high sensitivity of SARS-CoV-2 to IFN inhibition [4]. Indeed, the study of the association between disease severity and single-nucleotide polymorphisms (SNPs) in IFN-stimulated genes (ISGs) has shed light on the critical roles of certain ISGs in host susceptibility to virus infections [5,6].

The interferon-induced transmembrane protein (IFITM) family has been demonstrated to possess broad antiviral activity by blocking virus entry [7–14]. IFITM3 was reported to inhibit respiratory viruses, including influenza A virus [7–9,15]. IFITM3 knockout mice succumb to the challenge of the otherwise non-lethal influenza A virus [16,17]. In humans, SNP rs12252-C of IFITM3 has been reported to be associated with disease severities in patients infected with 2009 H1N1 virus, H7N9 virus, Hantaan virus, and HIV-1 [18–23]. This rs12252 C SNP was proposed to generate an IFITM3 splice variant (ENST00000526811), which may lead to truncation of the N-terminal 21 amino acids of IFITM3 and thus the loss of antiviral activity [16,24]. Another SNP rs34481144-A, which is located in the promoter of IFITM3 and is associated with lower expression of IFITM3, was also linked to the disease severity of influenza virus infection [25].

However, the possible role of IFITM3 in the recent SARS-CoV-2 pandemic is controversial [26,27]. Here, we demonstrate that IFITM3 inhibits SARS-CoV-2 infection by preventing virus entry, and the first 21 amino acids of IFITM3 are indispensable for anti-SARS-CoV-2 activity. The analysis of a Chinese COVID-19 patient cohort indicates that the rs12252 CC genotype of IFITM3 is associated with the risk of acquiring SARS-CoV-2 infection and the weakened neutralizing antibody response.

2. Materials and Methods

2.1. Plasmids

The IFITM3 cDNA was cloned into the pQCXIP vector, and the Δ1–21, YLAA, Y20F, and Y20A mutations were generated with a PCR-based mutagenesis method as described previously [24,28]. The pCAGGS-SARS-CoV-2-S expression plasmid was a gift from He Huang (IPB, Beijing, China). The TMPRSS2 expression plasmid was purchased from Origene (Cat. No. RC208677). Transfection of plasmid DNA into cells was performed with Lipofectamine 2000 (Thermo Fisher Scientific, MA, USA) in accordance with the manufacturer's instructions.

2.2. Pseudovirus Production and Infection

SARS-CoV-2 spike protein (S)-pseudoviruses were produced by transfecting HEK293T cells (CRL-2316) with psPAX2, pLenti-Luc/GFP, and pCAGGS-S. Forty-eight hours after transfection, the supernatant was collected and passed through a 0.45 μm filter. To determine viral infectivity, viruses were incubated with HEK293T-ACE2 cells for 48 h, and luciferase activities were measured using the Luciferase Assay System (Promega, WI, USA). Protein expression in cell lysates was detected by Western blotting.

2.3. β-Lactamase-Vpr Assay

BlaM-Vpr pseudoviral particle fusion assay was performed as previously described [29]. Briefly, cells were transfected with the following plasmids in a 10 cm dish, with 6 μg of psPAX2, 6 μg of BlaM-Vpr, and 6 μg of pCAGGS-S. Transfection was carried out using PEI. Forty-eight hours after transfection, the supernatant was collected and clarified with a 0.45 μm filter. Then, HEK293T-ACE2 cells were incubated with BlaM-Vpr pseudoviral particles for 8 h, before cells were washed with PBS and loaded with the CCF2-AM substrate (Invitrogen) for 2 h at 37 °C according to the manufacturer's instructions. Following substrate loading, cells were washed and transferred to FACS tubes for flow cytometry analysis.

2.4. Cell–Cell Fusion Assay

Cell–cell fusion assay was performed as previously described [30]. In brief, HeLa cells transiently co-expressing SARS-CoV-2 S protein and HIV-1 Tat-Flag were co-cultivated with HEK293T target cells expressing ACE2 and HIV-1 LTR-Luc for 40 h. Cell fusion was quantified with luciferase assay based on HIV-1 Tat transactivated expression of luciferase.

2.5. Western Blotting

Cells were lysed in RIPA buffer (25 mM Tris /HCl (pH 7.4), 150 mM NaCl, 1% NP-40, 0.25% sodium deoxycholate, 1 mM EDTA). After clarification by centrifugation, equal amounts of cell lysates quantified by BCA™ Protein Assay Kit (Thermo Fisher Scientific, MA, USA) were separated in SDS-PAGE and transferred to nitrocellulose membranes. Then, membranes were probed with the indicated antibodies, anti-HA antibody from Sigma-Aldrich (St. Louis, MO, USA) (1:4000, Cat. No. H6908), anti-Flag antibody from Sigma-Aldrich (1:4000, Cat. No. F3165), anti-actin antibody from Sigma-Aldrich (1:5000, Cat. No. A1978), anti-SARS-CoV-2 N antibody from Sino Biological (1:1000, Cat. No. 40143-R019), and anti-IFITM3 antibody from Proteintech (1:1000, Cat. No. 11714-1-AP), followed by IRDye™ secondary antibodies (Odyssey, Lincoln, NE, USA). The signals were collected with the Odyssey Infrared Imaging System (Li-Cor, Lincoln, NE, USA).

2.6. Subjects and Samples

Samples of 203 patients (with a median age of 51, 105 males, and 98 females) diagnosed with quantitative RT-PCR for SARS-CoV-2 infection were investigated in this study. All patients required hospitalization in JinYinTan Hospital or FangCang Hospital, Wuhan. Throat swabs and plasma samples of patients were collected for additional viral testing, cytokine and chemokine quantification, and DNA extraction. This study was approved by the Ethical Review Board of the Institute of Pathogen Biology, Chinese Academy of Medical Sciences and Peking Union Medical College. Consent was obtained from competent patients directly or from relatives/friends/welfare attorneys of incapacitated patients.

2.7. Definition of Mild and Severe Cases

According to the new coronavirus disease-19 (COVID-19)-related pneumonia diagnosis and treatment guideline issued by the National Health Commission, patients with mild clinical symptoms and negative lung imaging present were defined as mild cases. Patients were defined as severe cases who had pneumonia confirmed by chest imaging and had an oxygen saturation (SaO_2) of 94% or less while they were breathing ambient air or a ratio of the partial pressure of oxygen (PaO_2) to the fraction of inspired oxygen (FiO_2) (PaO_2:FiO_2) at or below 300 mm Hg.

2.8. Sequencing and Genotyping of rs12252

The region encompassing the human IFITM3 rs12252 sequences was amplified by PCR from DNA obtained from throat swabs. The amplification was performed using the following forward and reverse primers, 5′-GGAAACTGTTGAGAAACCGAA-3′ and 5′-CATACGCACCTTCACGGAGT-3′ (Beijing Institute of Genomics, Shenzhen, China). The products encompassing IFITM3 rs12252 were purified and Sanger sequenced on an Applied Biosystems 3730 × 1 DNA Analyzer (GATC Biotech, Konstanz, Germany). SNP genotypes were identified by MEGA-X. Homozygotes were called based on high, single-base peaks with high Phred quality scores, while heterozygotes were identified based on low, overlapping peaks of two bases with lower Phred quality scores relative to surrounding base calls. SNP genotype frequencies in our sequencing were compared with that of the Chinese Han population from the 1000 Genomes Project database.

2.9. Cytokines and Chemokines Quantification

Plasma cytokines and chemokines were measured using the Human Cytokine Standard 27-Plex Assays panel and the Bio-Plex 200 system (Bio-Rad, CA, USA) according to the manufacturer's instructions. Plasma samples from four healthy adults were used as controls for cross-comparison. These four healthy adults were tested negative for SARS-CoV-2 by RT-PCR and did not have respiratory symptoms or other health-related complaints before sample collection.

2.10. SARS-CoV-2 Viral RNA Analysis

SARS-CoV-2 viral RNA content was tested using validated primers and probes targeting the nucleocapsid (N) gene, and data were acquired with the Bio-Rad real-time PCR system [31].

2.11. Enzyme-Linked Immunosorbent Assay

Enzyme-linked immunosorbent assay (ELISA) was used for detecting IgM and IgG antibodies against the SARS-CoV-2 N, S, and receptor-binding domain (RBD) as previously described [32]. Briefly, the purified RBD, S, and N proteins of SARS-CoV-2 were used as coating antigens. Horseradish peroxidase (HRP)-conjugated goat anti-human IgG (Sigma-Aldrich, St. Louis, MO, USA) and Horseradish Peroxidase-conjugated AffiniPure Goat Anti-Human IgM (Jackson ImmunoResearch Inc., West Grove, PA, USA) were used as the second antibody in 1:60,000 dilutions. The optimal coating concentrations of RBD, S, and N antigen were 10 ng/well, 20 ng/well, and 10 ng/well. The optimal plasma dilutions were 1:400.

2.12. Neutralizing Antibody Detection

Among the 203 patient samples in this cohort, we had sufficient plasma samples of only 81 patients to measure the neutralizing antibody level. The neutralization assay was performed with Vero E6 cells. Serial two-fold dilutions of plasma samples (starting at 1:10) were pre-incubated with SARS-CoV-2 at 100 TCID50 (50% tissue culture infective doses) for 2 h and then added to Vero E6 cells. The virus/plasma mixtures were removed after 1 h, and fresh growth medium was added to each well. The cytopathic effects were evaluated 5 days after incubation at 37 °C. For each plasma dilution, 4 duplicate wells were used. The neutralizing antibody titers were calculated using the Reed and Muench method [33].

2.13. Statistical Analysis

Statistical analysis of genetic data was performed using the chi-square test with SPSS software (version 23.0, SPSS Inc., Chicago, IL, USA). The statistical significance of allele frequencies was determined with Fisher's exact test. Statistical analysis of cytokine and chemokine data and antibody data was performed using Student's t test or the Mann–Whitney test, and the correlation analysis was performed using Pearson or Spearman correlation analysis with SPSS software (version 23.0). * indicates a significant difference $p < 0.05$, ** indicates $p < 0.01$, *** indicates $p < 0.001$, and **** indicates $p < 0.0001$.

3. Results

3.1. IFITM3 Restricts SARS-CoV-2 Infection

We and others reported that many IFN-inducible genes, including IFITM3, were up-regulated in COVID-19 patients [34–36]. Given the broad antiviral activity of IFITM3 [7–11], we investigated whether IFITM3 inhibits SARS-CoV-2 infection. Since IFITM3 impedes virus entry, we measured the effect of IFITM3 on the infection of lentiviral particles pseudotyped with SARS-CoV-2 spike protein. IFITM3 overexpression diminished SARS-CoV-2 S-mediated infection by 2-fold (Figure 1A). Then, siRNAs targeting IFITM3 led to a 50% increase in the infection of SARS-CoV-2 S-pseudotyped virus (Figure 1B). The N-terminal 21 amino acids of IFITM3, which contain the endocytic sorting signal 20-YEML-23, were reported to regulate IFITM3 trafficking in cells and antiviral activity [14,24,28]. The N-terminal 21 amino acids, the tyrosine-based protein sorting signal 20-YEML-23 in particular, enable IFITM3 to undergo endocytosis and are essential for the antiviral function of IFITM3 [24,28]. We have thus generated the N-terminal 21 amino acids deletion mutant (Δ1–21), the endocytic sorting signal mutants YLAA, Y20F, and Y20A, as controls of non-antiviral IFITM3 mutants. Not surprisingly, the N-terminal 21 amino acids deletion mutant (Δ1–21), the endocytic sorting signal mutant YLAA, Y20F, and Y20A, all lost the ability to inhibit the infection of SARS-CoV-2 S pseudoviruses (Figure S1). Next, we examined the effect of IFITM3 on SARS-CoV-2 infection of Vero E6 and Huh 7 cells. IFITM3 transfected Vero E6 cells were infected with SARS-CoV-2. Viral RNA was quantitated by qRT-PCR, and a 5-fold decrease was observed in IFITM3-transfected cells at both 4 hpi and 8 hpi (Figure 1C). Consistent with its neutral effect on SARS-CoV-2 S pseudoviruses, the N-terminal 21 amino acids deletion mutant (Δ1–21) did not inhibit SARS-CoV-2 infection (Figure 1C). In IFITM3-knockdown Huh 7 cells, levels of SARS-CoV-2 viral RNA and N protein expression increased by 2-fold (Figure 1D). These data demonstrate the anti-SARS-CoV-2 activity of IFITM3.

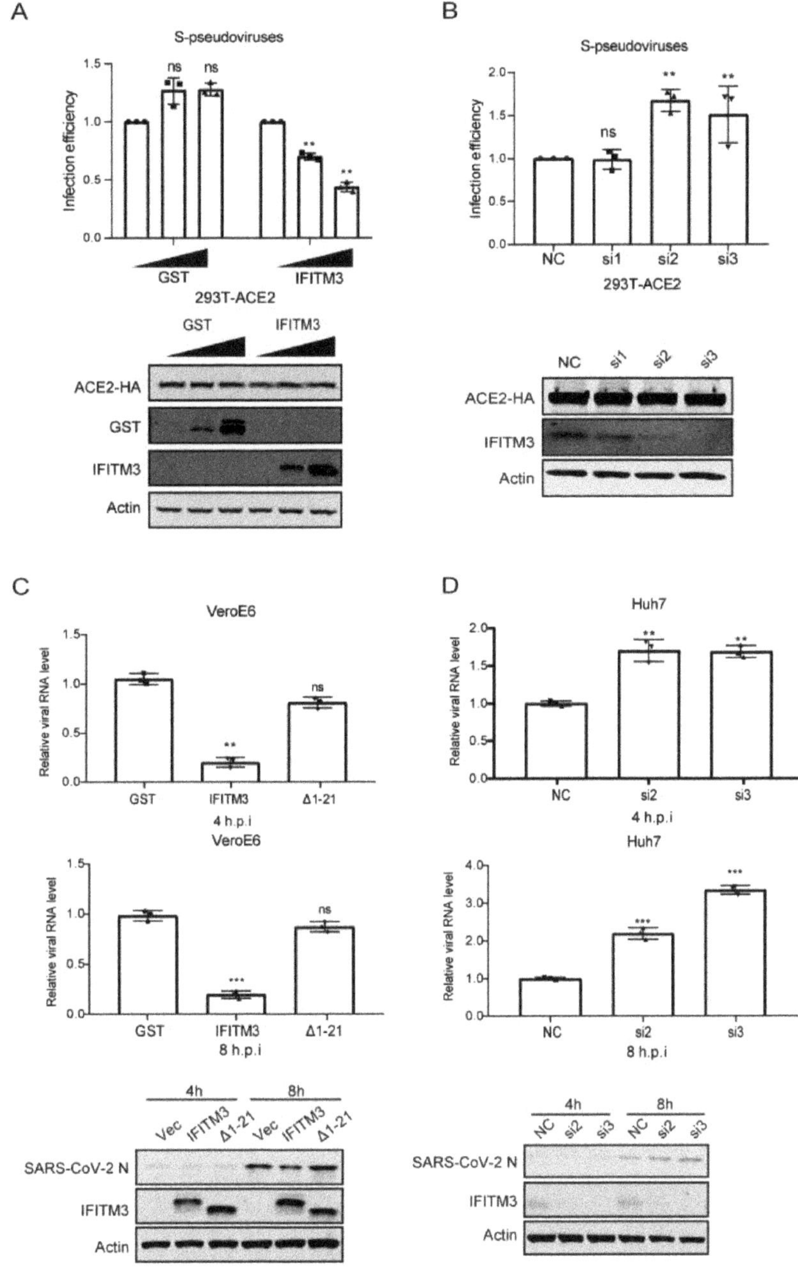

Figure 1. IFITM3 inhibits SARS-CoV-2 infection. (**A**) HEK293T-ACE2 cells were transfected with increasing amounts of GST or IFITM3 plasmid DNA and then infected with lentiviral reporter viruses pseudotyped with SARS-CoV-2 spike protein. Infection efficiency was determined by luciferase activities and normalized to that of vector DNA-transfected cells. Statistical significance was determined

with Student's *t* test. ns, not significant; **, $p < 0.01$. Expression of GST, IFITM3, ACE2, and actin was measured by Western blotting. (**B**) HEK293T-ACE2 cells were transfected with siRNAs targeting IFITM3 (si1, si2, si3) and then infected with lentiviral reporter viruses pseudotyped with SARS-CoV-2 spike protein. Infection efficiency was determined by luciferase activities and normalized to that of cells transfected with control siRNA (NC). Expression of IFITM3, ACE2, and actin was measured by Western blotting. (**C**) Vero E6 cells were transfected with GST, IFITM3, or Δ(1–21) IFITM3 and then infected with SARS-CoV-2 for 4 h or 8 h. Viral RNA level was determined by qRT-PCR, and expression of N, IFITM3, and actin was measured by Western blotting. (**D**) Huh7 cells were transfected with siRNAs targeting IFITM3 (si1, si2, si3) and then infected with SARS-CoV-2 for 4 h or 8 h. Viral RNA level was determined by qRT-PCR, and expression of N, IFITM3, and actin was measured by Western blotting. Data in the bar charts are presented as mean ± SD of three independent experiments. ns, not significant; NC, negative control; si1, IFITM3 siRNA1; si2, IFITM3 siRNA2; si3, IFITM3 siRNA3; h.p.i., hour post-infection. ns, not significant; **, $p < 0.01$; ***, $p < 0.001$.

3.2. IFITM3 Inhibits SARS-CoV-2 Entry

As IFITM proteins diminish the entry of different viruses, we next sought to examine the effect of IFITM3 on SARS-CoV-2 entry by BlaM-Vpr fusion assay. BlaM-Vpr is incorporated into SARS-CoV-2 S-pseudotyped lentiviral particles and delivered into target cells upon virus entry. Cleavage of CCF2-AM by BlaM-Vpr produces a fluorescence emission shift from green (520 nm) to blue (450 nm) (Figure 2A). As shown in Figure 2B,C, IFITM3 expression reduced SARS-CoV-2 S-mediated virus entry in a dose-dependent manner. Compared with wild-type IFITM3, IFITM3(Δ1–21) had minimal impact on virus entry (Figure 2D,E). These data support that IFITM3 inhibits SARS-CoV-2 infection by preventing virus entry.

We further investigated the effect of IFITM3 on SARS-CoV-2 S-induced membrane fusion in the cell–cell fusion assay. SARS-CoV-2 S and Tat transfected HeLa cells were co-cultured with HEK293T-ACE2 cells that expressed IFITM3 and Tat-inducible LTR-luciferase. After cell–cell fusion, Tat transactivates luciferase expression. Results of luciferase assay showed that IFITM3 inhibited SARS-CoV-2 S-mediated cell–cell fusion, while no effect was observed for IFITM3(Δ1–21) (Figure S2). Together, these data demonstrate that IFITM3 inhibits SARS-CoV-2 S-protein-mediated virion–cell fusion, as well as cell–cell fusion.

3.3. TMPRSS2 Attenuated the Inhibition of IFITM3

TMPRSS2 allows SARS-CoV-2 entry at the cell surface [37,38], and has been shown to render SARS-CoV-2 resistant to the inhibition of IFITM3 [26,39,40]. Indeed, when TMPRSS2 and IFITM3 were co-expressed in 293T-ACE2 cells, infection of SARS-CoV-2 S-pseudotyped viruses was less inhibited by IFITM3 compared to that in cells not expressing TMPRSS2 (Figure 3A). We further knocked down IFITM3 in TMPRSS2-positive Calu-3 cells, and did not observe any effect on infection of SARS-CoV-2 S-pseudotyped virus (Figure 3B). These data are consistent with previous reports that TMPRSS2 counters the inhibition of IFITM proteins [26,39,40].

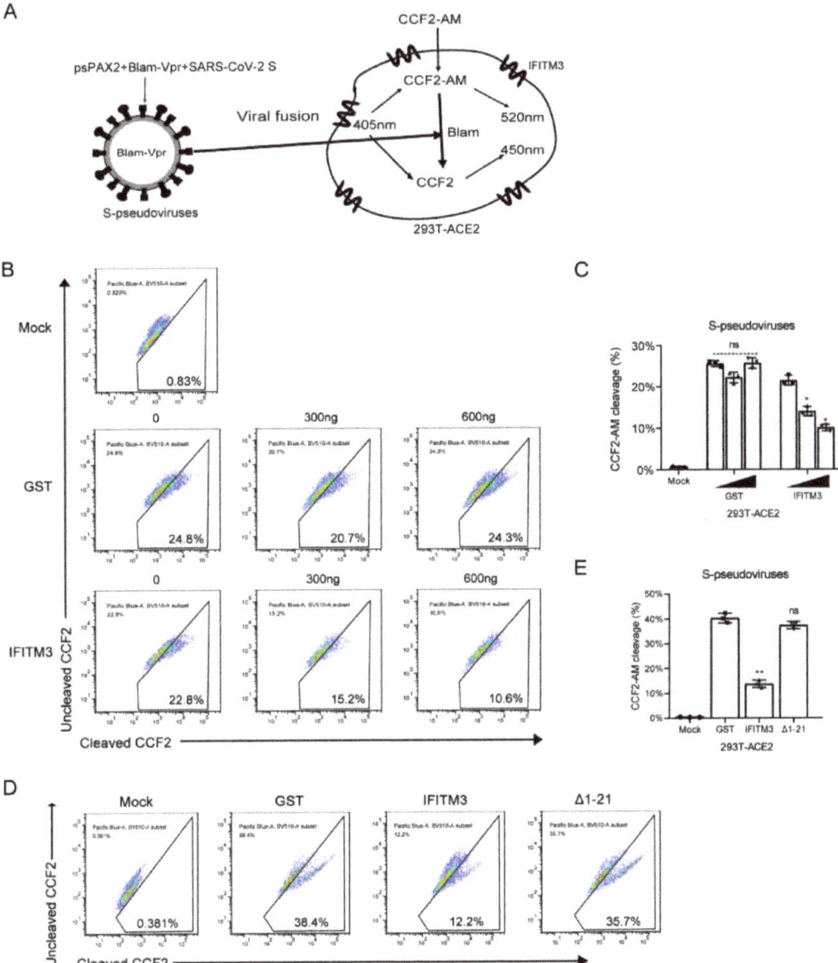

Figure 2. IFITM3 inhibits SARS-CoV-2 entry. (**A**) Illustration of the β-lactamase-Vpr assay. (**B**) HEK293T-ACE2 cells were transfected with increasing amounts of GST or IFITM3 and then infected with BlaM pseudovirus particles. BlaM activity was determined by flow cytometry. (**C**) Summary of the percentages of cells exhibiting CCF2-AM cleavage in (**B**). (**D**) HEK293T-ACE2 cells were transfected with IFITM3 or Δ(1–21) IFITM3 DNA, and then infected with BlaM pseudovirus particles. BlaM activity was determined by flow cytometry. (**E**) Summary of the percentages of cells exhibiting CCF2-AM cleavage in (**D**). Data in the bar charts are presented as mean ± SD of three independent experiments. ns, not significant. *, $p < 0.05$; **, $p < 0.01$.

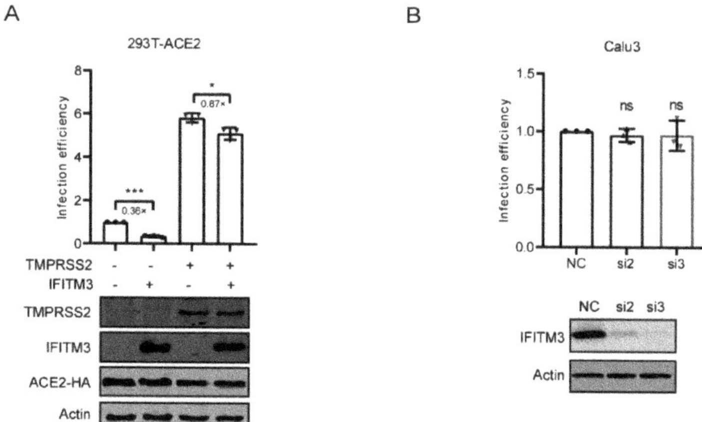

Figure 3. TMPRSS2 attenuated the inhibition of IFITM3. (**A**) HEK293T-ACE2 cells were transfected with IFITM3, TMPRSS2, or IFITM3 together with TMPRSS2 plasmid DNAs and then infected with lentiviral reporter viruses pseudotyped with SARS-CoV-2 spike protein. Infection efficiency was determined by luciferase activities and normalized to that of vector DNA-transfected cells. Expression of TMPRSS2, IFITM3, ACE2, and actin was measured by Western blotting. (**B**) Calu3 cells were transfected with siRNAs targeting IFITM3 (si2, si3) and then infected with lentiviral reporter viruses pseudotyped with SARS-CoV-2 spike protein. Infection efficiency was determined by luciferase activities and normalized to that of cells transfected with control siRNA (NC). Expression of IFITM3 and actin was measured by Western blotting. Data in the bar charts are presented as mean ± SD of three independent experiments. Statistical significance was determined with Student's *t* test. ns, not significant; *, $p < 0.05$; ***, $p < 0.001$; NC, negative control; si2, IFITM3 siRNA2; si3, IFITM3 siRNA3.

3.4. The rs12252 CC Genotype of IFITM3 Is Associated with the Risk of Acquiring SARS-CoV-2 Infection

Two SNPs of IFITM3, rs12252-C, and rs34481144-A (Figure 4A) were reported to be associated with the disease severity of several virus infections [18–23]. To investigate the possible association of these SNPs with SARS-CoV-2 infection, we sequenced the IFITM3 locus of 203 hospitalized COVID-19 patients, which included 133 mild cases (65.5%), 43 severe cases (21.2%), and 27 deaths (13.3%) (Figure 4B). Sequence analysis showed that 77 patients (37.9%) had the rs12252 CC genotype, 88 patients (43.4%) had the CT genotype, and 38 patients (18.7%) had the TT WT genotype (Figure 4B). Compared with the prevalence rate of 25.4% CC in the healthy Chinese Han population (from the 1000 genome project), the frequency of the CC genotype in 203 COVID-19 patients (37.9%) was significantly increased ($p = 0.04$) (Figure 4B,C), even though total C allele frequency did not show significant changes (Figure 4D). This result suggests that the CC genotype may increase the risk of SARS-CoV-2 infection. We further explored the correlation between CC genotype and disease severity, and found that the distributions of CC and CT/TT were 36.1% and 63.9% in mild cases, 41.9% and 58.1% in severe cases, 40.7% and 59.3% in deaths (Figure 4E,F). Although no statistically significant differences between these groups were found, the CC genotype in the severe and dead population tends to be over-presented. Next, the viral loads in throat swabs were determined, and the results showed no differences between CC and CT/TT genotypes in the entire patient cohort (Figure 4G), patients with mild symptoms (Figure 4H), or patients with severe symptoms (Figure 4I).

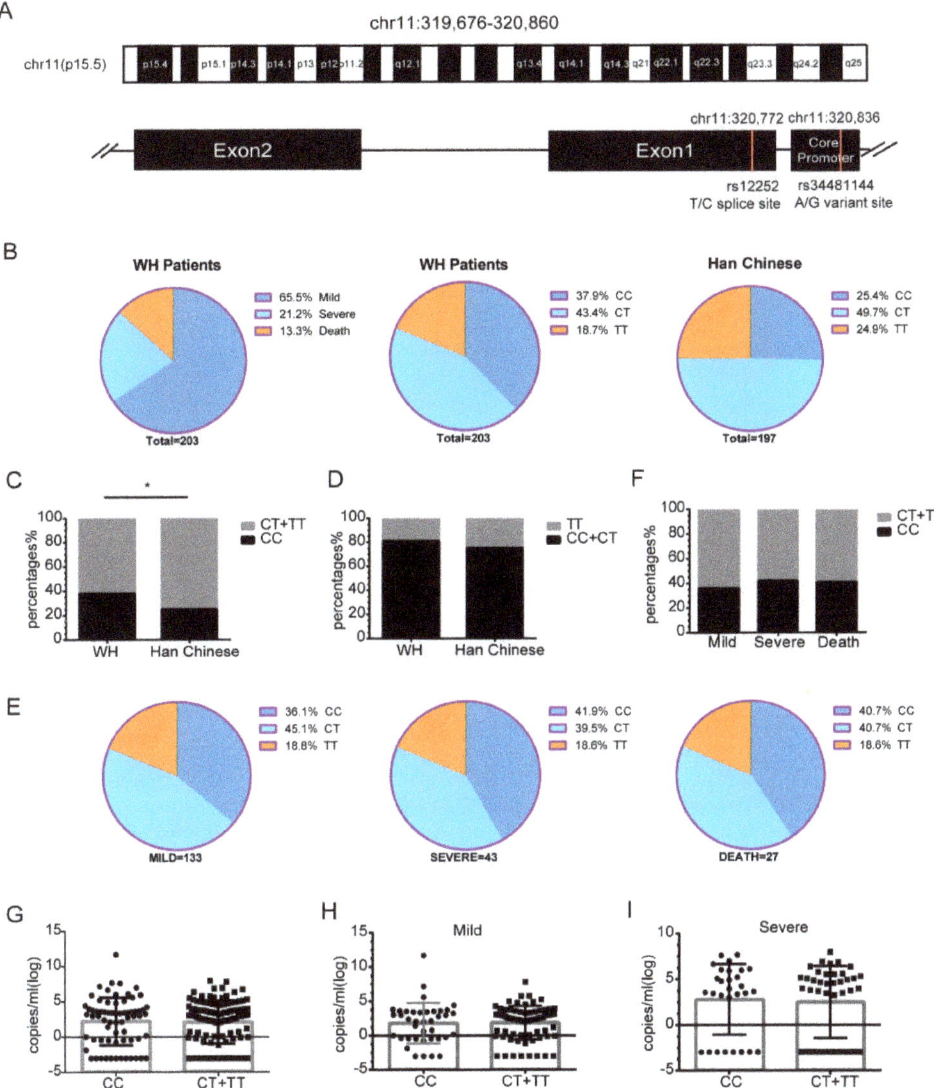

Figure 4. The rs12252 CC genotype of IFITM3 is associated with the risk of SARS-CoV-2 infection. (**A**) Illustration of SNPs rs12252 and rs34481144 of IFITM3. (**B**) Frequency of rs12252 alleles in our 203 COVID-19 patient cohort (WH patients), including 133 mild cases (65.5%), 43 severe cases (21.2%), and 27 deaths (13.3%). Han Chinese: healthy Chinese Han population (from the 1000 genome project). (**C**) The frequency of the CC genotype (37.9%) was increased ($p = 0.04$) in our COVID-19 patient cohort. (**D**) Frequency of the rs12252-C allele in our cohort and in the Chinese Han population (data from the 1000 genomes project). (**E**) Frequencies of the rs12252 alleles in mild, severe, and death cases. (**F**) Frequency of the CC genotype in mild, severe, and death cases. (**G**) Viral load in the throat swab samples from the CC and CT/TT carriers. (**H,I**) Viral load in the CC and CT/TT carriers from the mild group (**H**) and the severe group (**I**). WH patients, Wuhan patients; *, $p < 0.05$.

Furthermore, we investigated the association of another SNP rs34481144-A with SARS-CoV-2 infection. Only one patient in our cohort carries the AG genotype, which was consistent with 1000 genomes project data. Due to the limited number of the AG genotype in our patient cohort, we could not conclude the association of rs34481144 A with SARS-CoV-2 infection (Table S1).

3.5. Levels of Cytokines and Chemokines in COVID-19 Patients with CC or CT/TT Genotypes

Cytokine and chemokine levels in 21 patients (12 CC, 7 CT, and 2 TT genotypes) were determined. Compared with the healthy controls, plasma levels of IL-1β, IL1ra, IL-2, IL-4, IL-6, IL-8, IL-7, IL-9, IL-13, IL-15, IL-17, basic FGF, IP-10, MCP-1, MIP-1α, PDGF, VEGF, MIP-1β, TNF-α, and RANTES were significantly elevated in COVID-19-related pneumonia patients, whereas levels of IL-5, IL-10, IL-12, Eotaxin, G-CSF, GM-CSF, and IFN-γ were similar between the healthy control group and COVID-19 patients (Figure S3A). Further comparison of mild and severe cases showed that levels of MCP-1, MIP-1β, and MIP-1α were significantly higher in severe cases (Figure S3B). We also found that patients with the CC genotype had lower levels of IL-2, IL-17, and basic FGF than patients with the CT/TT genotype, but the level of RANTES was higher in CC patients, while levels of IL-1β, IL1ra, IL-4, IL-6, IL-8, IL-7, IL-9, IL-13, IL-15, IP-10, MCP-1, MIP-1α, PDGF, VEGF, MIP-1β, TNF-α, IL-5, IL-10, IL-12, Eotaxin, G-CSF, GM-CSF, and IFN-γ did not show a statistically significant difference between these two groups (Figure 5).

3.6. The rs12252 CC Genotype Might Be Associated with SARS-CoV-2 Neutralizing Antibody Positivity

We next measured SARS-CoV-2 neutralizing antibodies in 81 patients (35 CC, 32 CT, and 14 TT genotypes) two to three weeks after the onset of symptoms. We found that 18 of 35 patients with the CC genotype (51.4%) were positive for neutralizing antibodies, and 34 were positive (73.9%) in the 46 patients of the CT or TT genotype (Figure 6A). The neutralizing antibody positivity rate was significantly lower in the CC group (χ^2 = 4.356, p = 0.037), although no difference in neutralizing antibody titers in all positive cases was observed (Figure 6B). These data suggest that the rs12252 CC genotype is probably associated with poor neutralizing antibody production. In addition, we measured the levels of anti-N antibody, anti-S antibody, and anti-receptor-binding-domain (RBD) antibody. There were no differences in the levels of anti-N antibody or anti-S antibody between patients with CC and CT/TT genotypes, whereas patients with the CC genotype had a lower level of anti-RBD antibody IgG (Figure S4).

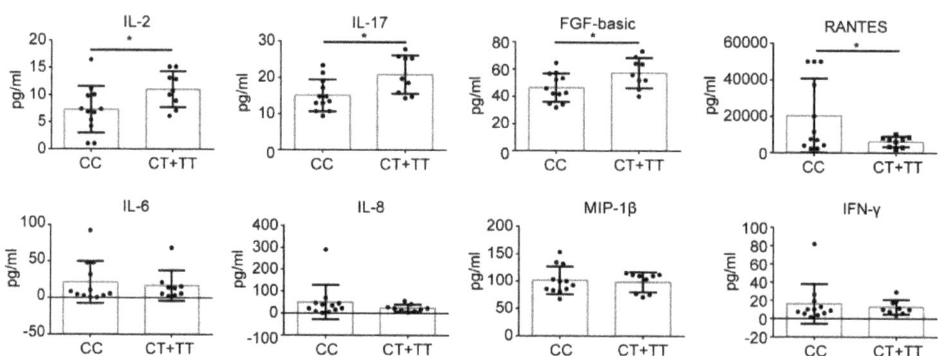

Figure 5. *Cont.*

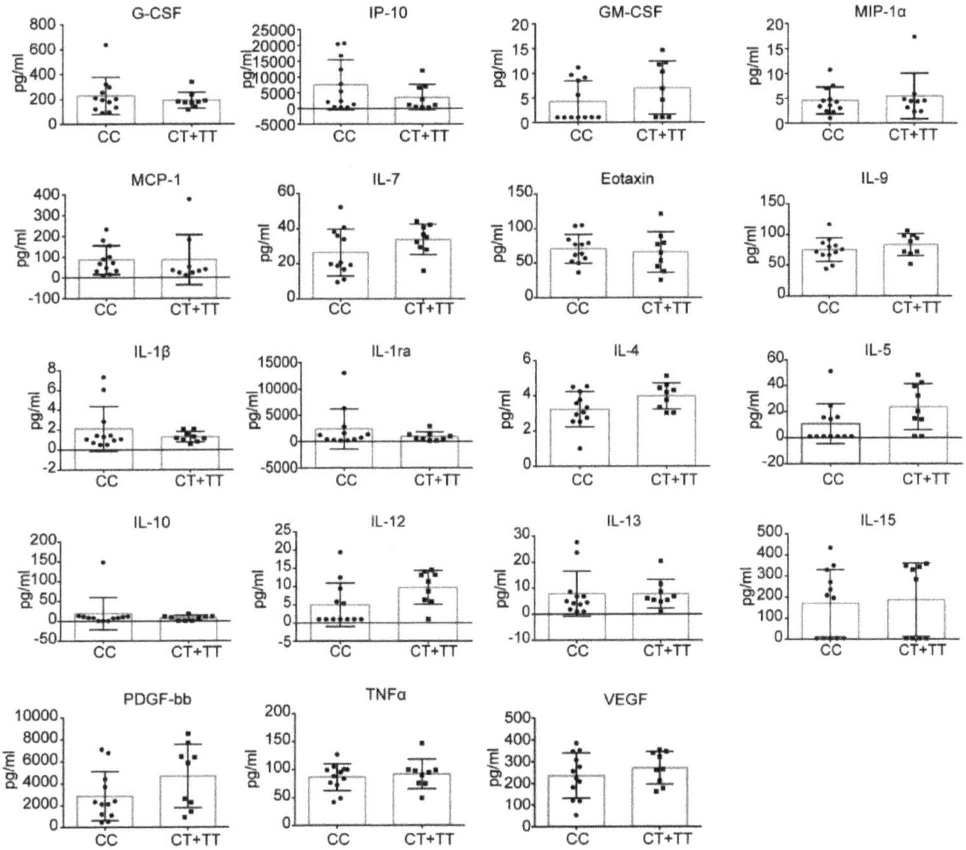

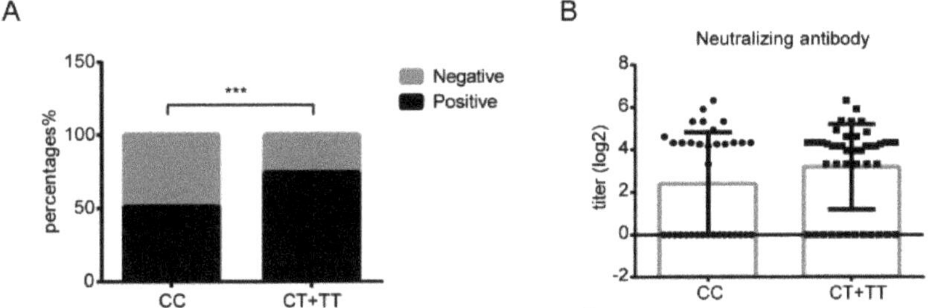

Figure 5. Levels of cytokines and chemokines in COVID-19 patients with the CC genotype or the CT/TT genotypes. Statistically significant differences were observed for IL-2 (p = 0.043), IL-17 (p = 0.013), FGF-basic (p = 0.034), and RANTES (p = 0.034). *, $p < 0.05$.

Figure 6. The rs12252 CC genotype is associated with lower positivity frequency of SARS-CoV-2 neutralizing antibody. (**A**) The positivity frequency of neutralizing antibodies in patients with different rs12252 genotypes in two to three weeks after symptom onset (p = 0.0008). (**B**) Neutralizing antibody titers in CC and CT/TT patients. ***, $p < 0.001$.

4. Discussion

IFITM3 has been reported to inhibit a number of pathogenic viruses in the families of orthomyxoviruses, flaviviruses, filoviruses, and coronaviruses [7–14]. Interestingly, IFITM3 was reported to restrict the entry mediated by the spike proteins of SARS-CoV, MERS-

CoV, hCoV-229E, and hCoV-NL63, yet it enhances the infection of hCoV-OC43 [8,9,15,41]. IFITM1, IFITM2, and IFITM3 were reported to inhibit SARS-CoV-2 infection, particularly in cells not expressing the TMPRSS2 protease, which allows SARS-CoV-2 entry at the plasma membrane [26,39,40,42,43]. In this study, we further showed that IFITM3 inhibited SARS-CoV-2 infection by preventing virus entry. The first 21 amino acids of IFITM3, which carry the endocytic sorting signal 20YEML23 and the Y20 phosphorylation site and have been shown essential for the restriction of VSV and influenza A virus, are also important for inhibiting SARS-CoV-2. In addition, results of BlaM-Vpr fusion assay and cell–cell fusion assay support that IFITM3 inhibited SARS-CoV-2 S-protein-mediated virion–cell fusion and cell–cell fusion, consistent with a previous report [40].

The rs12252 C allele of IFITM3 has been reported to be associated with the disease severity of several viral infections, including influenza virus, Hantaan virus, and HIV-1 [18–23]. We now show that our COVID-19 cohort has a significantly higher presentation of the rs12252 CC homozygous genotype compared to the general control population, which is in agreement with previous reports [44–46]. The rs12252 C allele was predicted to affect the splicing of IFITM3 RNA, resulting in the loss of the first 21 amino acids of IFITM3 and the expression of a truncated version of IFITM3 protein $\Delta(1–21)$, which did not inhibit influenza A virus (A/WSN/1933) [16]. Our study showed that this $\Delta(1–21)$ IFITM3 mutant also lost the ability to inhibit SARS-CoV-2 infection, which at least partially underlies the increased risk of acquiring SARS-CoV-2 for those carrying the rs12252 CC genotype. Given the current lack of strong evidence supporting the expression of $\Delta(1–21)$ IFITM3 in the rs12252 CC carriers [47,48], the molecular mechanisms behind the association of rs12252 CC with viral infections warrant further investigation. In any case, the rs12252 C may serve as one genetic marker for vulnerability to SARS-CoV-2 infection. IFITM3 SNP rs12252 has been reported to be associated with SARS-CoV-2 infection risk, COVID-19 disease outcome, and mortality rate in some cohorts [44–46,49–51], but not in others [52,53]. In this cohort of Chinese COVID-19 patients, we did not observe a statistically significant difference in the frequency of the CC genotype between the mild, severe, and fatal COVID-19 cases, nor the viral loads, but did find significantly higher presentations of the CC genotype in SARS-CoV-2 positive individuals, suggesting an association of IFITM3 with the acquisition of SARS-CoV-2 infection. These results together suggest that the ultimate effect of IFITM3 polymorphisms on SARS-CoV-2 infection and COVID-19 disease severity might depend on the genetic background of the studied populations.

IFITM3 SNP rs34481144 was reported to be associated with the risk for hospitalization after SARS-CoV-2 infection [50,51]. In this study, we also investigated the SNP rs34481144 at the IFITM3 core promoter, and found only one patient carrying the AG genotype in our cohort. This is consistent with 1000 genomes project data, i.e., Asian descents have a very low frequency of rs34481144 AA and AG genotypes, which precludes the further study of the possible association of this SNP with COVID-19 using our cohort.

COVID-19 is characterized by high levels of pro-inflammatory cytokines and chemokines, including IL-2, IL-7, IL-10, G-CSF, IP10, MIP-1A, MCP-1, and TNFα, especially in patients cared for in ICU [3]. Early studies have shown that increased levels of pro-inflammatory cytokines in plasma, such as MCP1, were associated with pulmonary inflammation and severe lung damage in SARS-CoV patients [54]. In our COVID-19 cohort, we observed increased levels of IL-2, IL-4, IL-15, IL17, and IP-10, which likely activate both T-helper-1 (Th1) and T-helper-2 (Th2) cells and lead to lung inflammation and damage, similar to what has been reported in SARS-CoV and MERS-CoV infections. In addition, we detected higher levels of RANTES and lower levels of IL-2, IL17, and basic FGF in CC carriers, which suggests that Th2 and T-helper-17 (Th17) cell functions might have been compromised in the CC patients. A detailed follow up of cytokine levels in COVID-19 patients at all stages of the disease will help to illustrate a more accurate map of host inflammatory responses to SARS-CoV-2 infection. Only 21 patients had sequential samples, and we therefore chose to evaluate the levels of cytokines and chemokines in these patients. Moreover, for the reason of strict blood sample control, we did not obtain sufficient samples to measure the levels of cytokines and chemokines in some patients. Due to the enormous pressures doctors and nurses faced in the early phase of the COVID-19 pandemic, we were unable to have regular periods of sample collection nor detailed clinical information, which limits the further interpretation of our data on cytokine and chemokine levels and association with IFITM3 polymorphisms in the context of clinical characteristics other than disease severity.

Polymorphisms in IFITM3 have been reported to modulate antibody responses to virus infection. Individuals with the rs12252 CC genotype have been shown to have higher pre-existing immunity to pdm09H1N1 than those with the TT genotype in a young people cohort [55]. Another study reported that the CC carriers presented lower seroconversion (SCR) and HI antibody titer 28 days after inoculation of the trivalent inactivated vaccine (TIV) [56]. In our study, we found that patients with the CC genotype had a lower positivity frequency of neutralizing antibody and lower levels of RBD antibody IgG than that in patients with CT/TT genotypes (Figure S4). These findings indicate that the rs12252 SNP in IFITM3 may modulate antibody responses, especially IgG production. A previous study showed that IFITM3 deficient mice had reduced the number of activated B cells and dysfunction of the germinal center to plasma/memory transition [56], which may in part explain the association of deficient antibody response with the rs12252 SNP in IFITM3.

In summary, our data demonstrate that IFITM3 is able to inhibit SARS-CoV-2 by impeding virus entry. We further observed the association of the rs12252 SNP in IFITM3 with a higher risk of SARS-CoV-2 acquisition, thus revealing one genetic factor underpinning the susceptibility to COVID-19.

Supplementary Materials: The following supporting information can be downloaded at: https://www.mdpi.com/article/10.3390/v14112553/s1, Figure S1: Deletion of first 21 aa of IFITM3 abolishes its antiviral activity; Figure S2: IFITM3 inhibits SARS-CoV-2 spike-protein-mediated cell–cell fusion; Figure S3: Levels of cytokines and chemokines in COVID-19 patients; Figure S4: Levels of anti-N, anti-S and anti-RBD antibodies in COVID-19 patients with the CC and CT/TT genotypes; Table S1: Genotype frequencies of IFITM3 rs34481144.

Author Contributions: L.R., F.G., J.W., F.X., G.W. and F.Z. conceived and designed the study. F.X., G.W., F.Z., Y.H. (Yu Huang), Z.F., S.M., Y.X., L.W., C.W. and Y.H. (Yamei Hu) carried out the experiments. S.C., C.L., L.R., F.G. and J.W. analyzed the data. F.G., C.L., G.W. and F.X. drafted the manuscript. All authors have read and agreed to the published version of the manuscript.

Funding: This work was supported by funds from the National Key Plan for Scientific Research and Development of China (2022YFE0203100 and 2020YFA0707602), the CAMS Innovation Fund for Medical Sciences (CIFMS 2021-1-I2M-038 and CIFMS 2022-I2M-JB-014), the Ministry of Science and Technology of China (2018ZX10301408-003), the National Natural Science Foundation of China (82271802, 82072288 and 32200718), and the Canadian Institutes of Health Research (CCI-132561).

Institutional Review Board Statement: The study was conducted in accordance with the Declaration of Helsinki, and approved by the Institutional Review Board of Wuhan Infectious Diseases Hospital (protocol KY-2020-02.02, January 2020).

Informed Consent Statement: Informed consent was obtained from all subjects involved in the study.

Data Availability Statement: The data presented in this study are available upon request from the corresponding author.

Conflicts of Interest: The authors declare no conflict of interest.

References

1. Zhu, N.; Zhang, D.; Wang, W.; Li, X.; Yang, B.; Song, J.; Zhao, X.; Huang, B.; Shi, W.; Lu, R.; et al. A Novel Coronavirus from Patients with Pneumonia in China, 2019. *N. Engl. J. Med.* **2020**, *382*, 727–733. [CrossRef]
2. Wu, F.; Zhao, S.; Yu, B.; Chen, Y.-M.; Wang, W.; Song, Z.-G.; Hu, Y.; Tao, Z.-W.; Tian, J.-H.; Pei, Y.-Y.; et al. A new coronavirus associated with human respiratory disease in China. *Nature* **2020**, *579*, 265–269. [CrossRef] [PubMed]
3. Huang, C.; Wang, Y.; Li, X.; Ren, L.; Zhao, J.; Hu, Y.; Zhang, L.; Fan, G.; Xu, J.; Gu, X.; et al. Clinical features of patients infected with 2019 novel coronavirus in Wuhan, China. *Lancet* **2020**, *395*, 497–506. [CrossRef]

4.   Lei, X.; Dong, X.; Ma, R.; Wang, W.; Xiao, X.; Tian, Z.; Wang, C.; Wang, Y.; Li, L.; Ren, L.; et al. Activation and evasion of type I interferon responses by SARS-CoV-2. *Nat. Commun.* **2020**, *11*, 3810. [CrossRef]

5.   Hijikata, M.; Ohta, Y.; Mishiro, S. Identification of a Single Nucleotide Polymorphism in the MxA Gene Promoter (G/T at nt −88) Correlated with the Response of Hepatitis C Patients to Interferon. *Intervirology* **2000**, *43*, 124–127. [CrossRef] [PubMed]

6.   Yakub, I.; Lillibridge, K.M.; Moran, A.; Gonzalez, O.Y.; Belmont, J.; Gibbs, R.A.; Tweardy, D.J. Single Nucleotide Polymorphisms in Genes for 2′-5′-Oligoadenylate Synthetase and RNase L in Patients Hospitalized with West Nile Virus Infection. *J. Infect. Dis.* **2005**, *192*, 1741–1748. [CrossRef] [PubMed]

7.   Brass, A.L.; Huang, I.-C.; Benita, Y.; John, S.P.; Krishnan, M.N.; Feeley, E.M.; Ryan, B.J.; Weyer, J.L.; van der Weyden, L.; Fikrig, E.; et al. The IFITM Proteins Mediate Cellular Resistance to Influenza a H1N1 Virus, West Nile Virus, and Dengue Virus. *Cell* **2009**, *139*, 1243–1254. [CrossRef]

8.   Huang, I.-C.; Bailey, C.C.; Weyer, J.L.; Radoshitzky, S.; Becker, M.M.; Chiang, J.J.; Brass, A.L.; Ahmed, A.A.; Chi, X.; Dong, L.; et al. Distinct Patterns of IFITM-Mediated Restriction of Filoviruses, SARS Coronavirus, and Influenza A Virus. *PLOS Pathog.* **2011**, *7*, e1001258. [CrossRef]

9.   Wrensch, F.; Winkler, M.; Pöhlmann, S. IFITM Proteins Inhibit Entry Driven by the MERS-Coronavirus Spike Protein: Evidence for Cholesterol-Independent Mechanisms. *Viruses* **2014**, *6*, 3683–3698. [CrossRef]

10.   Lu, J.; Pan, Q.; Rong, L.; He, W.; Liu, S.-L.; Liang, C. The IFITM Proteins Inhibit HIV-1 Infection. *J. Virol.* **2011**, *85*, 2126–2137. [CrossRef]

11.   Perreira, J.M.; Chin, C.R.; Feeley, E.M.; Brass, A.L. IFITMs Restrict the Replication of Multiple Pathogenic Viruses. *J. Mol. Biol.* **2013**, *425*, 4937–4955. [CrossRef]

12.   Yao, L.; Dong, H.; Zhu, H.; Nelson, D.; Liu, C.; Lambiase, L.; Li, X. Identification of the IFITM3 gene as an inhibitor of hepatitis C viral translation in a stable STAT1 cell line. *J. Viral Hepat.* **2011**, *18*, e523–e529. [CrossRef]

13.   Wrensch, F.; Karsten, C.B.; Gnirß, K.; Hoffmann, M.; Lu, K.; Takada, A.; Winkler, M.; Simmons, G.; Pöhlmann, S. Interferon-Induced Transmembrane Protein–Mediated Inhibition of Host Cell Entry of Ebolaviruses. *J. Infect. Dis.* **2015**, *212* (Suppl. 2), S210–S218. [CrossRef] [PubMed]

14.   Weidner, J.M.; Jiang, D.; Pan, X.-B.; Chang, J.; Block, T.M.; Guo, J.-T. Interferon-Induced Cell Membrane Proteins, IFITM3 and Tetherin, Inhibit Vesicular Stomatitis Virus Infection via Distinct Mechanisms. *J. Virol.* **2010**, *84*, 12646–12657. [CrossRef] [PubMed]

15.   Zhao, X.; Sehgal, M.; Hou, Z.; Cheng, J.; Shu, S.; Wu, S.; Guo, F.; Le Marchand, S.J.; Lin, H.; Chang, J.; et al. Identification of Residues Controlling Restriction versus Enhancing Activities of IFITM Proteins on Entry of Human Coronaviruses. *J. Virol.* **2018**, *92*, e01535-17. [CrossRef] [PubMed]

16.   Everitt, A.R.; Clare, S.; Pertel, T.; John, S.P.; Wash, R.S.; Smith, S.E.; Chin, C.R.; Feeley, E.M.; Sims, J.S.; Adams, D.J.; et al. IFITM3 restricts the morbidity and mortality associated with influenza. *Nature* **2012**, *484*, 519–523. [CrossRef] [PubMed]

17.   Bailey, C.C.; Huang, I.-C.; Kam, C.; Farzan, M. Ifitm3 Limits the Severity of Acute Influenza in Mice. *PLOS Pathog.* **2012**, *8*, e1002909. [CrossRef] [PubMed]

18.   Pan, Y.; Yang, P.; Dong, T.; Zhang, Y.; Shi, W.; Peng, X.; Cui, S.; Zhang, D.; Lu, G.; Liu, Y.; et al. IFITM3 Rs12252-C Variant Increases Potential Risk for Severe Influenza Virus Infection in Chinese Population. *Front. Cell. Infect. Microbiol.* **2017**, *7*, 294. [CrossRef]

19.   Lee, N.; Cao, B.; Ke, C.; Lu, H.; Hu, Y.; Tam, C.H.T.; Ma, R.C.W.; Guan, D.; Zhu, Z.; Li, H.; et al. IFITM3, TLR3, and CD55 Gene SNPs and Cumulative Genetic Risks for Severe Outcomes in Chinese Patients with H7N9/H1N1pdm09 Influenza. *J. Infect. Dis.* **2017**, *216*, 97–104. [CrossRef]

20.   Wang, Z.; Zhang, A.; Wan, Y.; Liu, X.; Qiu, C.; Xi, X.; Ren, Y.; Wang, J.; Dong, Y.; Bao, M.; et al. Early hypercytokinemia is associated with interferon-induced transmembrane protein-3 dysfunction and predictive of fatal H7N9 infection. *Proc. Natl. Acad. Sci. USA* **2014**, *111*, 769–774. [CrossRef]

21.   Zhang, Y.-H.; Zhao, Y.; Liu, N.; Peng, Y.-C.; Giannoulatou, E.; Jin, R.-H.; Yan, H.-P.; Wu, H.; Liu, J.-H.; Liu, N.; et al. Interferon-induced transmembrane protein-3 genetic variant rs12252-C is associated with severe influenza in Chinese individuals. *Nat. Commun.* **2013**, *4*, 1418. [CrossRef] [PubMed]

22.   Xu-Yang, Z.; Pei-Yu, B.; Chuan-Tao, Y.; Wei, Y.; Hong-Wei, M.; Kang, T.; Chun-Mei, Z.; Ying-Feng, L.; Xin, W.; Ping-Zhong, W.; et al. Interferon-Induced Transmembrane Protein 3 Inhibits Hantaan Virus Infection, and Its Single Nucleotide Polymorphism rs12252 Influences the Severity of Hemorrhagic Fever with Renal Syndrome. *Front. Immunol.* **2016**, *7*, 535. [CrossRef] [PubMed]

23.   Zhang, Y.; Makvandi-Nejad, S.; Qin, L.; Zhao, Y.; Zhang, T.; Wang, L.; Repapi, E.; Taylor, S.; McMichael, A.; Li, N.; et al. Interferon-induced transmembrane protein-3 rs12252-C is associated with rapid progression of acute HIV-1 infection in Chinese MSM cohort. *AIDS* **2015**, *29*, 889–894. [CrossRef] [PubMed]

24.   Jia, R.; Pan, Q.; Ding, S.; Rong, L.; Liu, S.-L.; Geng, Y.; Qiao, W.; Liang, C. The N-Terminal Region of IFITM3 Modulates Its Antiviral Activity by Regulating IFITM3 Cellular Localization. *J. Virol.* **2012**, *86*, 13697–13707. [CrossRef] [PubMed]

25.   Allen, E.K.; Randolph, A.G.; Bhangale, T.; Dogra, P.; Ohlson, M.; Oshansky, C.M.; Zamora, A.E.; Shannon, J.P.; Finkelstein, D.; Dressen, A.; et al. SNP-mediated disruption of CTCF binding at the IFITM3 promoter is associated with risk of severe influenza in humans. *Nat. Med.* **2017**, *23*, 975–983. [CrossRef]

26.   Shi, G.; Kenney, A.D.; Kudryashova, E.; Zani, A.; Zhang, L.; Lai, K.K.; Hall-Stoodley, L.; Robinson, R.T.; Kudryashov, D.S.; Compton, A.A.; et al. Opposing activities of IFITM proteins in SARS-CoV-2 infection. *EMBO J.* **2021**, *40*, e106501. [CrossRef]

27. Bozzo, C.P.; Nchioua, R.; Volcic, M.; Koepke, L.; Krüger, J.; Schütz, D.; Heller, S.; Stürzel, C.M.; Kmiec, D.; Conzelmann, C.; et al. IFITM proteins promote SARS-CoV-2 infection and are targets for virus inhibition in vitro. *Nat. Commun.* **2021**, *12*, 4584. [CrossRef]

28. Jia, R.; Xu, F.; Qian, J.; Yao, Y.; Miao, C.; Zheng, Y.-M.; Liu, S.-L.; Guo, F.; Geng, Y.; Qiao, W.; et al. Identification of an endocytic signal essential for the antiviral action of IFITM3. *Cell. Microbiol.* **2014**, *16*, 1080–1093. [CrossRef]

29. Tscherne, D.M.; Manicassamy, B.; García-Sastre, A. An enzymatic virus-like particle assay for sensitive detection of virus entry. *J. Virol. Methods* **2010**, *163*, 336–343. [CrossRef] [PubMed]

30. Su, B.; Wurtzer, S.; Rameix-Welti, M.-A.; Dwyer, D.; van der Werf, S.; Naffakh, N.; Clavel, F.; Labrosse, B. Enhancement of the Influenza A Hemagglutinin (HA)-Mediated Cell-Cell Fusion and Virus Entry by the Viral Neuraminidase (NA). *PLoS ONE* **2009**, *4*, e8495. [CrossRef] [PubMed]

31. Liu, Y.; Wang, Y.; Wang, X.; Xiao, Y.; Chen, L.; Guo, L.; Li, J.; Ren, L.; Wang, J. Development of two TaqMan real-time reverse transcription-PCR assays for the detection of severe acute respiratory syndrome coronavirus-2. *Biosaf. Health* **2020**, *2*, 232–237. [CrossRef] [PubMed]

32. Guo, L.; Ren, L.; Yang, S.; Xiao, M.; Chang, D.; Yang, F.; Cruz, C.S.D.; Wang, Y.; Wu, C.; Xiao, Y.; et al. Profiling Early Humoral Response to Diagnose Novel Coronavirus Disease (COVID-19). *Clin. Infect. Dis.* **2020**, *71*, 778–785. [CrossRef] [PubMed]

33. Matumoto, M. A note on some points of calculation method of LD50 by Reed and Muench. *Jpn. J. Exp. Med.* **1949**, *20*, 175–179. [PubMed]

34. Zhou, Z.; Ren, L.; Zhang, L.; Zhong, J.; Xiao, Y.; Jia, Z.; Guo, L.; Yang, J.; Wang, C.; Jiang, S.; et al. Heightened Innate Immune Responses in the Respiratory Tract of COVID-19 Patients. *Cell Host Microbe* **2020**, *27*, 883–890.e2. [CrossRef]

35. Hachim, M.; Al Heialy, S.; Hachim, I.Y.; Halwani, R.; Senok, A.C.; Maghazachi, A.; Hamid, Q. Interferon-Induced Transmembrane Protein (IFITM3) Is Upregulated Explicitly in SARS-CoV-2 Infected Lung Epithelial Cells. *Front. Immunol.* **2020**, *11*, 1372. [CrossRef]

36. Blanco-Melo, D.; Nilsson-Payant, B.E.; Liu, W.-C.; Uhl, S.; Hoagland, D.; Møller, R.; Jordan, T.X.; Oishi, K.; Panis, M.; Sachs, D.; et al. Imbalanced Host Response to SARS-CoV-2 Drives Development of COVID-19. *Cell* **2020**, *181*, 1036–1045.e9. [CrossRef] [PubMed]

37. Hoffmann, M.; Kleine-Weber, H.; Schroeder, S.; Krüger, N.; Herrler, T.; Erichsen, S.; Schiergens, T.S.; Herrler, G.; Wu, N.-H.; Nitsche, A.; et al. SARS-CoV-2 Cell Entry Depends on ACE2 and TMPRSS2 and Is Blocked by a Clinically Proven Protease Inhibitor. *Cell* **2020**, *181*, 271–280.e8. [CrossRef]

38. Koch, J.; Uckeley, Z.M.; Doldan, P.; Stanifer, M.; Boulant, S.; Lozach, P. TMPRSS2 expression dictates the entry route used by SARS-CoV-2 to infect host cells. *EMBO J.* **2021**, *40*, e107821. [CrossRef]

39. Peacock, T.P.; Goldhill, D.H.; Zhou, J.; Baillon, L.; Frise, R.; Swann, O.C.; Kugathasan, R.; Penn, R.; Brown, J.C.; Sanchez-David, R.Y.; et al. The furin cleavage site in the SARS-CoV-2 spike protein is required for transmission in ferrets. *Nat. Microbiol.* **2021**, *6*, 899–909. [CrossRef]

40. Buchrieser, J.; Dufloo, J.; Hubert, M.; Monel, B.; Planas, D.; Rajah, M.M.; Planchais, C.; Porrot, F.; Guivel-Benhassine, F.; Van der Werf, S.; et al. Syncytia formation by SARS-CoV-2-infected cells. *EMBO J.* **2020**, *39*, e106267. [CrossRef]

41. Zhao, X.; Guo, F.; Liu, F.; Cuconati, A.; Chang, J.; Block, T.M.; Guo, J.-T. Interferon induction of IFITM proteins promotes infection by human coronavirus OC43. *Proc. Natl. Acad. Sci. USA* **2014**, *111*, 6756–6761. [CrossRef] [PubMed]

42. Winstone, H.; Lista, M.J.; Reid, A.C.; Bouton, C.; Pickering, S.; Galao, R.P.; Kerridge, C.; Doores, K.J.; Swanson, C.M.; Neil, S.J.D. The Polybasic Cleavage Site in SARS-CoV-2 Spike Modulates Viral Sensitivity to Type I Interferon and IFITM2. *J. Virol.* **2021**, *95*, e02422-20. [CrossRef] [PubMed]

43. Zani, A.; Kenney, A.D.; Kawahara, J.; Eddy, A.C.; Wang, X.L.; Kc, M.; Lu, M.; Hemann, E.A.; Li, J.; Peeples, M.E.; et al. Interferon-induced transmembrane protein 3 (IFITM3) limits lethality of SARS-CoV-2 in mice. *bioRxiv* **2021**. [CrossRef]

44. Zhang, Y.; Qin, L.; Zhao, Y.; Zhang, P.; Xu, B.; Li, K.; Liang, L.; Zhang, C.; Dai, Y.; Feng, Y.; et al. Interferon-Induced Transmembrane Protein 3 Genetic Variant rs12252-C Associated with Disease Severity in Coronavirus Disease 2019. *J. Infect. Dis.* **2020**, *222*, 34–37. [CrossRef] [PubMed]

45. Gómez, J.; Albaiceta, G.M.; Cuesta-Llavona, E.; García-Clemente, M.; López-Larrea, C.; Amado-Rodríguez, L.; López-Alonso, I.; Melón, S.; Alvarez-Argüelles, M.E.; Gil-Peña, H.; et al. The Interferon-induced transmembrane protein 3 gene (IFITM3) rs12252 C variant is associated with COVID-19. *Cytokine* **2021**, *137*, 155354. [CrossRef]

46. Alghamdi, J.; Alaamery, M.; Barhoumi, T.; Rashid, M.; Alajmi, H.; Aljasser, N.; Alhendi, Y.; Alkhalaf, H.; Alqahtani, H.; Algablan, O.; et al. Interferon-induced transmembrane protein-3 genetic variant rs12252 is associated with COVID-19 mortality. *Genomics* **2021**, *113*, 1733–1741. [CrossRef]

47. Makvandi-Nejad, S.; Laurenson-Schafer, H.; Wang, L.; Wellington, D.; Zhao, Y.; Jin, B.; Qin, L.; Kite, K.; Moghadam, H.K.; Song, C.; et al. Lack of Truncated IFITM3 Transcripts in Cells Homozygous for the rs12252-C Variant That is Associated with Severe Influenza Infection. *J. Infect. Dis.* **2018**, *217*, 257–262. [CrossRef]

48. Randolph, A.G.; Yip, W.-K.; Allen, E.K.; Rosenberger, C.M.; Agan, A.A.; Ash, S.A.; Zhang, Y.; Bhangale, T.R.; Finkelstein, D.; Cvijanovich, N.Z.; et al. Evaluation of IFITM3 rs12252 Association with Severe Pediatric Influenza Infection. *J. Infect. Dis.* **2017**, *216*, 14–21. [CrossRef]

49. Ahmadi, I.; Afifipour, A.; Sakhaee, F.; Zamani, M.S.; Gheinari, F.M.; Anvari, E.; Fateh, A. Impact of interferon-induced transmembrane protein 3 gene rs12252 polymorphism on COVID-19 mortality. *Cytokine* **2022**, *157*, 155957. [CrossRef]

50. Cuesta-Llavona, E.; Albaiceta, G.M.; García-Clemente, M.; Duarte-Herrera, I.D.; Amado-Rodríguez, L.; Hermida-Valverde, T.; Enríquez-Rodriguez, A.I.; Hernández-González, C.; Melón, S.; Alvarez-Argüelles, M.E.; et al. Association between the interferon-induced transmembrane protein 3 gene (IFITM3) rs34481144/rs12252 haplotypes and COVID-19. *Curr. Res. Virol. Sci.* **2021**, *2*, 100016. [CrossRef]

51. Nikoloudis, D.; Kountouras, D.; Hiona, A. The frequency of combined IFITM3 haplotype involving the reference alleles of both rs12252 and rs34481144 is in line with COVID-19 standardized mortality ratio of ethnic groups in England. *PeerJ* **2020**, *8*, e10402. [CrossRef] [PubMed]

52. Schönfelder, K.; Breuckmann, K.; Elsner, C.; Dittmer, U.; Fistera, D.; Herbstreit, F.; Risse, J.; Schmidt, K.; Sutharsan, S.; Taube, C.; et al. The influence of IFITM3 polymorphisms on susceptibility to SARS-CoV-2 infection and severity of COVID-19. *Cytokine* **2021**, *142*, 155492. [CrossRef] [PubMed]

53. Dobrijevic, Z.; Robajac, D.; Gligorijevic, N.; Sunderic, M.; Penezic, A.; Miljus, G.; Nedic, O. The association of ACE1, ACE2, TMPRSS2, IFITM3 and VDR polymorphisms with COVID-19 severity: A systematic review and meta-analysis. *EXCLI J.* **2022**, *21*, 818–839. [CrossRef]

54. Wong, C.K.; Lam, C.W.K.; Wu, A.K.L.; Ip, W.K.; Lee, N.L.S.; Chan, I.H.S.; Lit, L.C.W.; Hui, D.S.C.; Chan, M.H.M.; Chung, S.S.C.; et al. Plasma inflammatory cytokines and chemokines in severe acute respiratory syndrome. *Clin. Exp. Immunol.* **2004**, *136*, 95–103. [CrossRef] [PubMed]

55. Qin, L.; Wang, D.; Li, D.; Zhao, Y.; Peng, Y.; Wellington, D.; Dai, Y.; Sun, H.; Sun, J.; Liu, G.; et al. High Level Antibody Response to Pandemic Influenza H1N1/09 Virus Is Associated With Interferon-Induced Transmembrane Protein-3 rs12252-CC in Young Adults. *Front. Cell. Infect. Microbiol.* **2018**, *8*, 134. [CrossRef]

56. Lei, N.; Li, Y.; Sun, Q.; Lu, J.; Zhou, J.; Li, Z.; Liu, L.; Guo, J.; Qin, K.; Wang, H.; et al. IFITM3 affects the level of antibody response after influenza vaccination. *Emerg. Microbes Infect.* **2020**, *9*, 976–987. [CrossRef]

MDPI

St. Alban-Anlage 66

4052 Basel

Switzerland

www.mdpi.com

Viruses Editorial Office

E-mail: viruses@mdpi.com

www.mdpi.com/journal/viruses